W9-CSP-209

3
Topics in Heterocyclic Chemistry

Series Editor: R. R. Gupta

Topics in Heterocyclic Chemistry
Series Editor: R. R. Gupta

Recently Published and Forthcoming Volumes

QSAR and Molecular Modeling Studies in Heterocyclic Drugs I

Volume Editor: Satya Prakash Gupta

With contributions by

R. Bahal · S. C. Basak · E. Benfenati · P. V. Bharatam
B. Bhhatarai · E. A. Castro · P. R. Duchowicz · R. Garg
M. M. Gromiha · B. D. Gute · S. Khanna · D. Mills · R. Natarajan
M. N. Ponnuswamy · K. Saraboji · S. M. M. Sony · A. A. Toropov

 Springer

The series *Topics in Heterocyclic Chemistry* presents critical reviews on "Heterocyclic Compounds" within topic-related volumes dealing with all aspects such as synthesis, reaction mechanisms, structure complexity, properties, reactivity, stability, fundamental and theoretical studies, biology, biomedical studies, pharmacological aspects, applications in material sciences, etc. Metabolism will be also included which will provide information useful in designing pharmacologically active agents. Pathways involving destruction of heterocyclic rings will also be dealt with so that synthesis of specifically functionalized non-heterocyclic molecules can be designed.

The overall scope is to cover topics dealing with most of the areas of current trends in heterocyclic chemistry which will suit to a larger heterocyclic community.

As a rule contributions are specially commissioned. The editors and publishers will, however, always be pleased to receive suggestions and supplementary information. Papers are accepted for *Topics in Heterocyclic Chemistry* in English.

In references *Topics in Heterocyclic Chemistry* is abbreviated *Top Heterocycl Chem* and is cited as a journal.

Springer WWW home page: springer.com
Visit the THC content at springerlink.com

Library of Congress Control Number: 2006926508

ISSN 1861-9282
ISBN-10 3-540-33378-9 Springer Berlin Heidelberg New York
ISBN-13 978-3-540-33378-4 Springer Berlin Heidelberg New York
DOI 10.1007/11577737

Springer is a part of Springer Science+Business Media

springer.com

© Springer-Verlag Berlin Heidelberg 2006
Printed in Germany

The use of registered names, trademarks, etc. in this publication does not imply, even in the absence of a specific statement, that such names are exempt from the relevant protective laws and regulations and therefore free for general use.

Cover design: *Design & Production* GmbH, Heidelberg
Typesetting and Production: LE-TEX Jelonek, Schmidt & Vöckler GbR, Leipzig

Printed on acid-free paper   02/3100 YL – 5 4 3 2 1 0

Series Editor

Prof. R. R. Gupta

10A, Vasundhara Colony
Lane No. 1, Tonk Road
Jaipur-302 018, India
rrg_vg@yahoo.co.in

Volume Editor

Prof. Dr. Satya Prakash Gupta

Department of Chemistry
Birla Institute of Technology and Science
Pilani-333 031, India
spg@bits-pilani.ac.in

Editorial Board

Prof. D. Enders

RWTH Aachen
Institut für Organische Chemie
D-52074, Aachen, Germany
enders@rwth-aachen.de

Prof. Steven V. Ley FRS

BP 1702 Professor
and Head of Organic Chemistry
University of Cambridge
Department of Chemistry
Lensfield Road
Cambridge, CB2 1EW, UK
svl1000@cam.ac.uk

Prof. G. Mehta FRS

Director
Department of Organic Chemistry
Indian Institute of Science
Bangalore- 560 012, India
gm@orgchem.iisc.ernet.in

Prof. A.I. Meyers

Emeritus Distinguished Professor of
Department of Chemistry
Colorado State University
Fort Collins, CO 80523-1872, USA
aimeyers@lamar.colostate.edu

Prof. K.C. Nicolaou

Chairman
Department of Chemistry
The Scripps Research Institute
10550 N. Torrey Pines Rd.
La Jolla, California 92037, USA
kcn@scripps.edu
and
Professor of Chemistry
Department of Chemistry and Biochemistry
University of California
San Diego, 9500 Gilman Drive
La Jolla, California 92093, USA

Topics in Heterocyclic Chemistry
Also Available Electronically

For all customers who have a standing order to Topics in Heterocyclic Chemistry, we offer the electronic version via SpringerLink free of charge. Please contact your librarian who can receive a password or free access to the full articles by registering at:

springerlink.com

If you do not have a subscription, you can still view the tables of contents of the volumes and the abstract of each article by going to the SpringerLink Homepage, clicking on "Browse by Online Libraries", then "Chemical Sciences", and finally choose Topics in Heterocyclic Chemistry.

You will find information about the

 – Editorial Board
 – Aims and Scope
 – Instructions for Authors
 – Sample Contribution

at springer.com using the search function.

Preface to the Series

Topics in Heterocyclic Chemistry presents critical accounts of heterocyclic compounds (cyclic compounds containing at least one heteroatom other than carbon in the ring) ranging from three members to supramolecules. More than 50% of billions of compounds listed in *Chemical Abstracts* are heterocyclic compounds. The branch of chemistry dealing with these heterocyclic compounds is called heterocyclic chemistry, which is the largest branch of chemistry and as such the chemical literature appearing every year as research papers and review articles is vast and can not be covered in a single volume.

This series in heterocyclic chemistry is being introduced to collectively make available critically and comprehensively reviewed literature scattered in various journals as papers and review articles. All sorts of heterocyclic compounds originating from synthesis, natural products, marine products, insects, etc. will be covered. Several heterocyclic compounds play a significant role in maintaining life. Blood constituent hemoglobin and purines as well as pyrimidines, the constituents of nucleic acid (DNA and RNA) are also heterocyclic compounds. Several amino acids, carbohydrates, vitamins, alkaloids, antibiotics, etc. are also heterocyclic compounds that are essential for life. Heterocyclic compounds are widely used in clinical practice as drugs, but all applications of heterocyclic medicines can not be discussed in detail. In addition to such applications, heterocyclic compounds also find several applications in the plastics industry, in photography as sensitizers and developers, and in dye industry as dyes, etc.

Each volume will be thematic, dealing with a specific and related subject that will cover fundamental, basic aspects including synthesis, isolation, purification, physical and chemical properties, stability and reactivity, reactions involving mechanisms, intra- and intermolecular transformations, intra- and intermolecular rearrangements, applications as medicinal agents, biological and biomedical studies, pharmacological aspects, applications in material science, and industrial and structural applications.

The synthesis of heterocyclic compounds using transition metals and using heterocyclic compounds as intermediates in the synthesis of other organic compounds will be an additional feature of each volume. Pathways involving the destruction of heterocyclic rings will also be dealt with so that the synthesis of specifically functionalized non-heterocyclic molecules can be designed. Each volume in this series will provide an overall picture of heterocyclic compounds critically and comprehensively evaluated based on five to ten years of literature. Graduates, research students and scientists in the fields of chemistry, pharmaceutical chemistry, medicinal chemistry, dyestuff chemistry, agrochemistry, etc. in universities, industry, and research organizations will find this series useful.

I express my sincere thanks to the Springer staff, especially to Dr. Marion Hertel, executive editor, chemistry, and Birgit Kollmar-Thoni, desk editor, chemistry, for their excellent collaboration during the establishment of this series and preparation of the volumes. I also thank my colleague Dr. Mahendra Kumar for providing valuable suggestions. I am also thankful to my wife Mrs. Vimla Gupta for her multifaceted cooperation.

Jaipur, 31 January 2006   R.R. Gupta

Preface

The series *Topics in Heterocyclic Chemistry* now devotes its two volumes, Vols. 3 and 4, to today's most fascinating area of medicinal chemistry: quantitative structure-activity relationships (QSAR) and molecular modeling, which has revolutionalized drug discovery in the present era. These two volumes together present some very timely and important reviews on QSAR and molecular modeling studies in heterocyclic drugs and are titled *QSAR and Molecular Modeling Studies in Heterocyclic Drugs I* and *QSAR and Molecular Modeling Studies in Heterocyclic Drugs II*. Since the pioneering work of Corwin Hansch from 1962–1964 that laid the foundations of QSAR by means of three important contributions: the combination of several physicochemical parameters in one equation, the definition of the lipophilicity parameter π, and the formulation of the parabolic model for nonlinear lipophilicity–activity relationships, the area of computer-aided drug design with the development of computer technology went through a revolutionary change from two-dimensional to three-dimensional and now to multi-dimensional QSAR. The QSAR and molecular modeling studies have drastically reduced the cost and the time involved in the drug design and development. With the objective that some timely in-depth reviews on such studies in heterocyclic drugs may be of great value to those involved in drug discovery, some leaders in the field were invited to contribute and the overwhelming response led to devote two volumes on the topic. Both volumes cover the excellent and novel articles of varied interest.

Volume 3 contains five articles. The first article by Castro et al. describes the application of flexible molecular descriptors in the QSAR study of heterocyclic drugs. In this article, the various formulations of optimal descriptors introduced by different authors during the last ten years are discussed for the special case of heterocyclic drugs. The second article by Basak et al. is on predicting pharmacological and toxicological activity of heterocyclic compounds using QSAR and molecular modeling. Heterocyclic compounds are important as drugs, toxicants, and agrochemicals. In this article, the authors report the QSAR modeling of pharmacological activity, insect repellency, and environmental toxicity for a few classes of heterocyclic compounds from their structure. Pharmacological activity of drugs depends mainly on the interaction with their biological targets, which have complex three-dimensional structure, and their molecular recognitions are guided by the nature of intermolecular interactions. In the third article, therefore, Ponnuswamy et al. present conformational aspects and interaction studies of different heterocyclic drugs. In the next article, Khanna et al. describe, in detail, *in silico* studies on PPARγ agonistic heterocyclic systems. Several heterocyclic derivatives like oxazolidinedione, thiazolidinedione, tetrazole, phenoxazine, etc., are being developed for the treatment of insulin resistance and type 2 diabetes mellitus. The heterocyclic head group in these systems binds to and activates peroxisome proliferator activated receptor γ (PPARγ), a nuclear receptor that regulates the expression of several genes involved in the metabolism. In this article, therefore, various molecular modeling studies have been described that are important in understanding the drug–receptor interactions, analyzing the important pharmacophore features, identifying new scaffolds, and understanding the electronic structure and reactivity of these heterocyclic systems. The final article in this volume, written by Garg and Bhhatarai, is on QSAR and molecular modeling studies of HIV protease inhibitors. HIV protease is one of the major viral targets for the development of new chemotherapeutics against AIDS. In this article, therefore, Garg and Bhhatarai have presented a detailed study on structure–activity relationship studies on many groups of HIV protease inhibitors, providing the excellent rationale to design potent and pharmaceutically important protease inhibitors.

There are six fascinating articles in Vol. 4. These six articles present QSAR and molecular modeling on six different classes of heterocyclic drugs. The first article by Hadjipavlou-Litina is related to thrombin and factor FXa inhibitors. Both thrombin and factor FXa are bound to and are enzymatically active in blood clots. Thus a QSAR study on them may be of great use to investigate potent antithrombotics or antocoagulants. Similarly, the second article by Hannongbua has reviewed structural information and drug–enzyme interaction of the non-nucleoside reverse transcriptase inhibitors based on quantum chemical approaches, providing the valuable guidelines to design and develop potent anti-HIV drugs. Reverse transcriptase is an important enzymatic target to inhibit the growth of human immunodeficiency virus of type 1 (HIV-1), which is the causative agent of AIDS. In the next article, Vračko has described a QSAR approach in study of mutagenicity of aromatic and heteroaromatic amines. These compounds are highly hazardous to the environment and can be carcinogenic and thus are the subject of both theoretical and experimental studies.

Cocaine is a widely abused heterocyclic drug and there is no available anti-cocaine therapeutic, but in the fourth article Zhan describes the state of the art of molecular modeling of the reaction mechanism for the hydrolysis of cocaine and the mechanism-based design of anti-cocaine therapeutics. Amongst the heterocyclic systems, thiazolidine is a biologically important scaffold known to be associated with several biological activities. Some of the prominent biological responses attributed to this skeleton are antiviral, antibacterial, antifungal, antihistaminic, hypoglycemic, and anti-inflammatory activities. In the fifth article, therefore, Prabhakar et al. have presented a very comprehensive review on the QSAR studies of diverse biological activities of the thiazolidines published during the past decade. This study may be of importance to explore the possibility if thiazolidine nucleus can be exploited to design the drugs for some other diseases. In the final article, however, Gupta has reviewed the QSAR studies on calcium channel blockers (CCBs). CCBs have potential therapeutic uses against several cardiovascular and non-cardiovascular diseases and the article throws light on how to design more effective CCBs that may be therapeutically useful.

Thus both these volumes of *Topics in Heterocyclic Chemistry* are unique and make interesting readings for all those involved, theoretically or experimentally, in design and development of drugs. As an editor of these volumes, I have greatly enjoyed reading the articles and hope all readers will too.

Pilani, March 2006   Satya Prakash Gupta

Contents

Contents of Volume 4

QSAR and Molecular Modeling Studies in Heterocyclic Drugs II

Volume Editor: Satya Prakash Gupta
ISBN: 3-540-33233-2

Top Heterocycl Chem (2006) 3: 1–38
DOI 10.1007/7081_028
© Springer-Verlag Berlin Heidelberg 2006
Published online: 30 March 2006

Applications of Flexible Molecular Descriptors in the QSPR–QSAR Study of Heterocyclic Drugs

Pablo R. Duchowicz[1] · Eduardo A. Castro[1] (✉) · Andrey A. Toropov[2] · Emilio Benfenati[2]

[1]INIFTA, Departamento de Química, Facultad de Ciencias Exactas,
Universidad Nacional de La Plata, Suc.4, C.C. 16, La Plata 1900, Buenos Aires, Argentina
castro@quimica.unlp.edu.ar

[2]Laboratory of Environmental Chemistry and Toxicology,
Instituto di Ricerche Farmacologiche "Mario Negri", Via Eritrea 62, 20157 Milan, Italy

Abstract Various formulations of optimal descriptors introduced by different authors during the last 10 years are discussed for the special case of heterocyclic drugs. Usually, the first application of a new defined flexible variable involves alkanes or other types of homogeneous molecular sets, without considering any more rigorous details of structural diversity of the molecules. This chapter reviews significant examples from the literature that are devoted to heterocyclic compounds, and explores physicochemical properties and biological activities, such as aqueous solubility, singlet excitation energy, carcinogenic potential, mutagenicity, and anti-HIV-1 activity. The aim of the review is to provide non-specialist readers with conceptual information to understand the methodology employed.

The first approximation of hierarchy for topological, physical, chemical, and biological features of the molecular structure, which is able to be the basis of the next generation of optimal descriptors, is proposed.

Keywords Aqueous solubility · Anti-HIV-1 activity · Carcinogenic potential · Mutagenicity · Optimal descriptors · QSPR–QSAR theory

Abbreviations

AA	Amino acid
AO	Atomic orbital
CW	Correlation weight
DCW	Flexible molecular descriptor based on optimization of CW
GAO	Graph of atomic orbitals
HFG	Hydrogen-filled graph
HSG	Hydrogen-suppressed graph
LCCI	Linear combinations of connectivity indices
LCXCT	Linear combinations of higher-level molecular connectivity terms
LGI_k	Local graph invariant calculated at v_k
MCT	Molecular connectivity theory
OCWLGI	Optimization of correlation weights of local graph invariants
pMV	Partial molar volume
QSPR–QSAR	Quantitative structure–property/activity relationships
SMILES	Simplified molecular input line entry system
A	Adjacency matrix
A^{modif}	Modified A
F	Fischer F ratio
M	Molar mass
n	Number of vertices of the graph
N	Number of molecules
P	Physicochemical property or biological activity
R	Correlation coefficient of the model
S	Standard deviation of the model
v_k	k-th vertex of the graph

1
Introduction

Knowledge is multifunctional. One of the most important applications of knowledge is to produce new knowledge. In spite of the existence of large databases on molecular structures and the continuous growth of numerical experimental data on physicochemical properties and biological activities, the problem of estimating the properties of substances that have not yet been tested could take place in a more accurate way, at least in the next few decades. During the last half century it has become common practice to employ topological, physical, chemical, and biological numerical characteristics, depending on the molecular structure, to predict the properties of substances that remain unknown for different reasons, such as because they are unstable, toxic, or simply that their measurement requires too much time. The field of natural sciences, which aims to construct mathematical models to search for regularities in data and permit their systematization, has been addressed by the quantitative structure–property/activity relationships theory (QSPR–QSAR).

There are several motivations for establishing and developing QSPR–QSAR studies. In terms of economic aspects, the design of QSPR–QSAR models gives a chance for the rational use of the available resources present in the laboratory or even a plant, and avoids performing expensive and unnecessary experimental determinations. With respect to moral aspects, the QSPR–QSAR models applied to toxicology have great importance in the virtual screening of the toxic potential of compounds before their synthesis, and thus represent an effective alternative that reduces animal testing in biological assays. Finally, from the theoretical point of view, the model can illuminate the mechanisms of physicochemical properties or biological activities of the compounds.

Although the pioneering studies in QSPR–QSAR theory were established by Wiener in 1947 [1–4], other mathematical models had been reported previously for the prediction of the properties of substances. For example, it is well known that simple additive schemes and group contribution methods were used before the first QSPR–QSAR analyses [5–7]. We begin by summarizing these approaches, simply for the fact that optimal descriptors have gained some insight from these particular methods.

1.1
The Simple Additive Scheme

The essence of this technique for modeling an investigated property (P) relies on performing sums of parameters (C_k) associated with a collection of defined atomic groups or fragments present in the molecule, solely by means of a classical understanding:

$$P = \sum_{k=1} C_k \qquad (1)$$

In general, P represents heat capacities [5], Gibbs free energies [5], entropies [6], thermodynamic [7] or technological [8] properties of polymers, and other types of physicochemical properties [9, 10]. Although it has been reported that the simple additive scheme gives reasonable predictions for the properties of interest [5–8], unsatisfactory predictions were also mentioned in the relevant literature [9, 10]. One of the main disadvantages of the additive approach is the impossibility of providing all possible molecular fragments for all the chemicals available. However, Bicerano [8] described a tool for modeling the properties of polymers by using molecular topological connectivity indices (which can be defined for any structure), instead of a preliminary determined collection of fragments. In fact, the presented version of the simple additive scheme is universal, and the approach can be successfully used for predicting the properties of any type of molecule (not only polymers) [11, 12]. It is also interesting to note that the simple additive scheme has been used in psychonomic research [13].

1.2
Group Contribution Methods

The group contribution technique employs a special type of mathematical function for each property analyzed [14–20], whose parameters are calculated with the contributions of defined fragments in the molecule. The choice of the functions is made on the basis of physical meaning and/or previous practices. For example, Iwai et al. [14] established a model for the vapor pressure of alkane isomers by resorting to the following formula

$$\ln(P_v M) = \left[\sum_{i=1} n_i(a_i + b_i)/T^* - c_i \ln T^* - d_i T^* \right] + Q \qquad (2)$$

where P_v is the saturated vapor pressure, $T^* = T[K]/100$, M is the molecular weight, n_i is the count of the i-th group in the structure, Q is a correction factor specific for each compound, and a_i, b_i, c_i, and d_i correspond to the contributions of the i-th group. The first set of parameters for use in Eq. 2 are CH_4, $-CH_3$, $>CH_2$, $>CH-$, and $>C<$; a second order of groups is used for definition of Q in Eq. 2: $>CH-CH<$, $-CH(CH_3)CH_3$, $CH(CH_3)CH_2CH_3$, etc. The numerical values for a_i, b_i, c_i, and d_i are selected in such a way as to produce the smallest differences between the experimental and predicted vapor pressures. The group contribution method is widely used in biochemical research [16, 21].

2
Optimal Descriptors and their Applications

According to Balaban [22] one can classify topological indices as belonging to the following classes:

1. First generation, based on integers
2. Second generation, based on real numbers
3. Third generation, based on matrices
4. Fourth generation, which includes stereochemical features of molecules

Another class should be added to this classification:

5. Fifth generation, based on numerical optimization

The pool of molecular descriptors has increased dramatically during the last few decades, and the problem of selecting optimal molecular descriptors is a current topic of interest to many researchers [23]. There are two main procedures available for choosing optimal variables in QSPR–QSAR studies: one of these consists of evaluating thousands of molecular descriptors to find out the best ones; the alternative relies on selecting a few adjustable descriptors and optimizing their variable parts, in order to make them specific to the property–activity under study. Most of the molecular descriptors available in the literature are of the rigid type, that is, they are characterized by fixed numerical values which are independent of the property under consideration. Hence, these descriptors can be computed once the bonding pattern and the geometry in the case of 3D structural indices of a molecule are known. Since the selection of an optimal reduced set of descriptors among thousands of them is not a trivial task, one has to face the ambiguities arising from the high correlation between variables. In contrast, a flexible descriptor does not present this kind of problem and it can lead to a simpler model, both from the understanding and from the statistical points of view. If we extend the concept of a molecular descriptor as a "variable" or "flexible" function depending on some variable parts, then we can optimize them for every property considered during the regression analysis.

QSPR–QSAR studies have been traditionally developed by selecting a priori an analytical model (typically linear, polynomial, or log-linear) to quantify the correlation between selected molecular indices and the property of interest, followed by regression analysis to determine the parameters of the model. Although the above approaches have proven to be useful in some cases, they have a number of limitations. In fact, the quantitative relationships between the molecular structure and physicochemical and biological properties can be rather complex and highly nonlinear, so that determining the optimal analytical form of the QSPR–QSAR model presents a real challenge. Moreover, regression analysis becomes complex and less reliable as the number of descriptors increases.

As many QSPR–QSAR studies require the computation of topological indices for molecules containing heteroatoms and multiple bonds, this leads to the employment of the weighted graph theory, which assigns a parameter (weight) to each vertex and edge in the molecular graphs [24, 25]. In this approach the computation of molecular matrices and topological indices is generally performed with various sets of fixed (constant) vertex and edge parameters. However, the use of variable (optimized) vertex and edge parameters can lead to improvements in the quantitative relationships, since the variable descriptors are flexible enough to adjust themselves to different atomic environment, and to distinguish among the relative roles of heteroatoms when considering different properties. From the theoretical perspective, flexible descriptors are able to correlate with end points of interest not only in the domain of the training sets, but also for external sets of compounds. The verification of this hypothesis is essential and constitutes a clue for the validity of the QSPR–QSAR analyses.

Most of the descriptors proposed up to the present are size dependent and can produce apparently good correlations when used for samples of compounds that include molecules of different size. This is the opposite situation to when the same descriptors are used to describe variation in properties among molecules of the same size, as they often show a limited ability to characterize well the molecular shape. Also, when the size of the molecules increases, additional structural elements have to be considered in addition to the connectivity and branching. This objective can be well accomplished by resorting to optimal molecular descriptors.

In general, several strategies have been proposed for optimizing a molecular descriptor:

1. By optimizing the functional form; i.e., one can consider powers in the different topological indices
2. By optimizing the diagonal entries of an adjacency matrix to differentiate between atoms of different types
3. By optimizing off-diagonal entries of an adjacency matrix to differentiate between different bond types

In the present chapter we will discuss several examples of studies involving each type of the aforementioned optimizations.

2.1
The Variable Vertex Connectivity Index

One of the most successful topological indices, the vertex connectivity index of first order ($^1\chi$), was proposed by Milan Randic in 1975 according to the following definition [26]:

$$^1\chi = 1/2 \sum_{ij} \left(\deg_i \deg_j \right)^{-0.5} \tag{3}$$

where $\deg_i$ represents the degree of vertex v_i (the number of neighbors of v_i), and the sum runs over all possible products of vertex degrees for edge end points. It has to be noted that the vertex characteristic $\deg_i$ can be obtained by summing the elements over row i or column i from the adjacency matrix (A). In the same year, Kier, Hall, Randic and coworkers extended $^1\chi$ for paths longer than edges of length L, and defined the index of order L ($^L\chi$) as [27–29]:

$$^L\chi = 1/2 \sum_{ijk...} \left(\deg_i \deg_j ... \deg_L \right)^{-0.5} \tag{4}$$

where the product of vertex degrees is calculated with vertices situated along the topological distance L. To take into account the nature of atoms symbolized by vertices, Kier and Hall proposed the use of valence vertex connectivity indices ($^L\chi^v$) [30], which are calculated in the same way as $^L\chi$ but using valence delta weights (δ_i^v) instead of $\deg_i$.

$$^L\chi^v = 1/2 \sum_{ijk...} \left(\delta_i^v \delta_j^v ... \delta_L^v \right)^{-0.5} \tag{5}$$

$$\delta_i^v = \left(Z_i^v - H_i \right) / \left(Z_i - Z_i^v - 1 \right) \tag{6}$$

Here, Z_i^v indicates the number of valence electrons in atom v_i, Z_i is its atomic number, and H_i is the number of hydrogen atoms attached to atom v_i. For example, in the case of aliphatic amines, $\delta_i^v = 3$ for the nitrogen in the primary amino group, in the secondary amino group $\delta_i^v = 4$, and in the tertiary amino group $\delta_i^v = 5$, and the corresponding weights for the carbon atom are 1, 2, 3, or 4, depending on the bonding environment of the atom. Table 1 indicates the δ_i^v values generally accepted for different atoms.

In 1991 Randic introduced the variable vertex connectivity index ($^L\chi^f$) [31] as an alternative approach to Kier and Hall's index for characterization of heterosystems in QSPR–QSAR studies. The main difference between both molecular descriptors is that the former index uses optimized vertex weights ($\deg_i^f$) and the latter uses fixed vertex weights ($\deg_i$):

$$^L\chi^f = 1/2 \sum_{ijk...} \left(\deg_i^f \deg_j^f ... \deg_L^f \right)^{-0.5} \tag{7}$$

This flexible descriptor can be obtained from a modified adjacency matrix (A^{modif}), where its main diagonal elements have been replaced with variable parameters that have to be optimized for each type of atom present in the molecule. For example, Fig. 1 shows the hydrogen-suppressed graph (HSG) [32, 33] for the heterocyclic amino acid tryptophan, and Table 2 gives its A^{modif} with the $\deg_i^f$ corresponding to each vertex. It should be noted that for clarity hydrogen atoms can be included for heteroatoms in the HSG. In C – O bonds an order of 1.5 was assumed, with a value of 1 for single and 2 for double bonds. Clearly, in the present case $^1\chi^f$ depends on the optimizable variables x, y, and z assigned to carbon, oxygen, and nitrogen atoms, respectively. The index can be calculated by adding the bond contributions for all

Table 1 Valence vertex degrees for different atoms

Li	Na	K	Rb	Cs	Be	Mg	Ca	Sr	Ba	Cl	Br	I	S
1	1/9	1/17	1/35	1/53	2	2/9	2/17	2/35	2/53	7/9	7/27	7/45	5/9

the bonds:

$$
\begin{aligned}
{}^{1}\chi^{f} = {}&3/(3+x) + 2/(4+x) + 1/[(3+x)(2+x)]^{0.5} \\
&+ 1/[(2+x)(4+x)]^{0.5} + 4/[(3+x)(4+x)]^{0.5} \\
&+ 2/[(4+x)(1.5+y)]^{0.5} + 2/[(3+x)(1+z)]^{0.5} \\
&+ 1/[(4+x)(2+z)]^{0.5} \\
= {}&f(x,y,z)
\end{aligned}
\tag{8}
$$

Hence, the variable connectivity index is a function of several variables and can assume numerical values only after particular values for x, y, and z are selected. In contrast, if we set $x = 0$, $y = 0$, and $z = 0$, we obtain the rigid index ${}^{1}\chi$ for tryptophan. The variable vertex connectivity index remained overlooked

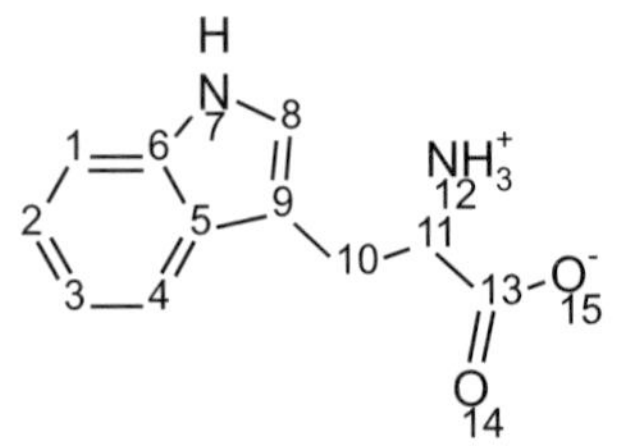

Fig. 1 Numbering of atoms in tryptophan

Table 2 Modified adjacency matrix of tryptophan with x, y, and z entries differentiating carbon, oxygen, and nitrogen atoms, respectively

	1	2	3	4	5	6	7	8	9	10	11	12	13	14	15	$\deg_i^f$
1	x	1.5	0	0	0	1.5	0	0	0	0	0	0	0	0	0	$3+x$
2	1.5	x	1.5	0	0	0	0	0	0	0	0	0	0	0	0	$3+x$
3	0	1.5	x	1.5	0	0	0	0	0	0	0	0	0	0	0	$3+x$
4	0	0	1.5	x	1.5	0	0	0	0	0	0	0	0	0	0	$3+x$
5	0	0	0	1.5	x	1.5	0	0	1	0	0	0	0	0	0	$4+x$
6	1.5	0	0	0	1.5	x	1	0	0	0	0	0	0	0	0	$4+x$
7	0	0	0	0	0	1	z	1	0	0	0	0	0	0	0	$2+z$
8	0	0	0	0	0	0	1	x	2	0	0	0	0	0	0	$3+x$
9	0	0	0	0	1	0	0	2	x	1	0	0	0	0	0	$4+x$
10	0	0	0	0	0	0	0	0	1	x	1	0	0	0	0	$2+x$
11	0	0	0	0	0	0	0	0	0	1	x	1	1	0	0	$3+x$
12	0	0	0	0	0	0	0	0	0	0	1	z	0	0	0	$1+z$
13	0	0	0	0	0	0	0	0	0	0	1	0	x	1.5	1.5	$4+x$
14	0	0	0	0	0	0	0	0	0	0	0	0	1.5	y	0	$1.5+y$
15	0	0	0	0	0	0	0	0	0	0	0	0	1.5	0	y	$1.5+y$

for several years, and so later Randic and coworkers published several studies revealing the advantages of this particular index [34–36].

2.1.1
QSPR Modeling of Partial Molar Volumes of 20 Amino Acids

Up to now, the characterization of amino acids by theoretical structural descriptors has not received wide attention. The study reported in [35] employs $^1\chi^f$ for predicting the partial molar volumes (pMV) of 17 amino acids (AA) that include some heterocyclic molecules, and appear listed in Table 3 together with the numerical values for $^1\chi$ and $^1\chi^f$. Three of the compounds have unknown values for the experimental property (isoleucine, threonine, lysine). This particular molecular set involves four optimizable parameters for each type of atom: x (carbon), y (oxygen), z (nitrogen), and w (sulfur). As a starting point in the search for the optimal values of the four parameters, it is assumed that all the variables have zero as the initial value. The simple

Table 3 Experimental and predicted partial molar volumes

AA	Exp. pMV	Pred. pMV [a]	Pred. pMV [b]
Glycine	43.25	43.499	43.384
Alanine	60.46	60.923	61.160
Valine	90.78	94.264	95.173
Leucine	107.75	109.478	110.695
Isoleucine	–	–	115.508
Methionine	105.35	100.477	101.511
Proline	82.83	81.106	81.749
Phenylalanine	121.48	112.063	113.331
Tryptophan	143.91	144.609	146.534
Serine	60.62	56.627	56.777
Threonine	–	–	87.397
Asparagine	78.0	83.756	84.453
Glutamine	93.9	107.551	–
Tyrosine	123.6	124.074	125.584
Cysteine	73.44	79.999	80.621
Lysine	–	–	95.016
Arginine	127.34	118.827	120.232
Histidine	98.79	100.695	101.734
Aspartic acid	73.83	68.551	68.941
Glutamic acid	85.88	84.712	85.429

[a] Including the outlier (glutamine), Eq. 10
[b] Excluding the outlier (glutamine), Eq. 11

connectivity index gives the following model:

$$pMV = 24.808 - 0.0265\ {}^1\chi$$
$$N = 20,\ R = 0.9168,\ S = 11.118,\ F = 79 \tag{9}$$

Now, in order to find the parameters for ${}^1\chi^f$, each variable can be altered one at a time until the minimum of S is reached. Proceeding this way, it is found that $x = -0.65$, $y = 5.25$, $z = 0.20$, and $w = 0.50$ and the improved relationship is:

$$pMV = 21.818 + 6.922\ {}^1\chi^f$$
$$N = 20,\ R = 0.9776,\ S = 5.961,\ F = 324 \tag{10}$$

It can be appreciated from this equation that S has been almost halved and F increased fourfold. In Fig. 2 is plotted the predicted pMV as a function of the experimental values. On the interpretation of Eq. 10, the numerical values found for the atomic parameters suggest that in AA compounds, carbon atoms have a greater role in explaining the pMV than sulfur or nitrogen, both of which have somewhat larger weights than oxygen. For some other properties the situation would change, as is the case for the solubility of alcohols, where oxygen play an important role. Equation 10 has glutamine with a residual that exceeds twice the value of S, which means that this AA is an outlier. By removing the compound from the set the following statistic is achieved:

$$pMV = 22.528 + 6.094\ {}^1\chi^f$$
$$N = 19,\ R = 0.9865,\ S = 4.727,\ F = 506 \tag{11}$$

The occurrence of unusual values of a given property under study for a compound (outliers) means either that the experimental data have errors or that the model used is unable to account for some specific structural features that are important for outliers and may be absent in other compounds. For example, if different kinds of oxygen atoms in AA are discriminated, such as

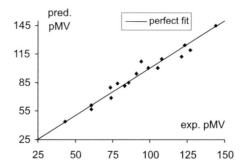

Fig. 2 Predicted partial molar volumes as a function of the experimental property

oxygen of the OH bond, oxygen of $C = O$, and oxygen of the COO^- group, one can expect further improvement in the regression analysis. However, in view of the limited amount of data analyzed here it is not possible to pursue this finer optimization any further. By applying Eq. 11 to predict the unknown isoleucine, threonine, and lysine, the predictions 115.508, 87.397, 95.016, respectively, are found.

2.2
Linear Combinations of Connectivity Indices and Higher-Level Connectivity Terms

From the molecular connectivity theory (MCT) [26–28, 30, 32, 33, 37–39] it is possible to define a medium-sized set composed of eight molecular connectivity indices (χ), which seems to be able to offer a satisfactory model of both many physicochemical properties and many classes of compounds, by means of optimal linear combinations of molecular connectivity indices (LCCI) [40–48]. This reduced set of molecular descriptors has allowed the successful estimation of the properties of natural amino acids, purine and pyrimidine bases, alkanes, organic phosphate derivatives, unsaturated organic compounds, inorganic salts, mixed classes of amino acids plus peptides, amino acids plus inorganic salts, amino acids plus purine and pyrimidine bases, and so on. The modeled properties include the pH at the isoelectric point, longitudinal relaxation time, the side-chain molecular volume, specific rotation, solubility, crystalline density, melting point, motor octane number, retention index for paper chromatography, enthalpy values, and hydration properties, etc.

The basic strategy behind the LCCI method is to employ optimal linear combinations of χ, chosen with the stepwise regression procedure or the full combinatorial technique [49, 50], which allow different classes of compounds to be modeled without recourse to empirical, quantum mechanical, or other kinds of molecular descriptors, which are frequently used in QSPR–QSAR studies to complement the topological indices in order to obtain better-quality models. The reason for choosing a limited set of eight connectivity indices is to keep under control the combinatorial problem that arises in the choice of the best combination of the connectivity indices. In some cases, however, the LCCI method requires an excessive number of indices to achieve satisfactory predictions, with the consequence of a loss of meaning of the corresponding LCCI. This fact led to the development of linear combinations (LCXCT) of nonlinear higher-level molecular connectivity terms ($X = f(\chi)$) or semiempirical terms ($X = f(\chi, P_{\exp})$), where $P_{\exp}$ is an experimental property different from the modeled property [46, 47, 50–52]. The passage from molecular connectivity indices to molecular connectivity terms represents a direct, easy, and general scheme to predict a wide variety of properties of interest for different classes of compounds. The medium-sized set of molecular connectivity indices $\{\chi\}$ is composed of the following eight molecular descriptors:

$$\{\chi\} = \left\{D, D^v, {}^0\chi, {}^0\chi^v, {}^1\chi, {}^1\chi^v, \chi_t, \chi_t^v\right\} \tag{12}$$

where the useful index χ_t was introduced by Needham et al. [53] and represents the total structure molecular connectivity index of a chemical graph over all n vertices:

$$\chi_t = (\delta_1\delta_2 \ldots \delta_n)^{-0.5} \tag{13}$$

By replacing δ_i with δ_i^v, the corresponding total valence molecular connectivity index (χ_t^v) is obtained. The sum delta (D) and valence sum delta (D^v) molecular connectivity indices are defined as follows, with the sum running over the vertices of the chemical graph:

$$D = \sum_{i=1} \delta_i \qquad D^v = \sum_{i=1} \delta_i^v \tag{14}$$

When considering outliers, these are often a result of association phenomena, either self-association or association with other types of molecules through different levels of noncovalent interactions (including hydrogen-bond interactions). The manifold of effects appear in classes made up of saturated, unsaturated, nonsubstituted, substituted, highly substituted, nonpolar, slightly polar, and highly polar compounds. Thus, one would try to give outliers different weights, that is to say, adjust them with different types of parameters. For this reason, the following medium-sized set of supraconnectivity indices (supraindices) was introduced, apart from the one given by Eq. 12:

$$\{(a\chi)^p\} = \left\{(aD)^p, (aD^v)^p, (a^0\chi)^p, (a^0\chi^v)^p, (a^1\chi)^p, (a^1\chi^v)^p, (a\chi_t)^p, (a\chi_t^v)^p\right\} \tag{15}$$

Here, a is an association factor that multiplies or divides the χ indices [41, 46, 49, 54]. The empirical parameter a can be the dielectric constant (ε), an ε-related parameter which can describe hydrogen bonds, and M. The dielectric constant has been selected to improve the modeling, since it is related to the noncovalent character of a compound and it is usually used to mimic the solvent behavior. For a class of highly heterogeneous solvents, it has been found that the best "ad hoc" ε-related parameters are $a_w \approx \varepsilon/15$, a_{OH}, and a_ε. When $\varepsilon/15 < 1$, $a_w = 1$ is assumed. The number 15 has been chosen as it represents the molar mass of a CH_3 radical. Hydrogen bonds in alcohols and acids contribute $a_w = 2$ whatever the value of $\varepsilon/15$, but for compounds with medium dielectric constant, like ethylene carbonate, $a_w = 3$ is preferred. Compounds with a very high ε value, like formamide, have $a_w = 7$ and the contribution due to the hydrogen bond is neglected, while for compounds with quite low ε value, like morpholine, $a_w = 1$ is preferred. The second ε-related parameter is $a_{OH} = 2 + \varepsilon/15$. When the number of alcohols in the data set is rather low, then $a_\varepsilon = a_w = \varepsilon/15$ is used instead of $a_w = 2$. In several studies it has been shown that the employment of the parameter $p = -1$

or $p = 2$ leads to an optimal model for some properties. For the case of χ_t and χ_t^v, these indices have to be divided by a, since their values decrease with increasing complexity of the chemical graph.

Assuming that the relationship between the physicochemical property and the subset of molecular connectivity indices $\{\chi\}$ or terms $\{X\}$ is linear, then the optimization problem to solve is:

$$P = |C.X| \tag{16}$$

where $P = [P_1, ..., P_n]$ is the column vector for the predicted property and $C = [c_1, ..., c_d]$ is the row vector of coefficients which are determined by the linear least-squares procedure. X is a $d \times n$ matrix that includes χ or X terms, and d is the number of variables involved. The bars in Eq. 16 stand for absolute values, to get rid of negative values with no physical meaning and simultaneously enhance the description of the property [49, 50].

In general, the molecular connectivity terms are obtained by a trial-and-error composition numerical procedure which is rather straightforward [46, 47, 51, 52]. Obviously, the fewer χ or X terms to be tried, the easier the trial-and-error search. Experience has shown that the most general form for X terms is:

$$X(\chi_i, \chi_j, \chi_k, \chi_l) = \frac{(\chi_i + b\chi_j)^p}{(c\chi_k + d\chi_l)^q} \tag{17}$$

where b, c, d, p, and q are optimization parameters. χ_i, χ_j, χ_k, and χ_l are normally taken from the set of Eq. 12, but they can also be taken from the set of Eq. 15. Kier and Hall have already suggested that the composition of χ indices into a single descriptor can give rise to improved descriptors [28]. The quality of the different combinations of indices is mainly controlled with two statistical parameters: (1) the quality factor, $Q = R/S$, and (2) the variance F (Fischer) ratio. In this way, the modeling is taken to be optimal when Q reaches a maximum together with F. For every LCCI or LCXCT equation, the fractional utility of each i index ($u_i = |c_i/s_i|$, i.e., the inverse of the fractional error), the vector associated to the fractional utilities (u), as well as the average fractional utility of the indices ($\langle u \rangle = (\sum u_i)/d$) of the found linear combinations have to be estimated [50]. The statistical parameter u_i would allow detection of the paradoxical situation of a LCCI with a good predictive power but with a poor utility at the level of some or all of its coefficients. This paradox can, in part, be removed by resorting to the use of orthogonal molecular connectivity indices.

Molecular connectivity terms can also be orthogonalized, generating orthogonal molecular connectivity terms that (1) reduce the intercorrelation among the X terms, (2) improve $\langle u \rangle$, (3) generate coefficients that are stable upon introduction of a new orthogonal index, and (4) detect dominant descriptors whenever X indices are poor descriptors [55–57]. When properties with negative and antithetical values are considered, like the specific rotations of L- and D-amino acids, Eq. 16 has to be corrected as:

$$P_{lUd} = C_{lUd}.X \quad \text{with} \quad C_d = - C_l \tag{18}$$

Consequently, once a subset is modeled the modeling of the other subset is straightforward.

2.2.1
QSPR Modeling of the Aqueous Solubility of Purine and Pyrimidine Bases

A preceding work [54] on a smaller set of bases, for which there was experimental evidence of association phenomena in solution, was the first in

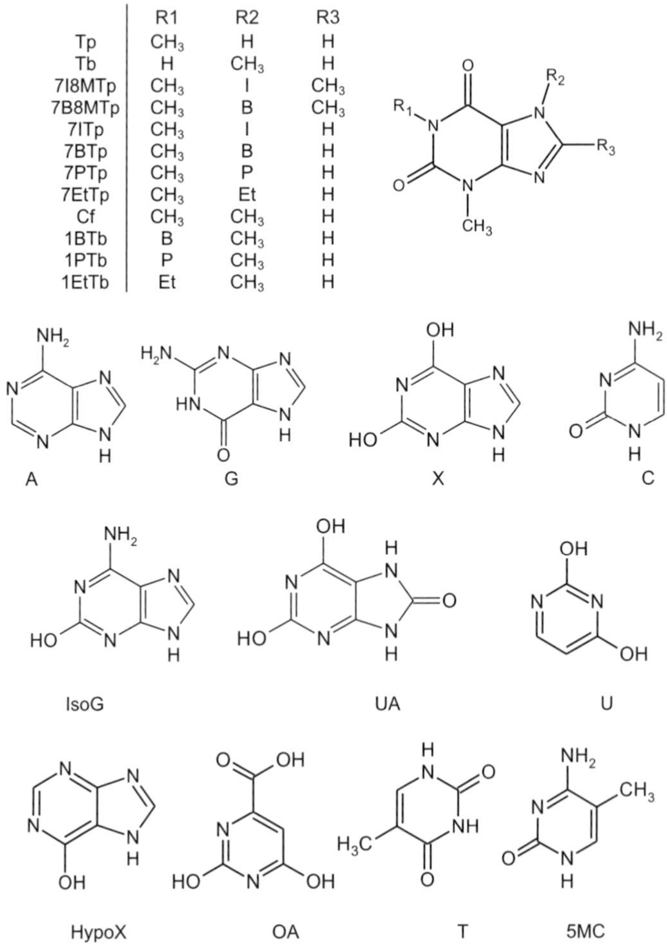

	R1	R2	R3
Tp	CH₃	H	H
Tb	H	CH₃	H
7I8MTp	CH₃	I	CH₃
7B8MTp	CH₃	B	CH₃
7ITp	CH₃	I	H
7BTp	CH₃	B	H
7PTp	CH₃	P	H
7EtTp	CH₃	Et	H
Cf	CH₃	CH₃	H
1BTb	B	CH₃	H
1PTb	P	CH₃	H
1EtTb	Et	CH₃	H

Fig. 3 The purine and pyrimidine bases modeled

Table 4 Experimental solubilities (at the indicated T ($°C$), in units of g.(100 ml of water)$^{-1}$) of 23 purine and pyrimidine bases

PP	Sol	T
7I8MTp	0.63	20
7B8MTp	0.45	20
7ITp	2.7	20
7BTp	0.37	30
1BTb	0.56	30
7PTp	23.11	30
1PTb	1.38	30
7EtTp	3.66	30
1EtTb	3.98	30
Cf	2.58	30
Tp	0.81	30
Tb	0.054	30
UA	0.002	20
OA	0.18	18
X	0.05	20
IsoG	0.006	25
G	0.004	40
HypoX	0.07	19
A	0.09	25
T	0.4	25
5MC	0.45	25
U	0.36	25
C	0.77	25

A = adenine, G = guanine, U = uracil, T = thymine, C = cytosine, OA = orotic acid, UA = uric acid, X = xanthine, M = methyl, Et = ethyl, P = propyl, B = butyl, I = isobutyl, Cf = caffeine, Tb = theobromine, Tp = theophylline

which the supramolecular indices were introduced. The modeling of aqueous solubility (Sol) for the purine and pyrimidine (PP) bases shown in Fig. 3 and Table 4 (see Table 5 for χ and M values) was achieved in 1996 [49], and the association values employed for five of the compounds were a(7PTp) = 4, a(1EtTb, 7EtTp, Cf) = 2, and a(7ITp) = 1.5. Figure 3 includes chemical elements such as boron (B), phosphorus (P), and iodine (I). It is found when using M that it is a very bad descriptor for this particular property of PP. The best single descriptor regression has the following statistics:

$$\left\{ \left(a^1 \chi \right)^2 \right\} : R = 0.9930, \ S = 5.700, \ F = 1553, \ Q = 0.176, \ \langle u \rangle = 22 \qquad (19)$$

Table 5 Connectivity indices values for 23 purines and pyrimidines

PP	D	D^v	$^0\chi$	$^0\chi^v$	$^1\chi$	$^1\chi^v$	χ_t	χ_t^v
7I8MTp	38	62	13.61036	11.38981	8.34111	5.97071	0.003564	8.51×10^{-5}
7B8MTp	38	62	13.44723	11.22667	8.48527	6.11486	0.003086	7.37×10^{-5}
7ITp	36	60	12.74012	10.46716	7.93043	5.53989	0.004365	9.82×10^{-5}
7BTp	36	60	12.57699	10.30402	8.07459	5.68405	0.00378	8.51×10^{-5}
1BTb	36	60	12.57699	10.30402	8.07459	5.68405	0.00378	8.51×10^{-5}
7PTp	34	58	11.86988	9.59691	7.57459	5.18405	0.005346	0.00012
1PTb	34	58	11.86988	9.59692	7.57459	5.18405	0.005346	0.00012
7EtTp	32	56	11.16277	8.88981	7.07459	4.68405	0.00756	0.00017
1EtTb	32	56	11.16277	8.88981	7.07459	4.68405	0.00756	0.00017
Cf	30	54	10.45567	8.1827	6.53658	4.10793	0.01069	0.00024
Tp	28	52	9.58542	7.23549	6.1259	3.71758	0.013095	0.000269
Tb	28	52	9.58542	7.23549	6.10906	3.7135	0.013095	0.000269
UA	26	54	8.71518	5.72474	5.6647	3.11237	0.01604	0.00013
OA	22	50	8.43072	5.24931	5.09222	2.66333	0.03928	0.00027
X	24	48	7.84493	5.34106	5.27086	2.92873	0.01964	0.00034
IsoG	24	46	7.84493	5.45738	5.27086	2.96049	0.01964	0.00043
G	24	46	7.84493	5.45738	5.27086	2.96049	0.01964	0.00043
HypoX	22	42	6.97469	4.95738	4.87701	2.74509	0.02406	0.00085
A	22	40	6.97469	5.07369	4.87701	2.77277	0.02406	0.00108
T	18	36	6.85337	4.89385	4.19838	2.4856	0.06804	0.00301
5MC	18	34	6.85337	5.01016	4.19838	2.51736	0.06804	0.0038
U	16	34	5.98313	3.9712	3.78769	2.06893	0.08333	0.00347
C	16	32	5.98313	4.08751	3.78769	2.1007	0.08333	0.00439

while the best two-descriptor LCCI is:

$$\left\{ \left(a^1\chi\right)^2, \left(\chi_t/a\right)^2 \right\} : R = 0.9970, \ S = 4.200, \tag{20}$$
$$F = 1445, \ Q = 0.240, \ \langle u \rangle = 22$$

Linear combinations with more than two descriptors show an unsatisfactory statistical behavior and a dramatically deteriorating utility vector. Thus, the modeling equation with two variables results in:

$$\text{Sol} = \left| 0.258 \left(a^1\chi\right)^2 + 1179 \left(\chi_t/a\right)^2 - 9.275 \right|$$
$$N = 23, \ R = 0.9970, \ S = 4.200, \ F = 1445 \tag{21}$$

Figure 4 compares the predictions for the 23 aqueous solubilities with the experimental data.

It is quite probable that other bases are expected to undergo some degree of self-association or association with the solvent in solution. An indirect an-

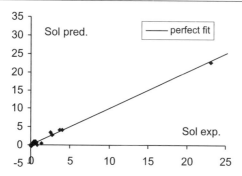

Fig. 4 Experimental and predicted aqueous solubilities for 23 purine and pyrimidine bases

swer to this topic can be given from the following cross-validation procedure: (1) excluding from the modeling the inferred value for 7ITp and modeling the $N = 22$ compounds with the same LCCI, we obtain a satisfactory statistical result: $Q = 0.234$, $F = 1372$, $R = 0.9970$; (2) excluding 7PTp as well, R starts to decrease consistently: $Q = 0.297$, $F = 120$, and $R = 0.9640$; (3) excluding 7ITp, 7PTp, Cf, 7EtTp, and 1EtTb, a poor modeling of the remaining $N = 18$ solubility points is obtained: $Q = 0.204$, $F = 4.8$, $R = 0.6240$. If, instead, we model only the 12 compounds from Tb to C (see Table 4) using the same LCCI, we obtain $Q = 0.896$, $F = 26$, $R = 0.9220$. The modeling of these 12 compounds can be improved further if the index $(\chi_t/a)^2$ is used alone; in fact, in this last case we obtain $Q = 0.944$, $F = 57$, $R = 0.9440$. Such an erratic behavior for the modeling of PP might be explained if it is assumed that more than five purines and pyrimidines undergo, to some extent, association phenomena in solution.

2.2.2
QSPR Modeling of the Aqueous Solubility of a Mixed Class of Purine and Pyrimidine Bases with Some Amino Acids

After the introduction of supramolecular connectivity indices, the modeling of the solubility was attempted and further refined for a mixed class of 43 AA plus PP [46, 47, 51, 52]. The training set of compounds was prepared by combining Tables 4 and 6, and the connectivity indices for the AA are shown in Table 7. The following new set of supraindices is defined:

$$\left\{ aD\chi_t^v, aD^v\chi_t^v, a^0\chi\chi_t^v, a^0\chi^v\chi_t^v, a^1\chi\chi_t^v, a^1\chi^v\chi_t^v, \chi_t a^{-1}, \chi_t^v a^{-1} \right\} \qquad (22)$$

Here, $a = 8$ for Pro, $a = 2$ for Ser, Hyp, and Arg, and $a = 1$ for the other AA. For purines and pyrimidines, instead, we have $a = 4$ for 7PTp, $a = 2$ for 7EtTb, EtTp, and Cf, $a = 1.5$ for 7ITp, and $a = 1$ for the remaining bases. To simplify things, the previous set can be modified as follows:

$$\left\{ {}^D S, {}^D S^v, {}^0 S, {}^0 S^v, {}^1 S, {}^1 S^v, S_t, S_t^v \right\} \qquad (23)$$

Table 6 Experimental values for aqueous solubilities of 20 amino acids (at 25 °C in units of g.(kg water)$^{-1}$)

AA	Sol
Gly	251
Ala	167
Ser	422
Val	58
Thr	97
Met	56
Pro	1622
Leu	23
Ile	34
Asn	25
Asp	5
Lys	6
Hyp	361
Gln	42
Glu	8.6
His	43
Arg	181
Phe	29
Tyr	0.5
Trp	12

When searching for an optimal X term, the overall best one for the 43 mixed compounds is:

$$X_{S3} = \left(^{D}S^{v}\right)^{0.3} - 0.9 \left(^{D}S\right)^{0.3} / \left(S_{t}^{v}\right)^{0.3} + 0.9 \left(S_{t}\right)^{0.3}$$
$$R = 0.9290, S = 97.000, F = 260, Q = 0.0096, \langle u \rangle = 14, u = (16, 13) \tag{24}$$

Descriptor X_{S3} is also good for both subclasses of compounds composed of AA and PP alone, according to the value of S.

AA: $R = 0.996, S = 35.000, F = 2070, Q = 0.029, \langle u \rangle = 38, u = (46, 31)$

PP: $R = 0.899, S = 22.000, F = 88, Q = 0.041, \langle u \rangle = 8.7, u = (9.4, 8.1)$ $\qquad$ (25)

2.2.3
QSPR Study of the Singlet Excitation Energy, Oscillator Strength, and Molar Absorption Coefficient of Five DNA-RNA Bases

In a recent work, five physicochemical properties of DNA-RNA bases were thoroughly analyzed: the first ($\Delta E1$) and second ($\Delta E2$) singlet excitation en-

Table 7 Connectivity indices values for 20 amino acids

AA	D	D^v	$^0\chi$	$^0\chi^v$	$^1\chi$	$^1\chi^v$	χ_t	χ_t^v
Gly	8	20	4.28446	2.63992	2.27006	1.18953	0.40825	0.03727
Ala	10	22	5.1547	3.51016	2.64273	1.62709	0.33333	0.03043
Ser	12	28	5.86181	3.66448	3.18074	1.77422	0.2357	0.00962
Val	14	26	6.73205	5.08751	3.55342	2.53777	0.19245	0.01757
Thr	14	30	6.73205	4.53473	3.55342	2.21862	0.19245	0.00786
Met	16	26.67	7.27602	6.14607	4.18074	4.04355	0.11785	0.01859
Pro	16	28	5.98313	4.55413	3.80453	2.76688	0.08333	0.00932
Leu	16	28	7.43916	5.79462	4.03658	3.02094	0.13608	0.01242
Ile	16	28	7.43916	5.79462	4.09142	3.07578	0.13608	0.01242
Asn	16	36	7.43916	4.70278	4.03658	2.30434	0.13608	0.00254
Asp	16	38	7.43916	4.57273	4.03658	2.23927	0.13608	0.00196
Lys	18	32	7.98313	5.91594	4.68074	3.36624	0.08333	0.00439
Hyp	18	34	6.85337	4.87159	4.19838	2.84158	0.06804	0.0034
Gln	18	38	8.14627	5.40997	4.53658	2.80434	0.09623	0.00179
Glu	18	40	8.14627	5.27984	4.53658	2.73927	0.09623	0.00139
His	22	42	8.26758	5.81918	5.19838	3.15529	0.03402	0.0008
Arg	22	42	9.56048	6.70883	5.53658	3.60022	0.04811	0.00078
Phe	24	42	8.97469	6.60402	5.69838	3.72222	0.02406	0.00069
Tyr	26	48	9.84493	6.97388	6.09222	3.85651	0.01964	0.00027
Trp	32	54	10.8365	8.10402	7.18154	4.71624	0.00567	0.00009

ergies, the first ($f1$) and second ($f2$) oscillator strengths of the first singlet excitation energy, and the molar absorption coefficient at 260 nm and pH = 7 (ε_{260}) (see Table 8) [47, 49, 51, 52]. The simulation of ε_{260} for nucleotides such as UMP, TMP, AMP, GMP, and CMP is carried out using the connectivity indices of U, T, A, G, and C only, simply because the only uncommon part of the nucleotides are these bases. The different kinds of optimal terms discovered are better molecular descriptors than M.

Up to now, the best X terms found by the trial-and-error procedure for f_1, f_2, and ε_{260} are:

$$X_{f_1} = {}^1\chi^v / \left({}^0\chi + 0.6\,{}^0\chi^v\right)$$
$$X_{f_2} = \chi_t^v / \left(\chi_t - 16\,{}^0\chi_t^v\right) \tag{26}$$
$$X_\varepsilon = {}^1\chi^v / \left({}^0\chi + 2\,{}^0\chi^v\right)$$

f_1: $R = 0.8900$, $S = 0.030$, $F = 12$, $Q = 29.7$, $\langle u \rangle = 3.1$

f_2: $R = 0.9200$, $S = 0.080$, $F = 17$, $Q = 11.5$, $\langle u \rangle = 3.3$

ε_{260}: $R = 0.9360$, $S = 1.200$, $F = 21$, $Q = 0.78$, $\langle u \rangle = 4.2$

Table 8 Experimental data of the nucleotide DNA-RNA bases [58]

Bases	$\varepsilon_{260}/1000$	$\Delta E1$	$\Delta E2$	$f1$	$f2$
A	15.4	4.75	5.99	0.28	0.54
G	11.7	4.49	5.03	0.2	0.27
U	9.9	4.81	6.11	0.18	0.3
T	9.2	4.67	5.94	0.18	0.37
C	7.5	4.61	6.26	0.13	0.72

The modeling of the singlet excitation energies $\Delta E1$ and $\Delta E2$ can be achieved with a unique term, behaving as an efficient descriptor for both properties:

$$X_{\Delta E} = \left[{}^0\chi / \left(\chi_t + 10^3 \chi_t^v \right) \right]^5 \tag{27}$$

$\Delta E1$: $R = 0.7900$, $S = 0.100$, $F = 5$, $Q = 7.9$, $\langle u \rangle = 55$

$\Delta E2$: $R = 0.9700$, $S = 0.100$, $F = 44$, $Q = 9.7$, $\langle u \rangle = 46$

As a final conclusion, the five properties can be described by a formally similar molecular connectivity term, while the two energies are modeled by the same term which is highly dependent on total molecular connectivity indices. Furthermore, every term is a mixing of χ and χ^v types of indices, and thus every property is then described as a function of δ and δ^v, in agreement with quantum chemistry [58].

2.3
Flexible Descriptors Obtained via Optimization of Correlation Weights of Local Graph Invariants

A somewhat different scheme for the design of optimal descriptors that is based on the hydrogen-filled graph (HFG) was suggested in [59]. According to the optimization of the correlation weights of local graphs invariants procedure (OCWLGI), a flexible descriptor results from the summation of the so-called correlation weights (CW), that is to say, special coefficients which produce the maximum R between a property/activity considered and the descriptor calculated with these CW. Each graph vertex v_k has two associated types of weights: first, weights for a given chemical element (that is, image of v_k in the graph, for example H, C, O, etc.); and second, weights depending on the numerical value of the local invariant calculated in v_k (i.e., 1, 2, 3, etc.). First of all, a general functional form of the descriptor has to be defined; one such expression is:

$$DCW_{HFG}\left(v_k, LGI_k\right) = \sum_{k=1}^{n} \left[CW(v_k) + CW\left(LGI_k\right) \right] \tag{28}$$

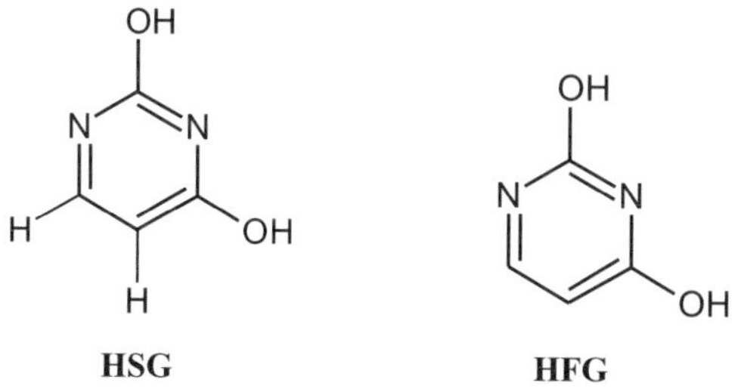

Fig. 5 Hydrogen-suppressed graph and hydrogen-filled graph for uracil

Table 9 Groups of atomic orbitals on some chemical elements

Chemical elements	Atomic orbitals
H	$1s^1$
C	$1s^2, 2s^2, 2p^2$
N	$1s^2, 2s^2, 2p^3$
O	$1s^2, 2s^2, 2p^4$

where v_k represents the chemical element in the HFG, LGI_k is the local graph invariant calculated at the v_k position, and $CW(v_k)$ and $CW(LGI_k)$ are the correlation weights of the mentioned graph invariants calculated with the Monte Carlo optimization method. These types of flexible molecular descriptors can also be based on the so-called graph of atomic orbitals (GAO) [60], with the idea of taking into account the structure of the atoms. An easy way to translate the HFG into the GAO is by means of the following steps:

1. Each atomic vertex in the HFG is replaced with a group of vertices of atomic orbitals (AO).
2. Two AO are defined as connected if, and only if, (a) they fall into groups of different atoms in the HFG and (b) these atoms are already connected in the HFG.

Figure 5 shows the HSG and HFG for uracil, while the GAO representation of the compound appears in Fig. 6. Table 9 indicates the different AO groups that are present in the heterocyclic molecule.

2.3.1
QSAR Study of Dihydrofolate Reductase Inhibition with Models Based on HFGs

An illustrative application of the OCWLGI procedure based on HFG is exemplified by the modeling of the inhibition of dihydrofolate reductase activity by 68 2,4-diamino-5-(substituted benzyl) pyrimidines [61]. Dihydrofolate reductase has been suggested as a target drug for malaria and cancer. The base

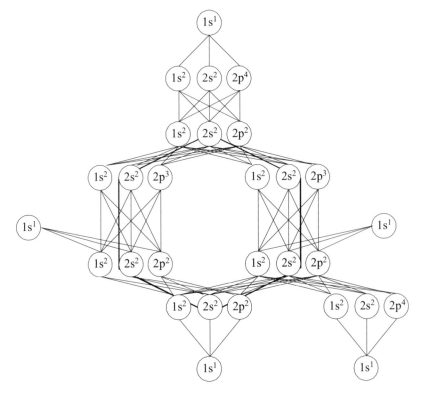

Fig. 6 Graph of atomic orbitals for uracil

structure for all the compounds is shown in Fig. 7, while the experimental activities for all the structures are listed in Table 10. In the present study, the LGI_k analyzed for designing the flexible descriptor are the nearest neighboring codes on v_k (NNC_k) [62] and the Morgan-extended connectivity indices on v_k of zero (^0EC_k), first (^1EC_k), and second (^2EC_k) order [63]. NNC_k is a function of the atomic composition of the vertex neighbors, and can be calculated as:

$$NNC_k = 100\,^tN_k + 10\,^CN_k + \,^HN_k \tag{29}$$

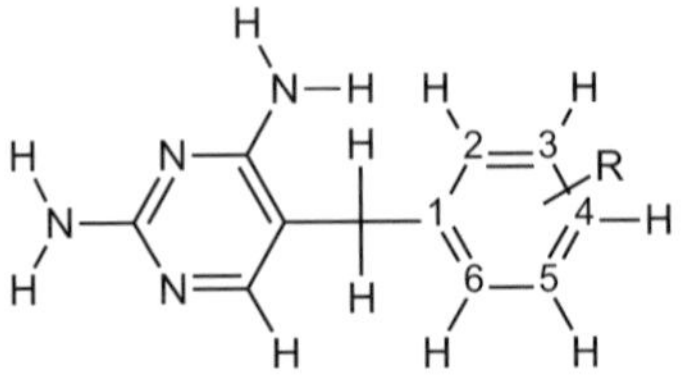

Fig. 7 Base structure for inhibitors of dihydrofolate reductase

Table 10 Experimental and predicted dihydrofolate reductase inhibitory activities for the training and test series

No.	R	DCW	Exp.	Pred.
Training set				
1	$4 - O(CH_2)_6CH_3$	34.817	6.100	6.351
2	$3 - F$	34.096	6.230	6.277
3	$3 - CH_2OH$	34.515	6.280	6.320
4	$4 - NH_2$	34.535	6.300	6.322
5	$3 - O(CH_2)_6CH_3$	34.817	6.390	6.351
6	$3 - OH$	35.553	6.470	6.427
7	$4 - CH_3$	39.690	6.480	6.853
8	$3 - OCH_2CONH_2$	36.885	6.570	6.564
9	$4 - OCF_3$	37.042	6.570	6.580
10	$3 - CH_2OCH_3$	38.728	6.590	6.754
11	$4 - OSO_2CH_3$	38.895	6.600	6.771
12	$3 - Cl$	37.792	6.650	6.658
13	$4 - N(CH_3)_2$	37.791	6.780	6.657
14	$3 - O(CH_2)_3CH_3$	38.048	6.820	6.684
15	$4 - Br$	40.312	6.820	6.917
16	$3 - O(CH_2)_5CH_3$	35.894	6.860	6.462
17	$4 - NHCOCH_3$	40.227	6.890	6.908
18	$4 - OCH_2C_6H_5$	39.902	6.890	6.875
19	$3 - OSO_2CH_3$	38.895	6.920	6.771
20	$4 - C_6H_5$	39.391	6.930	6.822
21	$3 - CF_3$	41.345	7.020	7.024
22	$3, 4 - OCH_2O$	42.467	7.130	7.139
23	$3, 5 - (OCH_3) - 2, 4 - O(CH_2)_7CH_3$	45.254	7.200	7.426
24	$3, 4 - (OCH_2CH_2OCH_3)_2$	44.469	7.220	7.345
25	$3 - I$	43.526	7.230	7.248
26	$3 - OCH_2CH_3, 4 - OCH_2C_6H_5$	46.094	7.350	7.513
27	$3, 5 - (OCH_2CH_3)_2, 4\text{-pyrryl}$	47.609	7.660	7.669
28	$3, 5 - (OCH_3)_2, 4 - N(CH_3)_2$	49.305	7.710	7.843
29	$3 - OCH_3, 4 - OCH_2CH_2OCH_3$	44.997	7.770	7.400
30	$3, 4, 5 - (CH_2CH_3)_3$	47.820	7.820	7.690
31	$3, 4, 5 - (OCH_3)_3$	51.280	8.080	8.047
32	$3, 5 - (OCH_3)_2, 4 - C(CH_3) = CH_2$	52.088	8.120	8.130
33	$3, 5 - (OCH_3)_2, 4 - Br$	51.825	8.180	8.103

(continued on next page)

where ^tN_k is the total number of neighbors of v_k in the HFG, ^CN_k is the number of neighbors of v_k that are images of carbon atoms, and ^HN_k is the number of neighbors of v_k that are images of hydrogen atoms. The Morgan-

Table 10 (continued)

No.	R	DCW	Exp.	Pred.
Test set				
1	$4 - O(CH_2)_5CH_3$	35.894	6.070	6.462
2	$- H$	34.010	6.180	6.268
3	$4 - NO_2$	36.219	6.200	6.496
4	$3 - O(CH_2)_7CH_3$	33.741	6.250	6.240
5	$3, 5 - (CH_2OH)_2$	35.020	6.310	6.372
6	$4 - F$	34.096	6.350	6.277
7	$4 - OCH_2CH_2OCH_3$	39.240	6.400	6.807
8	$4 - OH$	35.553	6.450	6.427
9	$4 - Cl$	37.792	6.450	6.658
10	$3, 4 - (OH)_2$	37.097	6.460	6.586
11	$3 - O(CH_2)_2OCH_3$	39.240	6.530	6.807
12	$3 - CH_2O(CH_2)_3CH_3$	37.009	6.550	6.577
13	$3 - CH_3$	39.690	6.700	6.853
14	$4 - OCH_3$	39.767	6.820	6.861
15	$3 - (OH), 4 - OCH_3$	41.310	6.840	7.020
16	$4 - O(CH_2)_3CH_3$	38.048	6.890	6.684
17	$3 - OCH_3$	39.767	6.930	6.861
18	$3 - Br$	40.312	6.960	6.917
19	$3 - NO_2, 4 - NHCOCH_3$	41.722	6.970	7.062
20	$3 - OCH_2C_6H_5$	39.902	6.990	6.875
21	$3, 5 - (CH_3)_2$	45.370	7.040	7.438
22	$3 - O(CH_2)_7CH_3, 4 - OCH_3$	39.497	7.160	6.833
23	$3, 5 - (OC_3H_7)_2$	44.239	7.410	7.322
24	$3 - OCH_3, 4 - OCH_2C_6H_5$	45.659	7.530	7.468
25	$3 - OCH_3, 4 - OH$	41.310	7.540	7.020
26	$3 - OCH_2C_6H_5, 4 - OCH_3$	45.659	7.660	7.468
27	$3, 5 - (OCH_2CH_3)_2$	46.393	7.690	7.543
28	$3 - OC_2H_5, 4 - OC_3H_7$	45.316	7.690	7.433
29	$3 - CF_3, 4 - OCH_3$	47.102	7.690	7.617
30	$3, 5 - (OCH_3)_2$	45.524	7.710	7.454
31	$3, 4 - (OCH_3)_2$	45.524	7.720	7.454
32	$3 - OSO_2CH_3, 4 - OCH_3$	44.652	7.800	7.364
33	$3 - OCH_3, 4 - OSO_2CH_3$	44.652	7.940	7.364
34	$3, 5 - (OCH_3)_2, 4 - SCH_3$	48.572	8.070	7.768
35	$3, 5 - (OCH_3)_2, 4 - O(CH_2)_2OCH_3$	50.753	8.350	7.993

extended connectivity of order m on v_k in the HFG is given as:

$$^{m}EC_k = \sum_{k,j}^{(m-1)} EC_j \tag{30}$$

Here, k,j represents an edge from the HFG, and it has to be noted that ^0EC_k coincides with deg_k.

A mathematical expression that defines the flexible variable was chosen as:

$$DCW_{HFG}\left(v_k, LGI_k\right) = \sum_{k=1}^{n}\left[CW\left(v_k\right)CW\left(LGI_k\right)\right] \tag{31}$$

Table 11 Correlation weights of local graph invariants for calculating the DCW_{HFG} (v_k, NNC_k)

LGI_k	CW_k
Chemical elements	
C	1.007
N	– 0.780
H	– 0.635
O	1.346
F	– 2.563
S	0.031
Cl	8.246
Br	15.612
I	25.012
NNC_k	
100	0.702
110	0.342
210	0.109
211	0.763
220	1.967
300	32.088
310	0.824
311	0.329
312	– 0.582
320	1.346
321	0.606
330	1.891
400	2.038
402	1.66
403	2.779
410	8.395
412	– 1.145
413	4.787
422	– 0.638
431	0.404

Table 12 Statistical characteristics for the models under consideration

Training set ($N = 33$)			Test set ($N = 35$)			Total set ($N = 68$)		
R	S	F	R	S	F	R	S	F
DCW_{HFG} (v_k, ^0EC_k), total number of optimization parameters is equal to 13								
0.6704	0.427	25	0.6248	0.506	21	0.6380	0.466	45
DCW_{HFG} (v_k, ^1EC_k), total number of optimization parameters is equal to 19								
0.9276	0.215	191	0.8474	0.357	84	0.8771	0.295	220
DCW_{HFG} (v_k, ^2EC_k), total number of optimization parameters is equal to 35								
0.9863	0.095	1106	0.8238	0.378	70	0.8938	0.277	262
DCW_{HFG} (v_k, NNC_k), total number of optimization parameters is equal to 29								
0.9630	0.155	395	0.9371	0.259	238	0.9410	0.213	511

The CW of LGI_k that maximize the correlation with the experimental activities were calculated with the Monte Carlo procedure and appear in Table 11. Table 12 displays the quality achieved for the different models when predicting the inhibitory activities (expressed as $\log(1/K)$) using the three LGI_k mentioned. In order to analyze the predictive ability of the molecular descriptors so defined, the total set of compounds was partitioned into a calibration and external test set. For doing this, the separation of compounds was carried out in such a manner that all local invariants which take place in all graphs under consideration are present in the training set. One can see that the best performance of the OCWLGI models have been obtained through DCW_{HFG} (v_k, NNC_k). The resulting equation is:

$$\log(1/K) = 0.103 \, DCW_{HFG} \, (v_k, NNC_k) + 2.765 \qquad (32)$$

$$N = 33, \quad R = 0.9630, \quad S = 0.155, \quad F = 395 \quad \text{(Training set)}$$
$$N = 35, \quad R = 0.9371, \quad S = 0.259, \quad F = 238 \quad \text{(Test set)}$$
$$N = 68, \quad R = 0.9410, \quad S = 0.213, \quad F = 511 \quad \text{(Total set)}$$

The best five-variable model of dihydrofolate reductase proposed in [64] for the 68 pyrimidine derivatives is statistically characterized by $R = 0.9300$, thus revealing the improvement that the relationship in Eq. 32 represents.

2.3.2
Comparison of QSARs of Anti-HIV-1 Potencies Based on HFG and GAO

There are at least two distinct classifications of human immunodeficiency virus, designated as HIV-1 and HIV-2, which differ basically in the nucleotide and AA sequences. Both of them can lead to the acquired immunodeficiency syndrome, although HIV-1 is predominant in this respect [65]. Reference [66] employed the OCWLGI technique to establish a QSAR model for 57 anti-

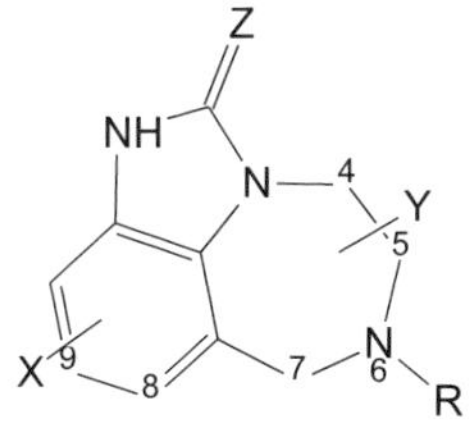

Fig. 8 TIBO and HEPT derivatives studied

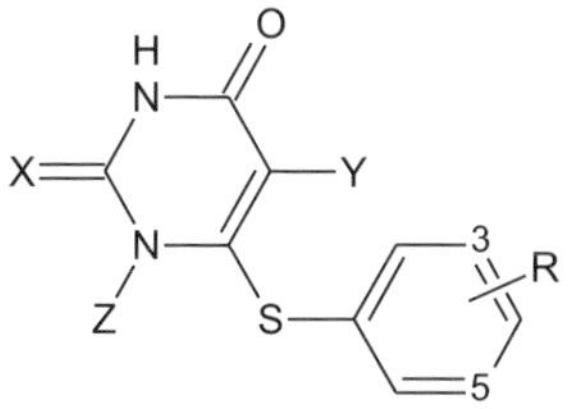

HIV-1 potencies of the two groups of reverse transcriptase inhibitors whose base structure is shown in Fig. 8 and listed in Table 13, commonly known as TIBO and HEPT derivatives. The activities are measured by the concentration of compound required to achieve 50% protection for MT-4 cells against the virus; for modeling reasons it is expressed as $\log(10^6 C_{50}^{-1})$. In the present analysis, both types of graphs based on HFG and GAO are used, including the Morgan-extended connectivities ^0EC_k, ^1EC_k, and ^2EC_k introduced previously. The flexible variable is defined as:

$$\mathrm{DCW}_{\mathrm{HFG/GAO}}\,(v_k, \mathrm{LGI}_k) = \sum_{k=1}^{n}[\mathrm{CW}(v_k) + \mathrm{CW}(\mathrm{LGI}_k)] \qquad (33)$$

In this equation, v_k is the image of a given chemical element in the HFG, and it also represents an image for a set of AO belonging to the element once considering the GAO representation. When partitioning the complete set of compounds into training and test series, Table 14 reveals that the best model incorporates $\mathrm{DCW}_{\mathrm{GAO}}\,(v_k, {}^0EC_k)$ as the optimal variable, producing the following model:

$$\log(10^6 C_{50}^{-1}) = 0.602\,\mathrm{DCW}_{\mathrm{GAO}}\,(v_k, {}^0EC_k) - 7.421 \qquad (34)$$

$N = 37$,	$R = 0.9437$,	$S = 0.508$,	$F = 285$	(Training set)
$N = 20$,	$R = 0.9413$,	$S = 0.540$,	$F = 140$	(Test set)
$N = 57$,	$R = 0.9409$,	$S = 0.515$,	$F = 425$	(Total set)

Table 15 indicates the CW_k for the calculation of $\mathrm{DCW}_{\mathrm{GAO}}\,(v_k, {}^0EC_k)$. It was reported [65] that the general interaction properties function approach (GIPF), which relies on quantum chemical parameters, was used for developing four-variable QSAR for the anti-HIV-1 potencies. However, the statistic obtained with this technique was $R = 0.9300$, $S = 0.597$ on the TIBO derivatives, and $R = 0.9390$, $S = 0.404$ on the HEPT derivatives. In other words, the results are practically equivalent.

Table 13 Training and test sets for TIBO and HEPT derivatives

No.	X	Y	Z	R	Type	DCW	Exp.	Pred.
Training set ($N = 37$)								
1	$8 - Br$	$5 - CH_3$	S	DMA[a]	TIBO	25.612	8.52	7.99
2	$8 - F$	$5 - CH_3$	S	DMA	TIBO	26.283	8.24	8.393
3	$8 - Cl$	$7 - CH_3$	S	DMA	TIBO	25.15	7.92	7.712
4	$8 - CH_3$	$5 - CH_3$	S	DMA	TIBO	24.513	7.87	7.328
5	$9 - Cl$	$5 - CH_3$	S	DMA	TIBO	25.15	7.47	7.712
6	$8 - Cl$	H	S	DMA	TIBO	24.502	7.34	7.322
7	$8 - I$	$5 - CH_3$	S	DMA	TIBO	25.393	7.32	7.858
8	$8 - CN$	$5 - CH_3$	S	DMA	TIBO	23.803	7.25	6.901
9	$8 - I$	$5 - CH_3$	O	DMA	TIBO	23.758	7.06	6.874
10	H	$5 - CH_3$	O	DMA	TIBO	22.329	7.01	6.014
11	$9 - Cl$	$7 - CH_3$	O	DMA	TIBO	23.614	6.8	6.788
12	$9 - CF_3$	$5 - CH_3$	S	DMA	TIBO	22.695	6.31	6.235
13	$8 - CH_3$	$5 - CH_3$	O	DMA	TIBO	22.977	6	6.404
14	$8 - CN$	$5 - CH_3$	O	DMA	TIBO	22.486	5.94	6.109
15	$9 - NO_2$	$5 - CH_3$	S	CPM [a]	TIBO	21.403	5.61	5.457
16	$10 - OCH_3$	$5 - CH_3$	S	DMA	TIBO	21.338	5.33	5.418
17	$9 - CF_3$	$5 - CH_3$	O	DMA	TIBO	21.159	5.23	5.31
18	H	$7 - CH_3$	O	DMA	TIBO	22.329	4.92	6.014
19	H	$5 - CH_3$	O	$CH_2C(C_2H_5) = CH_2$	TIBO	19.905	4.43	4.556
20	H	$5 - CH_3$	O	$CH_2CH_2CH = CH_2$	TIBO	19.257	4.3	4.166
21	H	$5 - CH_3$	O	C_3H_7	TIBO	18.542	4.22	3.736
22	H	$5 - CH_3$	O	$CH_2CH = CH_2$	TIBO	18.738	4.15	3.854
23	H	$4 - CH(CH_3)_2$	O	C_3H_7	TIBO	20.255	4.13	4.766
24	$8 - NH_2$	$5 - CH_3$	O	CPM	TIBO	17.894	3.07	3.346
25	O	C_2H_5	C_6H_5	$3,5 - (CH_3)_2$	HEPT	27.247	8.62	8.974
26	O	$CH(CH_3)_2$	CH_2OH	$3,5 - (CH_3)_2$	HEPT	25.418	8.48	7.873
27	O	C_2H_5	C_6H_5	H	HEPT	25.951	8.31	8.194
28	S	C_2H_5	CH_3	$3,5 - (CH_3)_2$	HEPT	26.433	8.25	8.484
29	O	C_2H_5	CH_3	$3,5 - (CH_3)_2$	HEPT	24.897	8.21	7.56
30	O	$CH(CH_3)_2$	CH_3	H	HEPT	24.795	8.09	7.498
31	S	$CH(CH_3)_2$	CH_3	H	HEPT	26.331	7.92	8.422
32	O	C_2H_5	CH_3	H	HEPT	23.601	7.66	6.78
33	S	C_2H_5	CH_3	H	HEPT	25.137	7.59	7.704
34	O	C_2H_5	CH_2OH	H	HEPT	22.928	6.92	6.375
35	O	CH_3	CH_2OH	$3,5 - (CH_3)_2$	HEPT	23.705	6.59	6.842
36	O	CH_3	CH_2OH	$3-CH_3$	HEPT	23.057	5.59	6.452
37	O	CH_3	C_2H_5	H	HEPT	23.601	5.52	6.78

(continued on next page)

Table 13 (continued)

No.	X	Y	Z	R	Type	DCW	Exp.	Pred.
Test set ($N = 20$)								
1	$8 - Cl$	$5 - CH_3$	S	DMA	TIBO	25.15	8.37	7.712
2	$9 - F$	$5 - CH_3$	S	DMA	TIBO	26.283	7.6	8.393
3	H	$5,7 - CH_3$	S	DMA	TIBO	24.513	7.38	7.328
4	H	$5 - CH_3$	S	DMA	TIBO	23.865	7.36	6.939
5	$8 - Br$	$5 - CH_3$	O	DMA	TIBO	24.076	7.33	7.066
6	H	$7 - CH_3$	S	DMA	TIBO	23.865	7.11	6.939
7	$8 - Cl$	$7 - CH_3$	O	DMA	TIBO	23.614	6.84	6.788
8	$9 - Cl$	H	S	DMA	TIBO	24.502	6.8	7.322
9	H	$7 - CH_3$	S	C_3H_7	TIBO	20.078	5.61	4.66
10	H	$5 - CH_3$	O	DMA	TIBO	22.329	5.48	6.014
11	$10 - OCH_3$	$5 - CH_3$	O	DMA	TIBO	19.802	5.18	4.494
12	$9 - NO_2$	$5 - CH_3$	O	CPM	TIBO	19.867	4.48	4.533
13	H	$5 - CH_3$	O	$CH_2(CH_2)_2CH_3$	TIBO	19.061	4	4.048
14	H	$5 - CH_3$	O	$CH_2CO(O)CH_3$	TIBO	16.516	3.07	2.517
15	O	$CH(CH_3)_2$	C_6H_5	H	HEPT	27.145	8.47	8.912
16	O	C_2H_5	CH_2OH	$3,5 - (CH_3)_2$	HEPT	24.224	7.8	7.155
17	O	$CH(CH_3)_2$	CH_2OH	H	HEPT	24.122	7.14	7.093
18	O	CH_3	C_6H_5	H	HEPT	25.432	7.03	7.881
19	O	CH_3	CH_3	H	HEPT	23.082	6.48	6.467
20	O	CH_3	CH_2OH	H	HEPT	22.409	5.19	6.062

[a] DMA = 3,3-dimethylallyl group; CPM = cyclopropylmethyl group

Table 14 Statistical characteristics for models obtained using different definitions of flexible descriptors

Training set ($N = 37$)			Test set ($N = 20$)			Total set ($N = 47$)		
R	S	F	R	S	F	R	S	F
DCW_{HFG} (v_k, ^0EC_k), total number of optimization parameters is equal to 13								
0.8674	0.765	106	0.9025	0.649	79	0.8781	0.720	185
DCW_{HFG} (v_k, ^1EC_k), total number of optimization parameters is equal to 21								
0.9138	0.624	177	0.8825	0.714	63	0.9023	0.651	241
DCW_{HFG} (v_k, ^2EC_k), total number of optimization parameters is equal to 36								
0.9496	0.482	321	0.9100	0.682	87	0.9325	0.554	367
DCW_{GAO} (v_k, ^0EC_k), total number of optimization parameters is equal to 31								
0.9437	0.508	285	0.9413	0.540	140	0.9409	0.515	425
DCW_{GAO} (v_k, ^1EC_k), total number of optimization parameters is equal to 47								
0.9636	0.411	455	0.9044	0.844	81	0.9309	0.592	357

Table 15 CWLGI on the DCW_{GAO} (v_k, ^0EC_k) optimal descriptor

LGI_k	CW_k
Atomic orbitals	
$1s^1$	0.475
$1s^2$	1.823
$2s^2$	2.071
$2p^2$	0.763
$2p^3$	5.028
$2p^4$	1.064
$2p^5$	3.544
$2p^6$	3.254
$3s^2$	4.518
$3p^4$	1.514
$3p^5$	2.138
$3p^6$	0.763
$3d^{10}$	0.987
$4s^2$	0.528
$4p^5$	0.309
$4p^6$	0.447
$4d^{10}$	0.350
$5s^2$	1.030
$5p^5$	1.243
^0EC_k	
3	2.279
4	1.009
5	0.387
6	0.266
7	0.712
8	0.556
9	0.434
10	0.612
11	2.951
12	4.685
14	0.810
17	0.700

2.3.3
Prediction of Mutagenicity with Models Depending on GAO

Another example employing OCWLGI based on GAO is provided by the modeling of the mutagenicity of 95 heteroaromatic amines, shown in Table 16 [60]. The mutagenic activities of these compounds in *Salmonella typhimurium* TA98 + S9 microsomal preparations are expressed as the mutation rate: ln (revertants mol^{-1}) = ln (R) [67]. In the present study the descriptor adopts the following functional form:

$$DCW_{GAO}(v_k, LGI_k) = \sum_{k=1}^{n} [CW(v_k)CW(LGI_k)] \tag{35}$$

and deg$_k$ is utilized as LGI$_k$; the groups of AO used for converting the HFG into GAO are indicated in Table 17. The best models founds are:

$$\ln(R) = 0.212\, DCW_{GAO}(v_k, deg_k) - 4.577 \tag{36}$$

$N = 47, R = 0.8739, S = 1.050, F = 145$ (Training set)

$N = 48, R = 0.8703, S = 0.860, F = 142$ (Test set)

$N = 95, R = 0.8706, S = 0.950, F = 291$ (Total set)

The performance of this equation compares favorably with the best mutagenicity model obtained by means of the similarity methods reported in [67], with statistics $N = 95, R = 0.8500, S = 1.020$.

2.3.4
Optimal Descriptors Calculated with SMILES

The simplified molecular input line entry system (SMILES) [68–71] is a compact and comfortable representation of the molecular structure from a chemical point of view. An increasing number of SMILES-based databases are gradually appearing on the internet, and thus it is interesting and important to search for suitable ways of using such a representation in QSPR–QSAR analyses. It has to be noted that the molecular graph contains details of the molecular architecture which is absent in SMILES. For instance, an extended connectivity of increasing order cannot be calculated directly from this notation.

High correlations between the octanol/water partition coefficient or aqueous solubility and molecular descriptors calculated with SMILES are reported in [72]; these descriptors are obtained with the so-called LINGO approach. The LINGO is a fragment of SMILES which contains q characters, defined by special rules. It is known that the presence of cycles in the SMILES structure is encoded by digits (1, 2, 3, etc.). However, an interesting feature of LINGO-based models is that information on cycles is neglected; for instance,

Table 16 Training and test sets for the mutagenicity of 95 heteroaromatic amines

No.	Compound name	DCW	Exp.	Pred.
Training set				
1	2-Bromo-7-aminofluorene	32.117	2.620	2.260
2	2-Methoxy-5-methylaniline	10.621	− 2.050	− 2.310
3	4-Ethoxyaniline	9.280	− 2.300	− 2.600
4	4-Aminofluorene	26.326	1.130	1.030
5	7-Aminofluoranthene	36.707	2.880	3.240
6	1,7-Diaminophenazine	26.451	0.750	1.060
7	4-Aminopyrene	36.707	3.160	3.240
8	2,4,5-Trimethylaniline	10.968	− 1.320	− 2.240
9	2,4-Dimethylaniline	8.912	− 2.220	− 2.680
10	3-Aminofluoranthene	36.707	3.310	3.240
11	2-Aminocarbazole	25.700	0.600	0.900
12	2-Hydroxy-7-aminofluorene	26.035	0.410	0.970
13	2,5-Dimethylaniline	8.912	− 2.400	− 2.680
14	2-Amino-4-methylphenol	6.565	− 2.100	− 3.180
15	4-Aminophenylsulfide	23.017	0.310	0.330
16	2,4-Dinitroaniline	12.127	− 2.000	− 1.990
17	2,4-Difluoroaniline	9.334	− 2.700	− 2.590
18	2-Aminofluoranthene	36.707	3.320	3.240
19	2-Amino-3′nitrobiphenyl	20.938	− 0.890	− 0.120
20	4,4′-Ethylenebis(aniline)	13.848	− 2.150	− 1.630
21	2-Aminophenanthrene	26.991	2.460	1.170
22	9-Aminophenanthrene	26.991	2.980	1.170
23	2-Aminopyrene	36.707	3.500	3.240
24	6-Aminoquinoline	14.782	− 2.670	− 1.430
25	1,6-Diaminophenazine	26.451	0.200	1.060
26	4-Aminophenyldisulfide	16.666	− 1.030	− 1.030
27	2,4-Diamino-*n*-butylbenzene	9.662	− 2.700	− 2.520
28	2-Aminobiphenyl	17.275	− 1.490	− 0.900
29	2-Chloroaniline	11.229	− 3.000	− 2.190
30	3-Amino-α,α,α-trifluorotoluene	17.070	− 0.800	− 0.940
31	3-Amino-4′-nitrobiphenyl	20.938	0.690	− 0.120
32	4-Bromoaniline	10.591	− 2.700	− 2.320
33	2-Amino-4-chlorophenol	10.938	− 3.000	− 2.250
34	3,3′-Dimethoxybenzidine	25.466	0.150	0.850
35	4-Cyclohexylaniline	10.565	− 1.240	− 2.330
36	4,4′-Methylene-bis(*o*-ethylaniline)	24.193	− 0.990	0.580
37	1-Amino-4-nitronaphthalene	19.559	− 1.770	− 0.410
38	4-Amino-3′-nitrobiphenyl	20.938	1.020	− 0.120
39	4,4′-Methylene-bis(*o*-fluoroaniline)	23.185	0.230	0.360

(continued on next page)

Table 16 (continued)

No.	Compound name	DCW	Exp.	Pred.
40	3-Aminoquinoline	14.782	− 3.140	− 1.430
41	4-Chloro-1,2-phenylenediamine	11.891	− 0.490	− 2.040
42	3-Aminophenanthrene	26.991	3.770	1.170
43	1-Aminoanthracene	26.991	1.180	1.170
44	1-Aminocarbazole	25.700	− 1.040	0.900
45	4-Aminocarbazole	25.700	− 1.420	0.900
46	4,4′-Methylenebis(o-isopropylaniline)	14.586	− 1.770	− 1.470
47	2,7-Diaminophenazine	26.451	3.970	1.060
Test set				
1	5-Aminoquinoline	14.782	− 2.000	− 1.430
2	1-Aminonaphthalene	15.896	− 0.600	− 1.190
3	2-Aminoanthracene	26.991	2.620	1.170
4	8-Aminoquinoline	14.782	− 1.140	− 1.430
5	2-Aminonaphthalene	15.896	− 0.670	− 1.190
6	3-Amino-3′-nitrobiphenyl	20.938	− 0.550	− 0.120
7	3-Aminofluorene	26.326	0.890	1.030
8	3,3′-Dichlorobenzidine	30.794	0.810	1.980
9	2,7-Diaminofluorene	26.988	0.480	1.170
10	2-Aminofluorene	26.326	1.930	1.030
11	2-Amino-4′-nitrobiphenyl	20.938	− 0.620	− 0.120
12	4-Aminobiphenyl	17.275	− 0.140	− 0.900
13	3-Methoxy-4-methylaniline	10.621	− 1.960	− 2.310
14	2-Amino-5-nitrophenol	8.172	− 2.520	− 2.840
15	2,2′-Diaminobiphenyl	17.937	− 1.520	− 0.760
16	1-Aminophenanthrene	26.991	2.380	1.170
17	4-Amino-2′-nitrobiphenyl	20.938	− 0.920	− 0.120
18	2-Aminophenazine	24.764	0.550	0.700
19	2,4-Diaminoisopropylbenzene	17.100	− 3.000	− 0.930
20	4,4′-Methylenedianiline	18.652	− 1.600	− 0.600
21	3,3′-Dimethylbenzidine	22.049	0.010	0.120
22	1-Aminofluoranthene	36.707	3.350	3.240
23	4-Chloroaniline	11.229	− 2.520	− 2.190
24	4-Fluoroaniline	7.067	− 3.320	− 3.070
25	3,3′-Diaminobiphenyl	17.937	− 1.300	− 0.760
26	2,6-Dichloro-1,4-phenylenediamine	18.320	− 0.690	− 0.670
27	2-Amino-7-acetamidofluorene	32.008	1.180	2.240
28	2,8-Diaminophenazine	26.451	1.120	1.060
29	4-Methoxy-2-methylaniline	10.621	− 3.000	− 2.310
30	3-Amino-2′-nitrobiphenyl	20.938	− 1.300	− 0.120
31	2,4′-Diaminobiphenyl	17.937	− 0.920	− 0.760

(continued on next page)

Table 16 (continued)

No.	Compound name	DCW	Exp.	Pred.
32	2-Bromo-4,6-dinitroaniline	17.918	– 0.540	– 0.760
33	4-Aminophenyl ether	19.646	– 1.140	– 0.390
34	1,9-Diaminophenazine	24.330	0.040	0.610
35	1-Aminofluorene	26.326	0.430	1.030
36	8-Aminofluoranthene	36.707	3.800	3.240
37	2-Amino-1-nitronaphthalene	19.559	– 1.170	– 0.410
38	4-Phenoxyaniline	18.984	0.380	– 0.530
39	2-Amino-7-nitrobiphenyl	29.989	3.000	1.810
40	Benzidine	17.937	– 0.390	– 0.760
41	4-Amino-4-nitrobiphenyl	20.938	1.040	– 0.120
42	1-Aminophenazine	25.789	– 0.010	0.920
43	4-Chloro-2-nitroaniline	14.892	– 2.220	– 1.410
44	3-Aminocarbazole	25.700	– 0.480	0.900
45	3,4′-Diaminobiphenyl	17.937	0.200	– 0.760
46	9-Aminoanthracene	26.991	0.870	1.170
47	6-Aminochrysene	38.086	1.830	3.540
48	1-Aminopyrene	36.707	1.430	3.240

c1ccccc1 and c2ccccc2 become c0ccccc0. This normalization reduces the number of possible LINGOs. Partial least squares (PLS) was employed as a tool for construction of the models, leading to quite good statistics: octanol/water partition coefficient: $N = 12\,831$, $R = 0.9644$, $S = 0.490$ (log unit); aqueous solubility: $N = 1309$, $R = 0.9539$, $S = 0.490$. Reference [73] describes another SMILES-based approach, where the optimal descriptors used are defined as:

$$\mathrm{DCW}_1(a, b, c, m)$$
$$= \left[a\mathrm{CW}(N_\mathrm{O}) + b\mathrm{CW}(N_\mathrm{DB}) + c\mathrm{CW}(\mathrm{d}N) + \sum_{k=1}^{n} \mathrm{CW}(s_k) \right]^m \tag{37}$$

$$\mathrm{DCW}_2(a, b, c, m)$$
$$= \left[a\mathrm{CW}(N_\mathrm{O}) + b\mathrm{CW}(N_\mathrm{DB}) + c\mathrm{CW}(\mathrm{d}N) + \sum_{k=1}^{n-1} \mathrm{CW}(ss_k) \right]^m \tag{38}$$

$$\mathrm{DCW}_3(a, b, c, m)$$
$$= \left[a\mathrm{CW}(N_\mathrm{O}) + b\mathrm{CW}(N_\mathrm{DB}) + c\mathrm{CW}(\mathrm{d}N) + \sum_{k=1}^{n-2} \mathrm{CW}(sss_k) \right]^m \tag{39}$$

Table 17 Groups of AO used for converting HFG into GAO

Atom	Group of atomic orbitals
H	$1s^1$
C	$1s^2, 2s^2, 2p^2$
N	$1s^2, 2s^2, 2p^3$
O	$1s^2, 2s^2, 2p^4$
F	$1s^2, 2s^2, 2p^5$
S	$1s^2, 2s^2, 2p^6, 3s^2, 3p^4$
Cl	$1s^2, 2s^2, 2p^6, 3s^2, 3p^5$
Br	$1s^2, 2s^2, 2p^6, 3s^2, 3p^6, 3d^{10}, 4s^2, 4p^5$

$$DCW_4(a, b, c, m)$$

$$= \left[aCW(N_O) + bCW(N_{DB}) + cCW(dN) + \sum_{k=1}^{n-3} CW(ssss_k) \right]^m \tag{40}$$

where N_O and N_{DB} denote the number of oxygen atoms and double bonds in the SMILES string, respectively, whereas $CW(N_O)$ and $CW(N_{DB})$ are their associated CW. Further, $dN = N_O - N_{DB} + 10$ and $CW(dN)$ is its correlation weight; s_k, ss_k, sss_k, and $ssss_k$ are, respectively, the one-, two-, three-, and four-symbol fragments of SMILES and $CW(s_k)$, $CW(ss_k)$, $CW(sss_k)$, $CW(ssss_k)$ are their CW. The numerical values of the CW were calculated as usual with the Monte Carlo optimization procedure. The object function is the correlation coefficient between DCW_i and the normal boiling points (NBP) of a training set of acyclic carbonyl compounds. As a final result one can obtain expressions of the form

$$NBP = C_0 + C_1 DCW_i \quad i = 1, 2, 3, 4 \tag{41}$$

for which the predictive ability must be validated by means of a test set. The best correlation takes place for the case of three-character SMILES fragments.

$$NBP = -193.687 + 94.144 \, DCW_3 \, (1, 1, 1, 0.5) \tag{42}$$

$$N = 100, \quad R = 0.9897, \quad S = 5.350, \quad F = 4673 \quad \text{(Training set)}$$

$$N = 100, \quad R = 0.9881, \quad S = 5.380, \quad F = 4055 \quad \text{(Test set)}$$

It is interesting to note that there is an analogy between increasing orders of the extended connectivities and a gradual increase in the number of SMILES characters in DCW_1, DCW_2, DCW_3, DCW_4.

3
Main Conclusions

Some of the optimal descriptors discussed in the present chapter are based on the additive principle. However, in contrast to the simple additive scheme, the optimized variables are obtained from general features that correspond simultaneously to all vertices of the molecular graph. Hence, uncomfortable situations arise quite frequently, where some increments are unavailable for specific structural fragments when predicting an end point of interest for a substance. Other subsets of optimal descriptors are calculated with special formulae, and some analogies between group contribution methods and such kinds of optimal descriptors can be established. However, in some situations dissimilarities appear, since the group methods can also incorporate additional physical and/or chemical information about the phenomena under modeling. In contrast, flexible descriptors rely solely on structural information as depicted by the QSPR–QSAR theory. An advantage of employing optimal descriptors is that by using extended connectivities of increasing orders as local graph invariants, rational and objective features of the molecular graphs can be rescued during the modeling process. At present, it is known that for some properties an increase of the order in the connectivities produces improvements of the quality of the models for both the training and test sets. It is to be noted, however, that OCWLGI based on a very high order can lead to overtraining, after which the OCWLGI model is characterized with high statistical characteristics for the training set and by unsatisfactory results for the test set.

Regarding the perspectives for optimal descriptors, the general ideas for their design can be improved in the following ways:

1. Involving new local and global topological graph invariants in the scheme of calculation of these descriptors (path of different lengths, valence shells, nearest-neighboring codes of increasing order, etc.)
2. Involving new physical and chemical features of molecular structure in the scheme of calculation of these descriptors (symmetry, chirality, quality and quantity of isomers, dynamic aspects of the molecular behavior, e.g., taking into account probabilities of different deformation of molecular systems [74])
3. Use of nonlinear correlation of optimal descriptors with end points of interest

Optimal descriptors are useful tools for QSPR–QSAR analyses. An advantage of molecular descriptors of such a kind is that they can be used to specifically address a given property/activity in a clear and transparent way. The different numerical values for the contributions of the various molecular fragments, such as extended connectivities of different orders, number of paths of length 2, 3, etc., valence shells, or presence of atomic orbitals of different kinds, can be used in searching for mechanistic interpretations of physical, chemical, and biological phenomena of interest.

References

1. Wiener H (1947) J Am Chem Soc 69:17
2. Wiener H (1947) J Am Chem Soc 69:2636
3. Wiener H (1948) J Phys Chem 52:425
4. Wiener H (1948) J Phys Chem 52:1082
5. Stull DR, Westrum EF Jr, Sinke GC (1969) The chemical thermodynamics of organic compounds. Wiley, New York
6. Charkin OP (1972) Russ Chem Bull (Historical Archive) 21:1386
7. Van Krevelen DW (1990) Properties of polymers: their correlation with chemical structure; their numerical estimation and prediction from additive group contributions. Elsevier, Amsterdam
8. Bicerano J (2002) Prediction of polymer properties, revisited and expanded, 3rd edn. Marcel Dekker, New York
9. Nikiforov VA, Karavan VS, Miltsov SA, Selivanov SL, Kolehmainen E, Wegelius E, Nissinen M (2003) ARKIVOC 6:191
   http://www.arkat-usa.org/ark/journal/2003/I06_Varvoglis/AV-744A/744A.pdf
10. Exner O, Bohm S (2001) Collect Czech Chem Commun 66:1623
11. Krenkel G, Castro EA, Toropov AA (2001) J Mol Struct Theochem 542:107
12. Duchowicz P, Castro EA, Toropov AA (2002) Comput Chem 26:327
13. Verguts T, Ameel E, Storms G (2004) Mem Cognit 32:379
14. Iwai Y, Yamanaga Sh, Arai Y (1999) Fluid Phase Equilib 163:1
15. Mazzobre MF, Roman MV, Morelle AF, Corti HR (2005) Carbohydr Res 340:1207
16. Ghafourian T, Barzegar-Jalali M (2002) Farmaco 57:565
17. Olsen E, Nielsen F (2001) Molecules 6:370 http://www.mdpi.org
18. Ren B (2002) Comput Chem 26:357
19. Thomsen M, Asmussen AG, Carlsen L (1999) Chemosphere 38:2613
20. Joback KG (2001) Fluid Phase Equilib 185:45
21. Casalenglo M, Sello G (2005) J Mol Struct Theochem 727:71
22. Balaban AT (1992) J Chem Inf Comput Sci 32:23
23. Randic M, Basak SC (1999) J Chem Inf Comput Sci 39:261
24. Barysz M, Jashari G, Lall RS, Srivastava VK, Trinajstic N (1983) In: King RB (ed) Chemical applications of topology and graph theory. Elsevier, Amsterdam
25. Ivanciuc O (2000) Rev Roum Chim 45:289
26. Randic M (1975) J Am Chem Soc 97:6609
27. Kier LB, Hall LH, Murray WJ, Randic MJ (1975) Pharm Sci 64:1971
28. Kier LB, Hall LH (1976) Molecular connectivity in chemistry and drug research. Academic, New York
29. Kier LB, Hall LH (1986) Molecular connectivity in structure-activity studies. Research Studies Press, Letchworth
30. Kier LB, Hall LH (1976) J Pharm Sci 65:1806
31. Randic M (1991) Chemometr Intell Lab Syst 10:213
32. Mekenian O, Bonchev D, Balaban AT (1984) Chem Phys Lett 109:85
33. Seybold PG, May M, Bagal UA (1987) J Chem Educ 64:575
34. Randic M, Dobrowolski JC (1998) Int J Quantum Chem 70:1209

35. Randic M, Mills D, Basak SC (2000) Int J Quantum Chem 80:1199
36. Randic M, Basak SC (2000) J Chem Inf Comput Sci 40:899
37. Randic M, Hansen PJ, Jurs PC (1988) J Chem Inf Comput Sci 28:60
38. Hansen PJ, Jurs PC (1988) J Chem Educ 65:574
39. Randic M (1991) J Mol Struct Theochem 233:45
40. Pogliani L (1992) J Pharm Sci 81:334
41. Pogliani L (1993) Comput Chem 17:283
42. Pogliani L (1993) J Phys Chem 97:6731
43. Pogliani L (1994) J Chem Inf Comput Sci 34:801
44. Pogliani L (1994) J Phys Chem 98:1494
45. Pogliani L (1997) Croat Chem Acta 70:803
46. Pogliani L (1997) Amino Acids 13:237
47. Pogliani L (1999) J Mol Struct Theochem 466:1
48. Pogliani L (2000) The concept of graph mass in molecular graph theory. A case in data reduction analysis. Nova, New York
49. Pogliani L (1996) J Chem Inf Comput Sci 36:1082
50. Pogliani L (1996) J Phys Chem 100:18065
51. Pogliani L (1997) Med Chem Res 7:380
52. Pogliani L (1999) J Chem Inf Comput Sci 39:104
53. Needham DE, Wei I, Seybold PG (1988) J Am Chem Soc 110:4186
54. Pogliani L (1995) J Phys Chem 99:925
55. Randic M (1991) New J Chem 15:517
56. Randic M (1991) J Chem Inf Comput Sci 31:311
57. Randic M (1994) Int J Quantum Chem Quantum Biol Symp 21:215
58. Ladik J, Appel K (1966) Theor Chim Acta 4:132
59. Toropov AA, Toropova AP (1998) Russ J Coord Chem 24:81
60. Toropov AA, Toropova AP (2001) J Mol Struct Theochem 538:287
61. Toropov AA, Nesterov IV, Nabiev OM (2003) J Mol Struct Theochem 622:269
62. Toropov AA, Toropova AP (2002) J Mol Struct Theochem 581:11
63. Toropov AA, Toropova AP (2002) J Mol Struct Theochem 578:129
64. Ivanciuc O, Ivanciuc T, Cabrol-Bass D (2002) J Mol Struct Theochem 582:39
65. Ganzalez OG, Murray JS, Peralta-Inga Z, Politzer P (2001) Int J Quantum Chem 83:115
66. Toropov AA, Toropova AP, Nesterov IV, Nabiev OM (2003) J Mol Struct Theochem 640:175
67. Basak SC, Grunwald GD (1995) J Chem Inf Comput Sci 35:366
68. Weininger D (1988) J Chem Inf Comput Sci 28:31
69. Weininger D, Weininger A, Weininger JL (1989) J Chem Inf Comput Sci 29:97
70. Weininger D (1990) J Chem Inf Comput Sci 30:237
71. http://www.daylight.com
72. Vidal D, Thormann M, Pons MJ (2005) J Chem Inf Model 45:386
73. Toropov AA, Toropova AP, Mukhamedzhanova DV, Gutman I (2005) Indian J Chem A 44:1545
74. Toropov AA, Toropova AP, Ismailov TT, Voropaeva NL, Ruban IN, Rashidova SSh (1996) Russ J Phys Chem 70:1081

Top Heterocycl Chem (2006) 3: 39–80
DOI 10.1007/7081_025
© Springer-Verlag Berlin Heidelberg 2006
Published online: 24 March 2006

Predicting Pharmacological and Toxicological Activity of Heterocyclic Compounds Using QSAR and Molecular Modeling

Subhash C. Basak (✉) · Denise Mills · Brian D. Gute · Ramanathan Natarajan

Natural Resources Research Institute, University of Minnesota Duluth,
5013 Miller Trunk Hwy., Duluth, MN 55811, USA
sbasak@nrri.umn.edu

"The most fundamental and lasting objective of synthesis
is not production of new compounds, but production of properties."
George S. Hammond [1]

"Ostensibly there is color, ostensibly odor, ostensibly bitterness,
but actually only atoms and the void"
Galen [2]

"... The ever-renewing force of your lila filled my heart.
I saw it in smiles, at its point of escape into the heart of beauty,
I saw it in shyness, at its point of hesitant switching to delight,
I felt the **Flux of Form**."
Rabindranath Tagore [3]

Abstract Heterocyclic compounds are important as drugs, toxicants, and agrochemicals. In this review, we report the QSAR modeling of pharmacological activity, insect repellency, and environmental toxicity for a few classes of heterocyclics from their structure. The calculated molecular descriptors fall into four classes: topostructural (TS), topochemical (TC), 3-dimensional or geometrical (3D), and quantum chemical (QC). The complexity and the computational time of the four classes of descriptors increase in the order QC > 3D > TC > TS. The HiQSAR approach utilizes these descriptor classes in a graduated manner, beginning with the most easily calculated TS descriptors. Three types of linear models are built, namely, principal components regression (PCR), partial least squares (PLS) regression, and ridge regression (RR), and the models with the best statistics (R_{cv}^2) are reported. With the number of descriptors being well over 300, we have used a descriptor trimming method, a modified Gram–Schmidt orthogonalization, to reduce the number of descriptors used in the linear models. Application of this methodology and the resulting improvement in the final models is discussed. In analyzing polychiral compounds, as in the case of piperidine amide mosquito repellents, we use a hierarchical molecular overlay method developed by us in order to differentiate the activities of the diastereomers. Finally, an important component of the molecular design process is the selection of analogs. We have used calculated molecular descriptors to develop similarity methods which are tailored to a particular property of interest. The results indicate that our tailored approach is superior to the widely used arbitrary molecular similarity approaches both in analog selection and in k-nearest neighbor (KNN) based estimation of properties of chemicals from their selected analogs.

Keywords Arbitrary molecular similarity · Hierarchical molecular overlay · Hierarchical quantitative structure–activity relationship (HiQSAR) · Ridge regression · Tailored molecular similarity

Abbreviations

QSAR	Quantitative structure–activity relationship
HiQSAR	Hierarchical quantitative structure–activity relationship
COX-2	Cyclooxygenase-2
TS	Topostructural
TC	Topochemical
3D	3-Dimensional
QC	Quantum chemical
RR	Ridge regression
PCR	Principal components regression
PLS	Partial least squares
PC	Principal component
PCA	Principal components analysis
OLS	Ordinary least squares
LOO	Leave-one-out
PRESS	Prediction sum of squares
SS_{Total}	Total sum of squares
$|t|$	Absolute value of the descriptor coefficient divided by its standard error
Ah	Aryl hydrocarbon
IC_{50}	Concentration required to achieve 50% inhibition
QSSAR	Quantitative stereochemical structure–activity relationship
MM2	Molecular mechanics force field

AM1   Austin model 1
HF   Hartree–Fock
DFT   Density functional theory
TI   Topological index (includes both topostructural and topochemical indices)
ED   Euclidean distance
kNN k-Nearest neighbors (where k, the number of neighbors selected, can equal any number ≥ 1, but is less than the total number of chemicals in the set)

1
Introduction

Heterocyclic molecules play a crucial role in health care and pharmaceutical drug design. The major constituents of many natural products used in ethnomedicine are heterocyclic chemicals. A review of the major category of drugs in use today would testify that a large number of those used in Western medical practice are heterocyclic molecules [4, 5]. Therefore, medicinal chemists, drug designers, and toxicologists remain keenly interested in the beneficial and deleterious effects of heterocyclic moieties in molecules. Although the major goal of drug research is to find new molecules that can cure diseases not controlled by the repertoire of currently available therapies, the toxic effects of drug molecules and their metabolites are important for determining the side effects of medication. A perusal of the literature will show that many potential drugs with good therapeutic activity never made it to the market because of unwanted toxicity. Many drugs have been withdrawn from the market after being in use for some time when reports of unacceptable side effects emerged from consumers.

2
Why QSAR and Molecular Modeling?

A molecule with a therapeutic activity of interest can be discovered by accidental finding (serendipity), rational design based on the effects of synthesized or natural molecules, side effects of well-established drug molecules, chemical analysis of extracts of substances used in ethnomedical practices, or analysis of samples of flora and fauna aimed at drug discovery, to name just a few methods [6]. A compound with potential therapeutic activity in a particular area is referred to as a "hit". Once a hit is found, the medicinal chemist goes into action and synthesizes a number of derivatives of the hit. Testing of the derivatives of the hit compound shows whether the structural class possesses, in general, the activity of interest. This phase helps us to concentrate on one, or a few, of the most promising compound(s), which are then used for the synthesis and testing of a large number of derivatives of the lead(s).

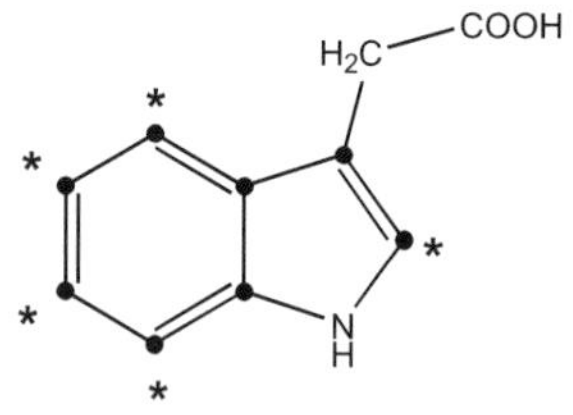

- 50 groups for each aromatic(*)
- 10 groups for esterification
- 10 groups for aliphatic C
- 10 groups for ring N

Total number of analogs
$= 50^5 \times 10 \times 10 \times 10$
$= 312.5$ billion analogs

Fig. 1 Combinatorial design, even for a small molecule, can result in an extremely large number of derivatives

This phase can be expensive, depending on how many derivatives one wishes to test and the cost of the bioassay. Synthesis and testing of hundreds of thousands of derivatives of a lead is not unusual according to the pharmaceutical drug discovery literature [7]. Figure 1 provides a hypothetical example illustrating that combinatorics, applied to even a small molecule, can result in a huge number of derivative compounds.

At the earlier stages of drug research, testing of a limited number of derivatives of the lead could result in fruitful drug discovery. More recently, with the identification of numerous toxicological and ecotoxicological effects of drugs and their metabolites, the cost of drug discovery has been calculated at a staggering $ 802 million [8]. At one time, the chemist would think of a synthesis or substitution scheme as follows: lead... methyl... ethyl... butyl... drug. Today's pessimistic situation is more accurately represented by: lead... methyl... ethyl... butyl... *futile!*

The astronomical cost of new drug discovery is fueled primarily by two factors: (a) Combinatoric explosion in the number of possible derivatives of a lead, and (b) Cost of the screening process. A chemical can be screened by three methods: *in vivo*, *in vitro*, and finally *in silico*. The potential therapeutic activity and toxicity of a chemical have usually been assessed in the pharmacological literature using a set of physicochemical properties and bioassay results. Table 1 gives a partial and ever expanding list of such properties.

Additionally, the Human Genome Project has catapulted us into the era of "omics" sciences, represented by genomics, proteomics, and metabolomics. Data from these sciences are thought to be useful in pharmaceutical drug design [9–12].

While *in vivo* and *in vitro* bioassays are costly, even experimental physicochemical properties such as those listed in Table 1 could be costly if the number of compounds to be evaluated is very large. That is why the recent trend

Table 1 Properties important for assessing therapeutic activity and toxicity

Physicochemical	Pharmacological/toxicological
Molar volume	Macromolecule level
Boiling point	: Receptor binding (KD)
Melting point	: Michaelis constant (Km)
Vapor pressure	: Inhibitor constant (Ki)
Water solubility	: DNA alkylation
Dissociation constant (pK_a)	: Unscheduled DNA synthesis
Partition coefficient	-omics response
: Air-water	: Pharmacogenomics/toxicogenomics
: Sediment-water	: Proteomics
Reactivity (electrophile, nucleophile,	: Metabolomics/metabonomics
or free radical)	Cell level
	: Salmonella mutagenicity
	: Mammalian cell transformation
	Organism level (acute)
	: Algae
	: Invertebrates
	: Fish
	: Birds
	: Mammals
	Organism level (chronic)
	: Bioconcentration
	: Carcinogenicity
	: Reproductive toxicity
	: Delayed neurotoxicity
	: Biodegradation
	Ecosystem level
	: Biodiversity degradation

Reprinted with permission; Source from [15]

is to use properties that are algorithmically derived, i.e., properties that can be calculated directly from molecular structure without the input of any other experimental data. Such properties can be calculated without any associated experimental error, they can be calculated quickly with the ever increasing speed of modern computers, and they can be calculated for any molecular structure, real or hypothetical. In contemporary combinatorial chemistry, large real or virtual libraries are created and evaluated with most of the chemicals having no experimentally determined data. *In silico* methods based on molecular structure alone constitute the first tier of a multitier evaluation of such libraries.

3
Structural Hierarchy: The Flux of Form

In the *in silico* approach, the essential aspects of the molecular structure deemed to be important for the particular bioactivity under investigation are represented by different types of molecular models. The ball and stick model, graph theoretical representations of structure using different types of simple weighted graphs [13–15], the 3D representation of molecules, and finally, the detailed characterization of molecular structure using quantum chemical Hamiltonians are all different distinct representations of molecular structure. When the organic chemist, biochemist, spectroscopist, and quantum chemist use the word "structure", they don't have the same image of molecular reality in mind. Some call this the *molecular structure conundrum* [16]. Corresponding to each representation (*model object*), there could be more than one way of quantifying the molecular structure or extraction of *molecular descriptor(s)* using different mathematical models [14, 17]. As a result, there is no unique way to represent and quantify molecular structure for the prediction of bioactivity or toxicity. In addition, it is not possible to know, *a priori*, which method will work best under a particular situation. Such information can be derived only by empirical means. A scheme of this multifarious mapping of chemical space (a set of chemical structures) into the set of real numbers R using experimental versus theoretical methods is depicted in Fig. 2.

The representation of molecular structure can be looked upon as a hierarchical process similar to the Russian doll, which contains a series of dolls hidden inside it. At the very elementary level, the structure of the model of an assembled entity, e.g., a molecule comprising atoms, can be represented by a graph $G = (V, E)$, where V represents a non-empty set and E is a binary relation defined on the set V. When V represents a set of atoms and E symbolizes the set of covalent (or any other) bonds, we have the simplest representation of molecular structure [13, 18]. Invariants derived from such graphs or matrices derived from them are the simplest structural descriptors of molecules.

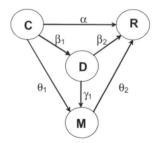

Fig. 2 Composition functions for structure–activity and property–activity relationships (C = chemicals, R = real numbers, D = structural descriptors, M = molecular properties) Reprinted with permission; source from [15]

They give us a hint of the "general topological form" of molecules without encoding information about the content of specific characters such as atom type, bonding patterns, electronic aspects of atoms, etc. Well-known indices including the Wiener index [19] and the simple connectivity index [20] fall in this category. Such indices, although holistic descriptors of molecular structure, often do not quantify interesting aspects of chemicals important for the chemical–biological interactions that lead to pharmacological and toxicological effects. We call these *topostructural* indices. Natural extension of the graph theoretic approach has led to the indices where both topological form (i.e., the distribution of adjacency and distance of atoms) and specific chemical attributes of atoms, bonds, lone pairs of atoms, etc. have been explicitly considered in the index development scheme. This led to the formulation of valence connectivity indices [21], electrotopological indices [22], and the information theoretic indices of neighborhood of symmetry developed by Basak and coworkers [23]. We termed such indices *topochemical* because they not only quantify the topology of the chemical species, but also specific chemical information. The addition of 3D and quantum chemical indices (both semiempirical and *ab initio*) to graph invariants provides a hierarchy of topostructural (TS), topochemical (TC), 3D, and quantum chemical (QC) molecular descriptors. The position of a parameter (e.g., QC indices) at a higher level of hierarchy does not necessarily mean that they are more important than those at some lower level. One important factor that distinguishes a parameter in a higher level of hierarchy from one situated at a lower level is that the former usually takes much more computational resources for calculation.

4
Hierarchical Structure–Activity Relationship (HiQSAR)

We formulated the scheme of hierarchical quantitative structure–activity relationship (HiQSAR) modeling where indices of increasing demand for computational resources are used in a graduated manner in QSAR development [24]. This was the result of the confluence of three different autonomous lines of research. First, since the 1970s, different classes of TS and TC indices have been developed for the characterization of molecular structure. Such indices were being applied in QSAR by different groups of researchers [20, 21, 25–30]. So, there was an interest in using the collection of these indices for QSAR instead of using only one class at a time. Secondly, in the mid 1980s, we [31] asked a more fundamental question: *What is the intrinsic dimensionality of molecular structure?* In other words, using computed descriptors, can we find a set of attributes that have intuitive appeal to the chemist and are, at the same time, orthogonal or minimally correlated? This was an extension of the three dimensional Cartesian coordinate system into n-dimensional ($n > 3$) space. We attempted to answer this question through the use of principal component an-

alysis (PCA) of a set of 90 topological indices calculated for a diverse set of 3692
chemicals. We could associate the first four orthogonal axes (first four PCs) as
representing molecular size, complexity, branching, and cyclicity. More recent

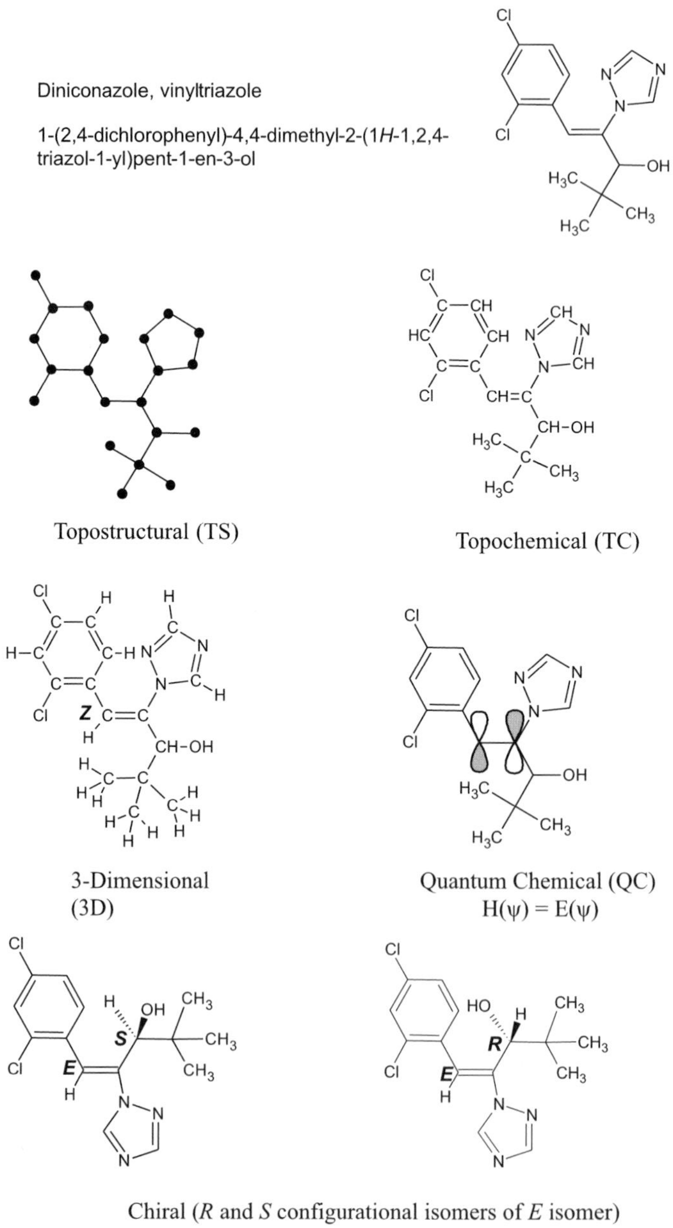

Diniconazole, vinyltriazole

1-(2,4-dichlorophenyl)-4,4-dimethyl-2-(1*H*-1,2,4-
triazol-1-yl)pent-1-en-3-ol

Topostructural (TS)

Topochemical (TC)

3-Dimensional
(3D)

Quantum Chemical (QC)
$H(\psi) = E(\psi)$

Chiral (*R* and *S* configurational isomers of *E* isomer)

Fig. 3 Hierarchy of molecular descriptors

studies by Basak et al. [32, 33] showed that the dimensionality of the structure space depends on how many and how specific indices are included in the set of independent variables. Thirdly, we wanted to explore whether we could use the originally calculated descriptors or variables transformed from them, e.g., the PCs, in a graduated manner. We thought these HiQSAR studies would give us some insight into the submolecular basis of the property under investigation, as indicated in a seminal paper by Klein [34]. Over the past 10 years or so, we have carried out HiQSAR modeling with numerous properties, including many heterocyclic compounds. In most of the cases, the combination of TS and TC indices resulted in the best QSARs; the addition of quantum chemical or 3D descriptors did very little in explaining variance not already handled by the TS + TC combination. As is well known, none of the four categories of indices, viz., TS, TC, 3D, and QC descriptors, can distinguish molecules with one or more chiral centers. More recently, to characterize such molecules, we have used two methods: (a) indices of chirality, and (b) overlay of structures derived by various techniques ranging from molecular mechanics to different levels of quantum chemistry. So, our current approach to HiQSAR consists of five classes of descriptors: (a) topostructural, (b) topochemical, (c) geometrical or 3D, (d) quantum chemical, and (e) chirality indices.

The hierarchical classification of molecular descriptors is exemplified in Fig. 3 with diniconazole. Vinyl triazoles are used extensively as agrochemicals. Diniconazole is 1-(2,4-dichlorophenyl)-4,4-dimethyl-2-(1H-1,2,4-triazol-1-yl)pent-1-en-3-ol and is capable of exhibiting both $E - Z$ and optical isomerism. Hence, each geometrical isomer has R and S enantiomers, and there are four isomeric compounds. The biological activity of diniconazole is concentrated only in the E isomers, and the optical antipodes of the E isomer have different biological activities. The (1E, 3R) isomer has fungicidal property while the (1E, 3S) isomer is a plant growth regulator. To model physicochemical properties of the isomers, one may use TS, TC, 3D, and QC descriptors, whereas to model the bioactivity, chiral indices are also needed.

In this chapter, we will discuss the utility of HiQSAR studies using these five classes of descriptors. We will also show the utility of molecular similarity and our recently developed tailored similarity approach in the clustering of chemical libraries and selection of analogs.

5
Materials and Methods

5.1
Calculation of Descriptors

Well over 300 structure-based chemodescriptors are typically calculated for use in our modeling studies using various software programs. From POLLY

Table 2 Symbols, definitions, and classification of calculated molecular descriptors

	Topostructural (TS)
I_D^W	Information index for the magnitudes of distances between all possible pairs of vertices of a graph
$\overline{I_D^W}$	Mean information index for the magnitude of distance
W	Wiener index = half-sum of the off-diagonal elements of the distance matrix of a graph
I^D	Degree complexity
H^V	Graph vertex complexity
H^D	Graph distance complexity
$\overline{IC}$	Information content of the distance matrix partitioned by frequency of occurrences of distance h
M_1	Zagreb group parameter = sum of square of degree over all vertices
M_2	Zagreb group parameter = sum of cross-product of degrees over all neighboring (connected) vertices
$^h\chi$	Path connectivity index of order $h = 0\text{--}10$
$^h\chi_C$	Cluster connectivity index of order $h = 3\text{--}6$
$^h\chi_{PC}$	Path-cluster connectivity index of order $h = 4\text{--}6$
$^h\chi_{Ch}$	Chain connectivity index of order $h = 3\text{--}10$
P_h	Number of paths of length $h = 0\text{--}10$
J	Balaban's index based on topological distance
nrings	Number of rings in a graph
ncirc	Number of circuits in a graph
DN^2S_y	Triplet index from distance matrix, square of graph order, and distance sum; operation $y = 1\text{--}5$
DN^21_y	Triplet index from distance matrix, square of graph order, and number 1; operation $y = 1\text{--}5$
$AS1_y$	Triplet index from adjacency matrix, distance sum, and number 1; operation $y = 1\text{--}5$
$DS1_y$	Triplet index from distance matrix, distance sum, and number 1; operation $y = 1\text{--}5$
ASN_y	Triplet index from adjacency matrix, distance sum, and graph order; operation $y = 1\text{--}5$
DSN_y	Triplet index from distance matrix, distance sum, and graph order; operation $y = 1\text{--}5$
DN^2N_y	Triplet index from distance matrix, square of graph order, and graph order; operation $y = 1\text{--}5$
ANS_y	Triplet index from adjacency matrix, graph order, and distance sum; operation $y = 1\text{--}5$
$AN1_y$	Triplet index from adjacency matrix, graph order, and number 1; operation $y = 1\text{--}5$
ANN_y	Triplet index from adjacency matrix, graph order, and graph order again; operation $y = 1\text{--}5$

Table 2 (continued)

ASV_y	Triplet index from adjacency matrix, distance sum, and vertex degree; operation $y = 1$–5
DSV_y	Triplet index from distance matrix, distance sum, and vertex degree; operation $y = 1$–5
ANV_y	Triplet index from adjacency matrix, graph order, and vertex degree; operation $y = 1$–5
	Topochemical (TC)
O	Order of neighborhood when IC_r reaches its maximum value for the hydrogen-filled graph
O_{orb}	Order of neighborhood when IC_r reaches its maximum value for the hydrogen-suppressed graph
I_{ORB}	Information content or complexity of the hydrogen-suppressed graph at its maximum neighborhood of vertices
IC_r	Mean information content or complexity of a graph based on the rth ($r = 0$–6) order neighborhood of vertices in a hydrogen-filled graph
SIC_r	Structural information content for rth ($r = 0$–6) order neighborhood of vertices in a hydrogen-filled graph
CIC_r	Complementary information content for rth ($r = 0$–6) order neighborhood of vertices in a hydrogen-filled graph
$^h\chi^b$	Bond path connectivity index of order $h = 0$–6
$^h\chi_C^b$	Bond cluster connectivity index of order $h = 3$–6
$^h\chi_{Ch}^b$	Bond chain connectivity index of order $h = 3$–6
$^h\chi_{PC}^b$	Bond path-cluster connectivity index of order $h = 4$–6
$^h\chi^v$	Valence path connectivity index of order $h = 0$–10
$^h\chi_C^v$	Valence cluster connectivity index of order $h = 3$–6
$^h\chi_{Ch}^v$	Valence chain connectivity index of order $h = 3$–10
$^h\chi_{PC}^v$	Valence path-cluster connectivity index of order $h = 4$–6
J^B	Balaban's index based on bond types
J^X	Balaban's index based on relative electronegativities
J^Y	Balaban's index based on relative covalent radii
AZV_y	Triplet index from adjacency matrix, atomic number, and vertex degree; operation $y = 1$–5
AZS_y	Triplet index from adjacency matrix, atomic number, and distance sum; operation $y = 1$–5
ASZ_y	Triplet index from adjacency matrix, distance sum, and atomic number; operation $y = 1$–5
AZN_y	Triplet index from adjacency matrix, atomic number, and graph order; operation $y = 1$–5
ANZ_y	Triplet index from adjacency matrix, graph order, and atomic number; operation $y = 1$–5
DSZ_y	Triplet index from distance matrix, distance sum, and atomic number; operation $y = 1$–5

Table 2 (continued)

DN^2Z_y	Triplet index from distance matrix, square of graph order, and atomic number; operation $y = 1-5$
nvx	Number of non-hydrogen atoms in a molecule
nelem	Number of elements in a molecule
fw	Molecular weight
si	Shannon information index
totop	Total topological index t
sumI	Sum of the intrinsic state values I
sumdelI	Sum of delta-I values
tets2	Total topological state index based on electrotopological state indices
phia	Flexibility index ($kp_1 * kp_2 / nvx$)
Idcbar	Bonchev–Trinajstić information index
IdC	Bonchev–Trinajstić information index
Wp	Wienerp
Pf	Plattf
Wt	Total Wiener number
knotp	Difference of chi-cluster-3 and path/cluster-4
knotpv	Valence difference of chi-cluster-3 and path/cluster-4
nclass	Number of classes of topologically (symmetry) equivalent graph vertices
NumHBd	Number of hydrogen bond donors
NumHBa	Number of hydrogen bond acceptors
SHCsats	E-State of C sp^3 bonded to other saturated C atoms
SHCsatu	E-State of C sp^3 bonded to unsaturated C atoms
SHvin	E-State of C atoms in the vinyl group, $= CH-$
SHtvin	E-State of C atoms in the terminal vinyl group, $= CH_2$
SHavin	E-State of C atoms in the vinyl group, $= CH-$, bonded to an aromatic C
SHarom	E-State of C sp^2 which are part of an aromatic system
SHHBd	Hydrogen bond donor index, sum of hydrogen E-state values for $-OH$, $= NH$, $-NH_2$, $-NH-$, $-SH$, and $\#CH$
SHwHBd	Weak hydrogen bond donor index, sum of $C-H$ hydrogen E-state values for hydrogen atoms on a C to which a F and/or Cl are also bonded
SHHBa	Hydrogen bond acceptor index, sum of the E-state values for $-OH$, $= NH$, $-NH_2$, $-NH-$, $>N-$, $-O-$, $-S-$, along with $-F$ and $-Cl$
Qv	General polarity descriptor
NHBint_y	Count of potential internal hydrogen bonders ($y = 2-10$)
SHBint_y	E-State descriptors of potential internal hydrogen bond strength ($y = 2-10$)
	Electrotopological state index values for atoms types: SHsOH, SHdNH, SHsSH, SHsNH2, SHssNH, SHtCH, SHother, SHCHnX, Hmax, Gmax, Hmin, Gmin, Hmaxpos, Hminneg, SsLi, SssBe, Sssss, Bem, SssBH, SsssB, SsssBm, SsCH3, SdCH2, SssCH2, StCH, SdsCH, SaaCH, SsssCH, SddC, StsC, SdssC, SaasC, SaaaC, SsssssC, SsNH3p, SsNH2, SssNH2p, SdNH, SssNH, SaaNH,

Table 2 (continued)

	StN, SsssNHp, SdsN, SaaN, SsssN, SddsN, SaasN, SsssssNp, SsOH, SdO, SssO, SaaO, SsF, SsSiH3, SsssSiH2, SsssSiH, SssssSi, SsPH2, SssPH, SsssP, SdsssP, SsssssP, SsSH, SdS, SssS, SaaS, SdssS, SddssS, SssssssS, SsCl, SsGeH3, SssGeH2, SsssGeH, SsssssGe, SsAsH2, SssAsH, SsssAs, SdsssAs, SssssssAs, SsSeH, SdSe, SssSe, SaaSe, SdssSe, SddssSe, SsBr, SsSnH3, SssSnH2, SsssSnH, SssssSn, SsI, SsPbH3, SssPbH2, SsssPbH, SssssPb

Geometrical (3D)/shape

kp_0	Kappa zero
$kp_1 - kp_3$	Kappa simple indices
$ka_1 - ka_3$	Kappa alpha indices
V_W	Van der Waals volume
^{3D}W	3D Wiener number based on the hydrogen-suppressed geometric distance matrix
$^{3D}W_H$	3D Wiener number based on the hydrogen-filled geometric distance matrix

Quantum chemical (QC)

E_{HOMO}	Energy of the highest occupied molecular orbital
E_{HOMO-1}	Energy of the second highest occupied molecular
E_{LUMO}	Energy of the lowest unoccupied molecular orbital
E_{LUMO+1}	Energy of the second lowest unoccupied molecular orbital
ΔH_f	Heat of formation
μ	Dipole moment

Reprinted with permission; source from [52]

v2.3 [35] and associated software, a set of 102 topological descriptors is available. An additional set of 100 topological descriptors is available from the Triplet [36] program. An extended set of connectivity indices, descriptors of polarity and hydrogen bonding, a large set of electrotopological state indices, as well as a small set of kappa shape indices are obtained using Molconn-Z v3.5 [37]. Additional geometrical descriptors were calculated for the aromatic and heteroaromatic amine data set (Sect. 5.3.2) using Sybyl v6.4 [38]. Austin Model 1 (AM1) semiempirical quantum chemical descriptors were obtained using MOPAC v6.00 [39], while *ab initio* quantum chemical descriptors at the STO-3G level were calculated using Gaussian 03W v6.0 [40]. A list of theoretical descriptors typically calculated for our modeling studies, along with brief descriptions and hierarchical classification, is provided in Table 2.

5.2
Statistical Methods

Prior to analyses, all calculated descriptors were transformed by $\ln (x + c)$, where x represents the original descriptor value and c is a constant added to avoid potential arithmetic error. In most cases, $c = 1$, as the original descriptor values are generally greater than $- 1$. A small number of descriptors, however, have minimum values less than or equal to $- 1$, in which case the constant added was the smallest natural number that would provide a positive sum for $(x + c)$.

For comparative purposes, three regression methodologies were used for the development of predictive QSAR models, namely ridge regression (RR) [41, 42], principal components regression (PCR) [43], and partial least squares (PLS) [44]. Each of these methodologies makes use of all available descriptors, as opposed to subset regression, and is useful when the number of descriptors is large with respect to the number of compounds and when the descriptors are highly intercorrelated. Formal comparisons have consistently shown that using a small subset of available descriptors is less effective than using alternative regression methods that retain all available descriptors, such as RR, PCR, and PLS [45, 46].

It should be strongly stated that ordinary least squares (OLS) regression is inappropriate when the number of descriptors is large with respect to the number of chemical compounds in the data set, and that the conventional R^2 metric is without value in this situation. Unlike R^2, which tends to *increase* upon the addition of any descriptor, the cross-validated R^2 tends to *decrease* upon the addition of irrelevant descriptors and is a reliable measure of model predictability [47]. Unlike OLS regression, the number of descriptors is not an issue with the regression methodologies used in the present analyses. RR, PCR, and PLS are appropriate methodologies when the number of descriptors exceeds the number of observations, and they are designed to utilize all available descriptors, as opposed to subset regression, in order to produce an unbiased model whose predictability is accurately reflected by the R^2_{cv}, regardless of the number of independent variables in the model. The distinction between these methods and OLS regression is an important one and cannot be overemphasized.

With respect to the applied regression methodologies, RR is similar to PCR in that the independent variables are transformed to their principal components (PCs). However, while PCR utilizes only a subset of the PCs, RR retains them all but downweights them based on their eigenvalues. With PLS, a subset of the PCs is also used, but the PCs are selected by considering both the independent and dependent variables. For each model developed, the cross-validated R^2 was obtained using the leave-one-out (LOO) approach and can be calculated as follows:

$$R^2_{cv} = 1 - \frac{PRESS}{SS_{Total}} \tag{1}$$

where PRESS is the prediction sum of squares and SS_{Total} is the total sum of squares. It is important to note that theoretic argument and empiric study have shown that the LOO cross-validation approach is *preferred* to the use of an "external test set" for small to moderately sized chemical databases [47]. For the sake of brevity, the highly parameterized models are not included in this chapter, but rather the cross-validated R^2 and PRESS statistic are reported for each model. Another useful statistical metric is the t-value associated with each model descriptor, defined as the descriptor coefficient divided by its standard error. Descriptors with large $|t|$-values are important in the predictive model and, as such, can be examined in order to gain some understanding of the nature of the property or activity of interest.

Predictive QSAR models have been developed using the hierarchical QSAR approach in which the four classes of calculated molecular descriptors, viz., TS, TC, 3D, and QC, have been utilized in a graduated manner. In this way, the relative contribution of each descriptor class can be examined by comparing the quality of the resulting models. An alternative approach involves first modestly thinning the entire pool of available descriptors, followed by regression analyses utilizing the reduced set of descriptors. While it is true that RR, PLS, and PCR can be applied without concern to the number of descriptors with respect to the number of observations, modest descriptor thinning can reduce computation time, improve model interpretability, and remove mere "noise". The descriptor thinning was accomplished using a modified Gram–Schmidt orthogonalization [48]. Although we used this algorithm for variable selection, it should be noted that it is typically used to stabilize regression coefficients for OLS regression. It is important to note that only modest thinning should be performed, allowing only the least correlated descriptors to be dropped, as excessive thinning leads to extreme bias in the predictor estimates [45].

5.3
QSAR Models

In previous studies, we have found our hierarchical QSAR approach to be successful in predicting properties, activities, and toxicities including:

- Mutagenicity of aromatic and heteroaromatic amines [24, 49–52]
- Complement–inhibitory activity of benzamidines [53]
- Vapor pressure [54, 55]
- Boiling point of structurally heterogeneous chemicals [56]
- Acute toxicity of benzene derivatives [57]
- Biological partition coefficients [58–60]
- Dermal penetration of polycyclic aromatic compounds [61]
- Toxicity of halocarbons [62]

In this section, we provide a few examples of our HiQSAR studies involving heterocyclic compounds, which are important both as drugs and toxicants.

5.3.1
Receptor Binding Affinity of Dibenzofurans

The aryl hydrocarbon (*Ah*) receptor is well documented in the field of toxicology, with the toxicity of certain classes of persistent pollutants, including dibenzofurans, being determined by *Ah* receptor interaction. We developed HiQSAR models [63] based on a set of dibenzofurans with *Ah* receptor binding potency values obtained from the literature (Fig. 4; Table 3) [64]. While the source data set was comprised of 34 compounds, we found two to be statistical outliers, namely 2,3,4-trCl and 1,2,4,6,8-peCl, and as such removed them from the study. Results reported here are associated with the resulting set of 32 compounds. Descriptors included TS, TC, 3D, and *ab initio* QC descriptors calculated at the STO-3G level. Statistical metrics for the predictive models are provided in Table 4. The TS + TC descriptors provide a high-quality RR model as evidenced by a R^2_{cv} value of 0.852. Note that the addition of the 3D and QC descriptors did not result in significant model improvement. The binding potency values as predicted by the TS + TC RR model are also provided in Table 3, along with the differences between the experimental and predicted values.

For comparative purposes, the same data set was analyzed using an alternative approach. Rather than employing the hierarchical method, we began with the entire pool of descriptors, viz., TS + TC + 3D + STO-3G, and performed variable thinning via the modified Gram–Schmidt algorithm. RR, PCR, and PLS regression analyses were performed using the reduced set of 72 descriptors. Again, a high-quality RR model was obtained with a R^2_{cv} value of 0.899. Comparing the descriptors with the highest $|t|$-values in this model and those in the best RR model obtained using the hierarchical approach, i.e., TS + TC, it was found that the same descriptors are important in both models (Table 5).

A perusal of the descriptors important for the correlation of *Ah* receptor binding affinity (Table 5) shows that electronic factors, as well as general molecular shape and size, play a dominant role. The SaaCH, SHarom, and sumDelI indices encode information about the hybridization and electronic states of atoms in the molecule. The $^4\chi^b$, $^4\chi$, Idcbar and P_4 descriptors are quantifiers of general molecular shape and size and are probably related to steric fit between the receptor and the ligand. In our PCA study involv-

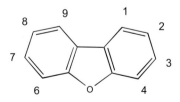

Fig. 4 Structure for the dibenzofuran data set provided in Table 3

Table 3 Experimental and cross-validated predicted *Ah* receptor binding affinity based on the TS + TC ridge regression model

No.	Chemical	Experimental pEC_{50}	Predicted pEC_{50}	Exp. – Pred.
1	2-Cl	3.553	3.169	0.384
2	3-Cl	4.377	4.199	0.178
3	4-Cl	3.000	3.692	– 0.692
4	2,3-diCl	5.326	4.964	0.362
5	2,6-diCl	3.609	4.279	– 0.670
6	2,8-diCl	3.590	4.251	– 0.661
7	1,2,7-trCl	6.347	5.646	0.701
8	1,3,6-trCl	5.357	4.705	0.652
9	1,3,8-trCl	4.071	5.330	– 1.259
10	2,3,8-trCl	6.000	6.394	– 0.394
11	1,2,3,6-teCl	6.456	6.480	– 0.024
12	1,2,3,7-teCl	6.959	7.066	– 0.107
13	1,2,4,8-teCl	5.000	4.715	0.285
14	2,3,4,6-teCl	6.456	7.321	– 0.865
15	2,3,4,7-teCl	7.602	7.496	0.106
16	2,3,4,8-teCl	6.699	6.976	– 0.277
17	2,3,6,8-teCl	6.658	6.008	0.650
18	2,3,7,8-teCl	7.387	7.139	0.248
19	1,2,3,4,8-peCl	6.921	6.293	0.628
20	1,2,3,7,8-peCl	7.128	7.213	– 0.085
21	1,2,3,7,9-peCl	6.398	5.724	0.674
22	1,2,4,6,7-peCl	7.169	6.135	1.035
23	1,2,4,7,8-peCl	5.886	6.607	– 0.720
24	1,2,4,7,9-peCl	4.699	4.937	– 0.238
25	1,3,4,7,8-peCl	6.699	6.513	0.186
26	2,3,4,7,8-peCl	7.824	7.479	0.345
27	2,3,4,7,9-peCl	6.699	6.509	0.190
28	1,2,3,4,7,8-heCl	6.638	6.802	– 0.164
29	1,2,3,6,7,8-heCl	6.569	7.124	– 0.555
30	1,2,4,6,7,8-heCl	5.081	5.672	– 0.591
31	2,3,4,6,7,8-heCl	7.328	7.019	0.309
32	Dibenzofuran	3.000	2.765	0.235

Reprinted with permission; source from [63]

ing 90 topological indices calculated for a structurally diverse set of 3692 molecules [13], we found such indices to be very strongly correlated with PC1 which, in turn, had a high correlation with size parameters such as molecular weight and number of vertices. The triplet index, ANV_1, is a composite of

Table 4 Summary statistics for predictive *Ah* receptor binding affinity models

Model type	RR		PCR		PLS	
	$R^2_{c.v.}$	PRESS	$R^2_{c.v.}$	PRESS	$R^2_{c.v.}$	PRESS
TS	0.731	16.9	0.690	19.4	0.701	18.7
TS + TC	0.852	9.27	0.683	19.9	0.836	10.3
TS + TC + 3D	0.852	9.27	0.683	19.9	0.837	10.2
TS + TC + 3D + QC	0.862	8.62	0.595	25.4	0.862	8.67
TS	0.731	16.9	0.690	19.4	0.701	18.7
TC	0.820	11.3	0.694	19.1	0.749	15.7
3D	0.508	30.8	0.523	29.9	0.419	36.4
QC	0.544	28.6	0.458	33.9	0.501	31.3

Reprinted with permission; source from [63]

Table 5 Descriptors important for the prediction of *Ah* receptor binding affinity of dibenzofurans, identified in both the hierarchical TS + TC RR model and the RR model developed following Gram–Schmidt variable thinning

Descriptor	Definition
SaaCH	Sum of E-states for CH with two aromatic bonds
SHarom	Sum of E-states for C sp^2 which are part of an aromatic system
$^4\chi^b$	Bond path connectivity index of order 4
$^4\chi$	Path connectivity index of order 4
P_4	Number of paths of length 4
sumDelI	Sum of delta-I values
Idcbar	Bonchev–Trinajstic information index
ANV_1	Triplet index from adjacency index, graph order, and vertex degree; operation = summation

the degree of adjacency and valency of atoms. This could be looked upon as a quantifier of molecular shape in the graph theoretical sense. Therefore, electronic, shape and size factors are most important in determining *Ah* receptor binding potency of dibenzofurans.

5.3.2
Mutagenicity of Aromatic and Heteroaromatic Amines

The set of 95 aromatic and heteroaromatic amines used to study mutagenic potency was obtained from Debnath et al. [65]. The mutagenic activity in

Table 6 Observed, fitted, and cross-validated predicted mutagenic potency for 95 aromatic and heteroaromatic amines based on the ridge regression TS + TC model

No.	Compound	$\ln(R)$			
		Obs.	Fitted	C.V.	Obs–C.V.
1	2-Bromo-7-aminofluorene	2.62	2.30	1.99	0.63
2	2-Methoxy-5-methylaniline (p-cresidine)	− 2.05	− 2.63	− 2.82	0.77
3	5-Aminoquinoline	− 2.00	− 1.88	− 1.85	− 0.15
4	4-Ethoxyaniline (p-phenetidine)	− 2.30	− 3.00	− 3.26	0.96
5	1-Aminonaphthalene	− 0.60	− 0.61	− 0.61	0.01
6	4-Aminofluorene	1.13	1.14	1.14	− 0.01
7	2-Aminoanthracene	2.62	1.53	1.34	1.28
8	7-Aminofluoranthene	2.88	3.06	3.10	− 0.22
9	8-Aminoquinoline	− 1.14	− 1.68	− 1.76	0.62
10	1,7-Diaminophenazine	0.75	0.85	0.88	− 0.13
11	2-Aminonaphthalene	− 0.67	− 0.39	− 0.32	− 0.35
12	4-Aminopyrene	3.16	2.75	2.57	0.59
13	3-Amino-3′-nitrobiphenyl	− 0.55	− 0.08	0.07	− 0.62
14	2,4,5-Trimethylaniline	− 1.32	− 1.23	− 1.18	− 0.14
15	3-Aminofluorene	0.89	1.18	1.25	− 0.36
16	3,3′-Dichlorobenzidine	0.81	0.21	− 0.17	0.98
17	2,4-Dimethylaniline (2,4-xylidine)	− 2.22	− 2.18	− 2.17	− 0.05
18	2,7-Diaminofluorene	0.48	0.92	1.11	− 0.63
19	3-Aminofluoranthene	3.31	2.83	2.65	0.66
20	2-Aminofluorene	1.93	1.18	1.03	0.90
21	2-Amino-4′-nitrobiphenyl	− 0.62	− 0.34	− 0.27	− 0.35
22	4-Aminobiphenyl	− 0.14	− 0.29	− 0.34	0.20
23	3-Methoxy-4-methylaniline (o-cresidine)	− 1.96	− 2.65	− 2.84	0.88
24	2-Aminocarbazole	0.60	0.68	0.69	− 0.09
25	2-Amino-5-nitrophenol	− 2.52	− 2.72	− 2.86	0.34
26	2,2′-Diaminobiphenyl	− 1.52	− 0.44	− 0.15	− 1.37
27	2-Hydroxy-7-aminofluorene	0.41	0.95	1.13	− 0.72
28	1-Aminophenanthrene	2.38	1.39	1.19	1.19
29	2,5-Dimethylaniline (2,5-xylidine)	− 2.40	− 2.08	− 1.98	− 0.42
30	4-Amino-2′-nitrobiphenyl	− 0.92	− 0.29	− 0.05	− 0.87
31	2-Amino-4-methylphenol	− 2.10	− 2.97	− 4.05	1.95
32	2-Aminophenazine	0.55	1.23	1.54	− 0.99
33	4, 4′-Diaminophenyl sulfide	0.31	− 0.30	− 0.62	0.93
34	2,4-Dinitroaniline	− 2.00	− 2.03	− 2.07	0.07
35	2,4-Diaminoisopropylbenzene	− 3.00	− 2.71	− 2.55	− 0.45
36	2,4-Difluoroaniline	− 2.70	− 2.54	− 2.43	− 0.27

Table 6 (continued)

No.	Compound	ln(R)			
		Obs.	Fitted	C.V.	Obs-C.V.
37	4,4′-Methylenedianiline	− 1.60	− 1.12	− 0.96	− 0.64
38	3,3′-Dimethylbenzidine	0.01	0.19	0.28	− 0.27
39	2-Aminofluoranthene	3.23	3.26	3.27	− 0.04
40	2-Amino-3′-nitrobiphenyl	− 0.89	− 0.69	− 0.64	− 0.25
41	1-Aminofluoranthene	3.35	3.06	3.00	0.35
42	4,4′-Ethylenebis (aniline)	− 2.15	− 1.87	− 1.10	− 1.05
43	4-Chloroaniline	− 2.52	− 2.58	− 2.60	0.08
44	2-Aminophenanthrene	2.46	1.57	1.45	1.01
45	4-Fluoroaniline	− 3.32	− 3.28	− 3.26	− 0.06
46	9-Aminophenanthrene	2.98	1.52	1.36	1.62
47	3,3′-Diaminobiphenyl	− 1.30	− 1.17	− 1.14	− 0.16
48	2-Aminopyrene	3.50	2.92	2.69	0.81
49	2,6-Dichloro-1,4-phenylenediamine	− 0.69	− 0.89	− 1.02	0.33
50	2-Amino-7-acetamidofluorene	1.18	1.01	0.37	0.81
51	2,8-Diaminophenazine	1.12	0.88	0.77	0.35
52	6-Aminoquinoline	− 2.67	− 1.70	− 1.52	− 1.15
53	4-Methoxy-2-methylaniline (m-cresidine)	− 3.00	− 2.73	− 2.65	− 0.35
54	3-Amino-2′-nitrobiphenyl	− 1.30	− 0.60	− 0.38	− 0.92
55	2,4′-Diamino-biphenyl	− 0.92	− 0.77	− 0.72	− 0.20
56	1,6-Diaminophenazine	0.20	− 0.12	− 0.34	0.54
57	4-Aminophenyldisulfide	− 1.03	− 0.35	0.91	− 1.94
58	2-Bromo-4,6-dinitroaniline	− 0.54	− 0.78	− 1.06	0.52
59	2,4-Diamino-n-butylbenzene	− 2.70	− 2.50	− 2.36	− 0.34
60	4-Aminophenyl ether	− 1.14	− 0.70	− 0.45	− 0.69
61	2-Aminobiphenyl	− 1.49	− 0.44	− 0.27	− 1.22
62	1,9-Diaminophenazine	0.04	0.02	0.00	0.04
63	1-Aminofluorene	0.43	1.08	1.25	− 0.82
64	8-Aminofluoranthene	3.80	3.19	3.05	0.75
65	2-Chloroaniline	− 3.00	− 2.66	− 2.43	− 0.57
66	2-Amino-$\alpha\alpha\alpha$-trifluorotoluene	− 0.80	− 0.97	− 1.88	1.08
67	2-Amino-1-nitronaphthalene	− 1.17	− 0.55	− 0.25	− 0.92
68	3-Amino-4′-nitrobiphenyl	0.69	0.48	0.43	0.26
69	4-Bromoaniline	− 2.70	− 2.40	− 1.48	− 1.22
70	2-Amino-4-chlorophenol	− 3.00	− 2.35	− 1.57	− 1.43
71	3,3′-Dimethoxybenzidine	0.15	0.59	1.11	− 0.96
72	4-Cyclohexylaniline	− 1.24	− 1.42	− 1.75	0.51
73	4-Phenoxyaniline	0.38	− 0.40	− 0.72	1.10

Table 6 (continued)

No.	Compound	ln(R) Obs.	Fitted	C.V.	Obs-C.V.
74	4,4′-Methylenebis(o-ethylaniline)	− 0.99	− 0.86	− 0.78	− 0.21
75	2-Amino-7-nitrofluorene	3.00	2.24	1.87	1.13
76	Benzidine	− 0.39	0.02	0.20	− 0.59
77	1-Amino-4-nitronaphthalene	− 1.77	− 0.73	− 0.23	− 1.54
78	4-Amino-3′-nitrobiphenyl	1.02	0.56	0.45	0.57
79	4-Amino-4′-nitrobiphenyl	1.04	− 0.26	− 0.80	1.84
80	1-Aminophenazine	− 0.01	1.00	1.32	− 1.33
81	4,4′-Methylenebis (o-fluoroaniline)	0.23	0.25	0.28	− 0.05
82	4-Chloro-2-nitroaniline	− 2.22	− 2.15	− 2.12	− 0.10
83	3-Aminoquinoline	− 3.14	− 1.51	− 1.26	− 1.88
84	3-Aminocarbazole	− 0.48	− 0.52	− 0.55	0.07
85	4-Chloro-1,2-phenylenediamine	− 0.49	− 1.69	− 2.21	1.72
86	3-Aminophenanthrene	3.77	1.57	1.36	2.41
87	3,4′-Diaminobiphenyl	0.20	− 0.27	− 0.46	0.66
88	1-Aminoanthracene	1.18	1.39	1.42	− 0.24
89	1-Aminocarbazole	− 1.04	− 0.71	− 0.54	− 0.50
90	9-Aminoanthracene	0.87	1.05	1.08	− 0.21
91	4-Aminocarbazole	− 1.42	− 0.72	− 0.32	− 1.10
92	6-Aminochrysene	1.83	3.03	3.35	− 1.52
93	1-Aminopyrene	1.43	3.17	3.72	− 2.29
94	4-4′-Methylenebis (o-isopropylaniline)	− 1.77	− 1.18	− 0.34	− 1.43
95	2,7-Diaminophenazine	3.97	1.70	0.87	3.10

Reprinted with permission; source from [66]

S. typhimurium TA98 + S9 microsomal preparation is expressed as the natural logarithm of R, where R is the number of revertants per nanomole. The compounds analyzed in this study [66] and their experimentally measured mutation rates are listed in Table 6. TS, TC, 3D, and QC descriptors were calculated, with the latter obtained at the AM1 semiempirical level. The regression results are summarized in Table 7. Generally, the results obtained using the ridge regression method are superior to those obtained using either principal components regression or partial least squares. The model statistics improve upon the addition of TC descriptors to the TS set, with R^2_{cv} increasing from 0.667 to 0.748; however, the addition of 3D and QC descriptors does not result in model improvement. The TC model is the best model developed using a single class of descriptors, with a R^2_{cv} of 0.734. The model developed using QC descriptors alone is quite poor.

Table 7 Summary of regression results for the prediction of mutagenicity in terms of $\ln(R)$, where R = the number of revertants per nanomole, for the set of 95 aromatic and heteroaromatic amines

Model type	RR		PCR		PLS	
	$R^2_{c.v.}$	PRESS	$R^2_{c.v.}$	PRESS	$R^2_{c.v.}$	PRESS
TS	0.667	116	0.631	128	0.634	127
TS + TC	0.748	87.6	0.642	124	0.705	103
TS + TC + 3D	0.751	86.8	0.646	123	0.702	104
TS + TC + 3D + QC	0.749	87.3	0.647	123	0.683	110
TS	0.667	116	0.631	128	0.634	127
TC	0.734	92.7	0.658	119	0.692	107
3D	0.578	147	0.577	147	0.601	139
QC	0.430	198	0.411	205	0.397	210

Reprinted with permission; source from [66]

Table 8 Descriptors important for the prediction of mutagenicity of aromatic and heteroaromatic amines

Descriptor	Definition
P_2	Number of paths of length 2
M_1	Zagreb group parameter = sum of square of degree over all vertices
M_2	Zagreb group parameter = sum of cross-product of degrees over all neighboring (connected) vertices
NHbint9	Number of potential internal hydrogen bonds of path length 9
AZN_4	Triplet index from adjacency matrix, atomic number, and graph order; operation = sum of inverse square root of cross-product over edges
P_5	Number of paths of length 5
P_6	Number of paths of length 6
ANS_4	Triplet index from adjacency matrix, graph order, and distance sum; operation = sum of inverse square root of cross-product over edges
IC_1	Mean information content or complexity of a graph based on the 1st order neighborhood of vertices in a hydrogen-filled graph
$^6\chi$	Path connectivity index of order 6

In order to identify the descriptors most important in the TS + TC RR model, they were ranked according to $|t|$-value. The ten descriptors with the highest values are reported in Table 8. It is clear that the most important molecular descriptors for the prediction of mutagenicity fall into following three classes:

Table 9 Predictive models for mutagenic potency in *S. typhimurium* TA98 + S9 based on the set of 95 aromatic and heteroaromatic amines, in order of increasing R^2

Descriptors	n	R^2	s	Investigator(s)	Refs.
Graphs of atomic orbitals	95	0.758	0.95	Toropov and Toropova (2001)	[68]
Electrotopological state indices	95	0.767	0.979	Cash (2001)	[69]
Topological, geometric, and quantum chemical	95	0.790	0.92	Basak et al. (1998)	[24]
Expanded set of topological, geometric, and quantum chemical	95	0.794	0.912	Basak et al. (2001)	[50]
Topological and geometric	95	0.797	0.91	Basak et al. (1997)	[51]
Log P, ϵ_{HOMO}, ϵ_{LUMO}, I_L	88	0.806	0.860	Debnath et al. (1992)	[65]
Expanded set of topological, geometric, and quantum chemical + electrotopological	95	0.821	0.84	Basak et al. (2001)	[52]
# rings, γ-polarizability, HASA1(SCF/AM1), HDSA (SCF/AM1), E_{tot} (C – C), E_{tot} (C – N)	95	0.834	0.811	Maran et al. (1999)	[70]

Reprinted with permission; source from [52]

1. Indices which are related to the shape and size of molecular graphs, including P_2, M_1, M_2, P_5, P_6, ANS_4 and $^6\chi$
2. Descriptors which indicate the presence and neighborhoods of heteroatoms, such as NHbint9
3. Indices which represent the degree of heterogeneity of atomic neighborhoods as coded by IC_1

In essence, a combination of steric and electronic factors dominate the set of most important descriptors.

It is interesting to note that in their correlation of mutagenicity for this set of compounds using log P and calculated electronic descriptors such as HOMO energy and LUMO energy, Debnath et al. [65] found that log P explained most of the variance in the data whereas electronic factors had a minor role. Our research showed that a combination of topological indices which quantify molecular size and shape, together with hydrogen bonding descriptors, can predict hydrophobicity very well [67]. So, it is not surprising that a combination of size, shape, and electronic heterogeneity descriptors is capable of correlating Ames' mutagenicity of aromatic and heteroaromatic amines.

Table 9 summarizes previous QSAR studies by our group [24, 50–52] and other researchers [65, 68–70] for the prediction of mutagenicity for the set of 95 aromatic and heteroaromatic amines. It should be noted, however, that the conventional R^2, rather than the R^2_{cv}, is reported in the previous studies.

5.3.3
Inhibition of COX-2 by Imidazoles

A series of 1,2-diarylimidazoles was synthesized and found to contain highly potent and selective inhibitors of the human cyclooxygenase-2 (COX-2) enzyme [71]. IC_{50} values for a set of 85 such compounds amenable to QSAR modeling (Fig. 5; Table 10) were obtained from a study by Garg et al. [72]. In a recent study (Basak and Mills, unpublished results), we calculated TS, TC, 3D, and QC descriptors, with the latter obtained at the *ab initio* STO-3G level. The modified Gram–Schmidt orthogonalization procedure was applied in the context of predictor thinning to the entire pool of descriptors, resulting in the retention of 77 indices. Subsequent RR, PCR, and PLS regression modeling with this set of descriptors yielded R^2_{cv} values of 0.798, 0.154, and – 0.146, respectively. The RR model is highly predictive. A negative R^2_{cv} value, such as that associated with the PLS model, indicates an extremely poor model (see Eq. 1). Table 10 includes experimentally determined activity values as well as cross-validated predicted values based on the RR model.

The descriptors of the RR model were ranked according to $|t|$-value, and those with the highest values are reported in Table 11. The most influential indices in the correlation can be divided into three groups:

1. Hydrogen bonding descriptors (NHbint5, NHbint9)
2. Electronic descriptors (SHssNH) which describe specific E-states
3. Indices that quantify the electronic character of the molecule, such as O_{orb} and a few of the triplet indices

Since COX-2 inhibitors must fit into the cavity of the enzyme surface, the internal hydrogen bonding parameters might also reflect the conformational restriction that is required for the enzyme inhibitory action of these molecules.

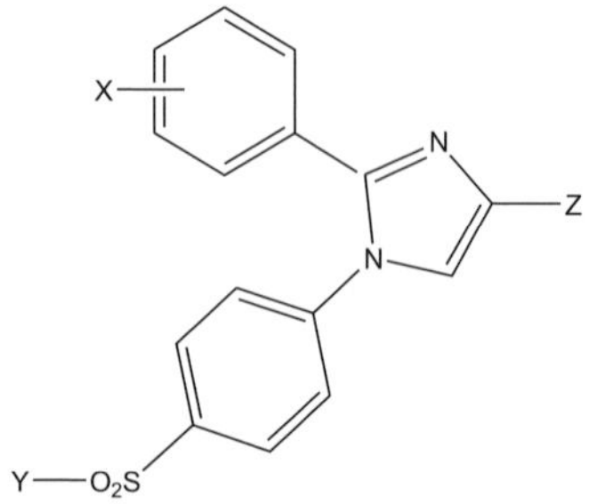

Fig. 5 Structure for the cyclooxygenase-2 (COX-2) data set provided in Table 8

Table 10 IC_{50} data for the inhibition of COX-2

No.	Substituent X	Y	Z	Log $1/C$ Expt.	Calc.	Δ
1	4-Cl	Me	CF_3	6.96	7.03	– 0.07
2	4-F	Me	CF_3	7.00	6.90	0.10
3	H	Me	CF_3	6.92	6.82	0.10
4	4-Me	Me	CF_3	6.80	7.07	– 0.27
5	4-OMe	Me	CF_3	6.24	6.57	– 0.33
6	4-NHMe	Me	CF_3	5.83	5.83	0.00
7	4-NMe$_2$	Me	CF_3	6.16	5.51	0.65
8	4-SMe	Me	CF_3	6.80	6.53	0.27
9	4– SO$_2$Me	Me	CF_3	5.24	5.19	0.05
10	4-Cl	NH$_2$	CF_3	8.00	7.98	0.02
11	4-F	NH$_2$	CF_3	8.00	7.81	0.19
12	H	NH$_2$	CF_3	7.40	7.48	– 0.08
13	4-Me	NH$_2$	CF_3	7.40	7.55	– 0.15
14	3-Cl	Me	CF_3	7.22	7.23	– 0.01
15	3-F	Me	CF_3	6.92	7.08	– 0.16
16	3-Br	Me	CF_3	7.10	7.31	– 0.21
17	3-Me	Me	CF_3	7.22	6.85	0.37
18	3-CF$_3$	Me	CF_3	6.68	6.20	0.48
19	3-OMe	Me	CF_3	6.46	5.96	0.50
20	3-SMe	Me	CF_3	6.46	6.63	– 0.17
21	3-CH$_2$OMe	Me	CF_3	4.17	5.07	– 0.90
22	3-NMe$_2$	Me	CF_3	5.50	5.44	0.06
23	3-NHMe	Me	CF_3	6.04	5.39	0.65
24	3-NH$_2$	Me	CF_3	5.23	6.15	– 0.92
25	3-NO$_2$	Me	CF_3	6.24	6.18	0.06
26	3-Cl	NH$_2$	CF_3	8.10	7.83	0.27
27	3-F	NH$_2$	CF_3	7.52	7.85	– 0.33
28	3-Br	NH$_2$	CF_3	8.16	7.92	0.24
29	3-Me	NH$_2$	CF_3	7.52	7.47	0.05
30	2-Cl	Me	CF_3	6.05	6.57	– 0.52
31	2-F	Me	CF_3	6.40	6.18	0.22
32	2-Me	Me	CF_3	6.10	5.93	0.17
33	2-OMe	Me	CF_3	4.00	4.85	– 0.85
34	2-F	NH$_2$	CF_3	7.00	6.90	0.10
35	2-Me	NH$_2$	CF_3	6.70	6.50	0.20
36	3-F-4-OMe	Me	CF_3	6.82	6.60	0.22
37	3-Cl-4-OMe	Me	CF_3	6.89	6.62	0.27
38	3-Cl-4-SMe	Me	CF_3	7.40	7.23	0.17
39	3-Cl-4-NMe$_2$	Me	CF_3	6.50	6.78	– 0.28

Table 10 (continued)

No.	Substituent X	Y	Z	Log $1/C$ Expt.	Calc.	Δ
40	3-F-4-NMe$_2$	Me	CF$_3$	6.48	6.74	− 0.26
41	3-Cl-4-NHMe	Me	CF$_3$	6.18	6.19	− 0.01
42	3-Cl-4-Me	Me	CF$_3$	7.52	7.39	0.13
43	3-F-4-Me	Me	CF$_3$	6.96	7.11	− 0.15
44	3-Me-4-F	Me	CF$_3$	6.77	7.00	− 0.23
45	3-Me-4-Cl	Me	CF$_3$	7.05	7.38	− 0.33
46	3-OMe-4-Cl	Me	CF$_3$	6.60	6.60	0.00
47	3-NMe$_2$-4-Cl	Me	CF$_3$	5.98	5.76	0.22
48	3,4-OCH$_2$O	Me	CF$_3$	6.77	7.38	− 0.61
49	3,4-F$_2$	Me	CF$_3$	6.92	6.61	0.31
50	3,4-Me$_2$	Me	CF$_3$	6.48	6.75	− 0.27
51	3-Me-5-Cl	Me	CF$_3$	7.10	7.06	0.04
52	3-Me-5-F	Me	CF$_3$	6.96	6.73	0.23
53	3-OMe-5-Cl	Me	CF$_3$	6.02	6.03	− 0.01
54	3,5-Cl$_2$	Me	CF$_3$	6.77	6.80	− 0.03
55	3-F-4-OMe	NH$_2$	CF$_3$	7.52	7.57	− 0.05
56	3-Cl-4-OMe	NH$_2$	CF$_3$	7.70	7.68	0.02
57	3-Br-4-OMe	NH$_2$	CF$_3$	7.52	7.73	− 0.21
58	3-Cl-4-SMe	NH$_2$	CF$_3$	8.00	8.31	− 0.31
59	3-Cl-4-Me	NH$_2$	CF$_3$	8.52	8.03	0.49
60	3-OMe-4-Cl	NH$_2$	CF$_3$	7.70	7.45	0.25
61	3,4-F$_2$	NH$_2$	CF$_3$	7.52	7.89	− 0.37
62	3-Me-5-Cl	NH$_2$	CF$_3$	7.40	7.77	− 0.37
63	3-Me-5-F	NH$_2$	CF$_3$	7.52	7.44	0.08
64	3-OMe-5-F	NH$_2$	CF$_3$	6.34	6.46	− 0.12
65	3,5-F$_2$-4-OMe	Me	CF$_3$	6.77	6.71	0.06
66	3,5-Cl$_2$-4-OMe	Me	CF$_3$	6.85	6.88	− 0.03
67	3,5-Br$_2$-4-OMe	Me	CF$_3$	7.05	6.81	0.24
68	3,5-Me$_2$-4-OMe	Me	CF$_3$	6.14	6.17	− 0.03
69	2,5-Me$_2$-4-OMe	Me	CF$_3$	4.91	4.72	0.19
70	3,5-Cl$_2$-4-NMe$_2$	Me	CF$_3$	6.85	6.35	0.50
71	3,5-F$_2$-4-OMe	NH$_2$	CF$_3$	7.52	7.65	− 0.13
72	4-Cl	Me	Me	6.62	7.79	− 1.17
73	4-Cl	Me	CF$_3$	6.96	7.03	− 0.07
74	4-Cl	Me	CHF$_2$	6.22	6.75	− 0.53
75	4-Cl	Me	CH$_2$F	6.39	5.84	0.55
76	4-Cl	Me	CHO	5.80	5.60	0.20
77	4-Cl	Me	CN	6.64	6.79	− 0.15
78	4-Cl	Me	COOC$_2$H$_5$	5.24	5.92	− 0.68

Table 10 (continued)

No.	Substituent X	Y	Z	Log $1/C$ Expt.	Calc.	Δ
79	4-Cl	Me	C_6H_5	6.62	8.04	– 1.42
80	4-Cl	Me	$CH_2OC_6H_4$-4-Cl	7.52	7.09	0.43
81	4-Cl	Me	$CH_2SC_6H_4$-4-Cl	7.30	7.43	– 0.13
82	4-Cl	Me	CH_2OMe	5.43	5.12	0.31
83	4-Cl	Me	CH_2OH	5.08	5.99	– 0.91
84	4-Cl	Me	CH_2SMe	6.50	6.23	0.27
85	4-Cl	Me	CH_2CN	5.81	5.66	0.15

Table 11 Descriptors important for the prediction of COX-2 inhibition of 1,2-diaryl-imidazoles

Descriptor	Definition
NHbint5	Number of potential internal hydrogen bonds of path length 5
NHbint9	Number of potential internal hydrogen bonds of path length 9
J	Balaban's index based on topological distance
DSN_2	Triplet index from distance matrix, distance sum, and graph order; operation = summation of squares
SHssNH	Sum of atom-type H E-states, NH with two single bonds
O_{orb}	Order of neighborhood when IC_r reaches its maximum value for the hydrogen-suppressed graph
ASN_2	Triplet index from adjacency matrix, distance sum, and graph order; operation = summation of squares
DN^2N_1	Triplet index from distance matrix, square of graph order, and graph order; operation = summation
DN^2I_3	Triplet index from distance matrix, square of graph order, and number 1; operation = Summation of square roots
DN^2N_5	Triplet index from distance matrix, square of graph order, and graph order; operation = product

The independent variables used in the model developed by Garg et al. [72] for the set of 1,2-diarylimidazoles included a calculated octanal:water partition coefficient, Verloop's sterimol parameter representing the length of the X substituents at the 2-position, McGowan's volume, and an indicator variable indicating NH_2 or Me for substituent Y. Modeling results included a q^2 value of 0.802; however, their model was based on 83 compounds, with two compounds removed as outliers.

5.4
Molecular Similarity and Tailored Similarity

Similarity is an old, intuitive concept that is frequently used to compare two or more objects. Like beauty, the perceived similarities (or differences) between any two objects depend largely on the person making the comparison. Which aspects are emphasized and which are ignored or minimized depends largely on personal perception or bias. Likewise, it is intuitively held that similar things behave or function in similar manners. On a broader scope, this "structure property similarity principle" includes the notion that similar "structural organizations" of objects leads to similar observable properties. Like general notions of similarity, molecular similarity is a largely intuitive concept and has been employed for years. In the realms of chemistry, biology, and toxicology, the natural extension of this structure–property similarity principle is that atoms, molecules, and macromolecules with similar structures will have similar physicochemical, biological, and toxicological properties. Chemical design and synthesis have long relied on the structure property similarity principle, but even so molecular similarity remains a largely intuitive, rather than quantitative, concept [74]. Xenoestrogens are an obvious recent example of this principle in action. The structural similarities between estrogen and a wide variety of hormonally active xenobiotics are the basis for endocrine disruption. Since the estrogen receptor cannot distinguish between these chemicals, it responds to a perceived overabundance of estrogen in the body [75].

One practical application of molecular similarity is the selection of analogs. Once a lead structure with interesting properties is identified, it is often useful to find similar chemicals with analogous properties. In contemporary drug discovery research, scientists usually search various proprietary and public domain databases for chemical analogs. Analogs can be selected based on the researcher's intuitive notion of chemical similarity, their similarity with respect to measured properties or calculated molecular descriptors, or their similarity based on an identified pharmacophore model. Since most of the chemicals in both public and proprietary databases have little experimental property data, similarity methods based on calculated properties, molecular descriptors, or pharmacophores are generally used for analog selection.

5.4.1
Construction of Molecular Similarity Spaces

There is no single consensus or method for characterizing and quantifying molecular similarity. In most cases, measures of molecular similarity are defined by the individual practitioner based on personal experience or chemical intuition. If the researcher selects n different quantifiable attributes for the molecules under investigation, the molecules can be looked upon as points in an n-dimensional space. Once such a similarity space is constructed, some sort of distance function such as Euclidean distance (Eq. 2) can be used to measure the distance between any two chemicals. The distance serves as a measure of the similarity or dissimilarity of any pair of molecules:

$$ED_{ij} = \left[\sum_{k=1}^{n} \left(D_{ik} - D_{jk} \right)^2 \right]^{\frac{1}{2}}$$

(2)

Difficulties in assessing molecular similarity arise from two major factors: (1) the selection of appropriate attributes for developing the space, and (2) the relevance of the attributes or descriptors to the property under investigation. Many practitioners of molecular similarity have their own favorite measures, but in many cases the axes selected are multicollinear, encoding essentially the same information multiple times. One solution to this problem is the use of principal components analysis (PCA) [76] to create completely orthogonal axes (principle components or PCs) derived from the original attributes [13, 77–88]. A more serious concern is whether or not the attributes, whether they are the original attributes or principle components, are relevant to the property of interest. While the use of PCA removes concerns about multicollinearity, the PCs are constructed in such a manner that they best characterize the descriptor space without any regard to the property of interest. So, once again, it becomes a question of selecting the best set of descriptors for estimating the property under investigation. One potential solution to this issue is the use of the tailored similarity method, which selects descriptors based on their relevance in predicting the property of interest [88–91].

5.4.2
The Tailored Approach to Molecular Similarity

User-defined or arbitrary molecular similarity methods perform reasonably well in narrow, well-defined situations, but the relationships between structural attributes and biomedicinal or toxicological properties are not always neatly defined, and human intuition often fails. Similarity methods based on objectively defined relationships are needed, rather than those derived from subjective or intuitive notions. In a multivariate space, this should be accomplished using robust statistical methods. The tailored similarity method makes use of a large number of molecular descriptors [88–91]. These descriptors are analyzed via ridge regression, modeling the property of interest and, based on the results of ridge regression, a small number of independent variables with high $|t|$-values are selected as the axes of the similarity space. In this way, we select variables that are strongly related to the property of interest, instead of subjectively selecting a group of descriptors based on the overall dimensionality of the descriptor set.

Our most recent tailored similarity study [89] examined the effects of tailoring on the estimation of mutagenicity for a set of 95 aromatic and heteroaromatic amines. In this study we utilized a much larger set of topological indices than have been used in many of our older studies, incorporating both the Triplet indices and a large set of descriptors from Molconn-Z v3.5. Three distinct similarity spaces were constructed, though two were "overlapping" spaces. The overlapping spaces were derived using principal components analysis on the set of 267 topological indices. The PCA created 20 completely orthogonal components with eigenvalues greater than one. These 20 PCs were used as the axes for the first similarity space.

The second similarity space was constructed by selecting the index most correlated with each PC. One of the arguments against using PCA to reduce the number of variables for modeling is that PCs, as linear combinations of the indices, are not easily interpretable. So, by selecting a single TI from each PC, we have a set of somewhat more easily interpretable set of axes to use in modeling.

Finally, the third set of indices was selected based on a ridge regression model using all 267 indices to model mutagenicity. From the modeling results, $|t|$-values were extracted and the 20 top indices were selected as axes for developing the similarity space. A summary of the correlation coefficients for

Table 12 Summary of the correlation coefficients (R) and standard errors ($s.e.$) for the three similarity methods

k	R			$s.e.$		
	PC	TI/PC	RR	PC	TI/PC	RR
1	0.749	0.717	0.850	1.406	1.471	1.051
2	0.783	0.752	0.871	1.242	1.319	0.957
3	0.790	0.755	0.879	1.194	1.278	0.929
4	0.804	0.761	0.877	1.149	1.254	0.927
5	0.786	0.754	0.878	1.192	1.271	0.923
6	0.758	0.760	0.872	1.255	1.251	0.943
7	0.756	0.754	0.862	1.263	1.269	0.976
8	0.757	0.717	0.857	1.265	1.348	0.992
9	0.763	0.720	0.861	1.258	1.343	0.981
10	0.760	0.701	0.850	1.276	1.380	1.016
15	0.719	0.634	0.854	1.395	1.505	1.014
20	0.753	0.555	0.855	1.444	1.631	1.031
25	0.761	0.509	0.849	1.519	1.685	1.069

estimating mutagenicity from the three similarity spaces for varying numbers of neighbors are presented in Table 12.

As can be easily observed from the table, tailoring the selected set of indices significantly improved the estimative power of the model, resulting in roughly a 10% increase to the correlation coefficient. These results, as with all of the results we have seen from tailored similarity spaces, are promising and we believe that tailored spaces will be very useful both in drug discovery and toxicological research.

5.5
Molecular Similarity and Analog Selection

As mentioned earlier, many times it is of interest to researchers to be able to select a set of analogs for a chemical of interest from a large, diverse

Probe Compound

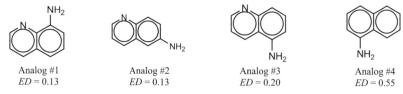

Analogs from a PC-based similarity space

Analog #1
ED = 0.13

Analog #2
ED = 0.13

Analog #3
ED = 0.20

Analog #4
ED = 0.55

Analogs from a TI-based similarity space derived from PCs

Analog #1
ED = 0.03

Analog #2
ED = 0.09

Analog #3
ED = 0.09

Analog #4
ED = 0.57

Analogs from a tailored TI-based similarity space

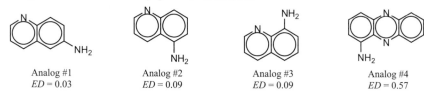

Analog #1
ED = 0.08

Analog #2
ED = 0.16

Analog #3
ED = 0.18

Analog #4
ED = 0.52

Fig. 6 Four nearest neighbors (analogs) for a heteroaromatic probe compound selected from a principle component-based similarity space, a topological index-based similarity space, and a tailored topological index space, respectively

Probe Compound

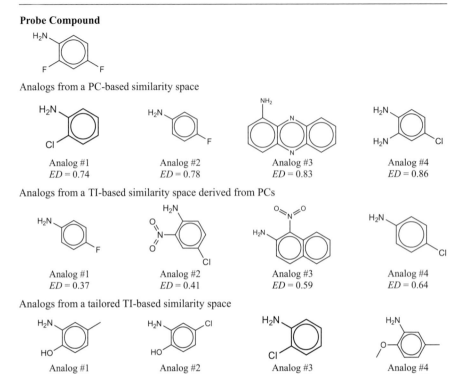

Fig. 7 Four nearest neighbors (analogs) for an aromatic probe compound selected from a principle component-based similarity space, a topological index-based similarity space, and a tailored topological index space, respectively

data set based on similarity spaces derived solely from calculated descriptors of molecular structure. In a number of publications, we have shown the utility of quantitative molecular similarity analysis techniques in the selection of analogs based on various molecular descriptor spaces [13, 76, 77, 80, 91, 92]. Figures 6 and 7 each present three sets of analogs selected for two different "chemicals of interest" from the set of 95 aromatic and heteroaromatic amines discussed above. Each figure shows analogs selected from the two non-tailored spaces and from the tailored space. Figure 6 presents the four nearest neighbors (or analogs) selected for one heteroaromatic compound from the set of 95 molecules, while Fig. 7 presents four selected analogs for one aromatic compound. As can be seen from the figures, while there is overlap between the analogs selected, various methods order the analogs differently and some surprising (non-intuitive) analogs are selected.

5.6
Hierarchical Molecular Overlay

5.6.1
Polychiral Diastereoisomerism: An Alternative Approach

Several pharmaceutical industries are introducing new enantiopure drugs (racemic switch), and the worldwide annual sales of enantiomeric drugs had exceeded 100 billion dollars [93]. A racemic switch is the redevelopment, in single-isomer form, of a chiral drug that was originally approved for marketing as a racemate. Though this is considered a management strategy to increase the patent life, it reduces the use of the less potent or the inactive component, which may affect non-target biological systems. This phenomenon can be explained with the example of a heterocyclic drug, glitazones, used to treat type II diabetes. Glitazones are 5-substituted 1,3-thiazolidine-2,4-diones (Fig. 8), with C-5 as the chiral center, and the drug has a pair of enantiomers. Due to the presence of the carbonyl group at the C-4 position, the compound undergoes spontaneous racemization and is sold only as the racemate. One of the glitozones, troglitazone (Rezulin), has a second chiral center and was marketed as a mixture of four stereoisomers. It was removed from the market by the US Food and Drug Administration on March 21, 2001 [94] because of fatal liver damage observed in a small number of patients who took troglitazone. The liver cells are believed to be affected by the toxic effects of the two stereoisomers arising out of the second chiral center in the chromane part of molecule (Fig. 8). Similar situations arise

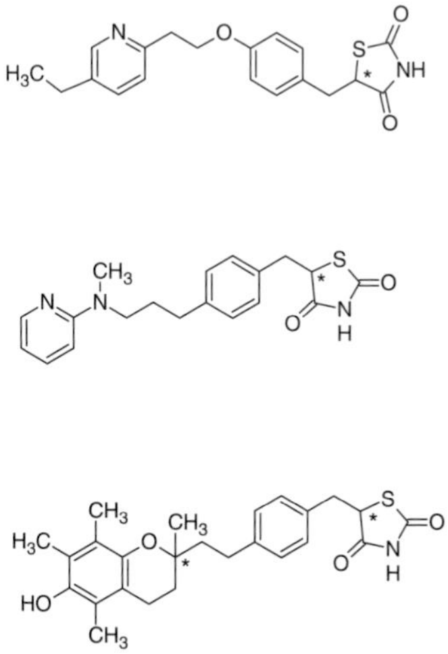

Fig. 8 Glitazones (chiral centers are marked with an *asterisk*): A pioglitazone (Actos), B rosiglitazone (Avandia), C troglitazone (Rezulin)

in the case of agrochemicals where different types of biological activity are observed between stereoisomers. Vinyl triazoles are used extensively as agrochemicals. Diniconazole is 1-(2,4-dichlorophenyl)-4,4-dimethyl-2-(1H-1,2,4-triazol-1-yl)pent-1-en-3-ol (structure is shown in Fig. 3) and is capable of exhibiting both E – Z and optical isomerism. Hence, each geometrical isomer has R and S enantiomers, and there are four isomeric compounds. Any biological activity of diniconazole is concentrated only in the E-isomers, and the Z-isomers are inactive. The optical antipodes of the E-isomer have different biological activities. ($1E, 3R$) isomer exhibits fungicidal activity while the ($1E, 3S$) isomer is a poor fungicide but a good plant growth regulator [95]. These examples and several other similar cases highlight the importance of the use of enantiopure compounds. Conventional QSAR modeling using simple computed molecular descriptors or physicochemical properties fails in handling compounds that exhibit polychiral diastereomerism. In such situations, a quantitative stereochemical structure–activity relationship modeling (QSSAR) approach is necessary. In order to utilize the QSSAR approach, we need descriptors that encode the stereochemical features of a molecule. Though Schultz et al. [96] developed a set of descriptors that could take into account stereoisomerism, they were found to have limitations in application to QSSAR of repellency of diastereomers of piperidine amides. The indices developed based on the Schultz formulation always have the highest value for a diastereoisomer that has the R-configuration in all chiral centers. For example, in the case of a compound with two asymmetric carbons, there will be four diastereomers namely, RR, RS, SR, and SS. Among these, RR has the highest value while SS has the lowest value of the Schultz-type indices, and these difficulties arise due to the treatment of chirality on a binary scale (+ 1 and – 1). Instead of this approach, a relative measure of chirality must be considered, which measures chirality on a continuous scale [97]. Indices that encode the relative chirality of diastereomers have not been developed. In the absence of such an approach, comparing the entire spatial configuration of atoms in molecules offers an alternative and very useful methodology.

5.6.2
Overlay of Mosquito Repellents

In our molecular overlay approach, optimized geometries of the two structures to be compared are overlaid, and the root mean square distance (RMSD) between the atoms is considered as the measure of similarity between the two overlaid structures. Two or more structures can be overlaid with respect to the atoms specified by the user, and the RMSD can be measured for any user-defined set of atoms, backbone atoms or all of the atoms including hydrogen. Due to this advantage, the approach could be used to overlay even structurally dissimilar molecules, as long as they have a common motif. One can zero in on a putative pharmacophore/biophore by overlaying different parts of the structures and correlating the efficacy of the various stereoisomers and the order of goodness of overlay measured in terms of RMSD. The use of molecular overlay to order the repellency of the diastereoisomers of two piperidine derivatives namely, AI3-37220 (1-(cyclohex-3-ene-1-ylcarbonyl)-2-methylpiperidine), and Picaridin (2-(2-hydroxyethyl)-1-piperidine carboxylic acid-1-methylpropyl ester) is explained here.

Both Picaridin and AI3-37220 have two asymmetric centers (Fig. 9), and the four diastereoisomers (*RR*, *RS*, *SR*, and *SS*) of each compound are known to have differing degrees of mosquito-repellent activity according to quantitative behavioral assays conducted at the US Department of Agriculture [98]. These compounds were tested as alternatives to DEET (*N,N*-diethyl-3-methylbenzamide), the most popular topical mosquito repellent, which has been on the market for nearly five decades. The order of repellency of the diastereomers of 220 is: *SS* > *RS* > *SR* > *RR*. In the case of Picaridin, the order is: *RS* > *RR* > *SR* > *SS*. The initial molecular overlay study [99] was carried out using molecular mechanics (MM2) optimized geometries of the diastereoisomers. It was clear from the study that most active compounds, namely Picaridin*RS*, 220*SS*, and DEET, have very similar structural motifs, which lead to a high degree of matching of the relevant parts of the molecule. In sharp contrast to this is the stereochemical dissimilarity between these three active structures, on the one hand, and the collection of less active diastereoisomers of 220 and Picaridin (Fig. 10).

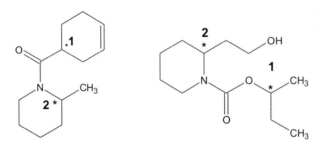

Fig. 9 Structures of AI3-37220 and Picaridin

Fig. 10 Similarity among the most active compounds (220*SS*, Picaridin*RS*, and DEET) and dissimilarity of the least active compounds (220*RR* and Picaridin*SS*) from the standard (DEET)

Though the active and inactive diastereoisomers were distinguished, the proper ordering of the eight diastereoisomers and DEET was not achieved using the overlay of MM2-optimized structures. The limitation in the molecular mechanics (MM2) geometry optimization was considered to be one of the reasons for this. In order to overcome this handicap, the lowest energy conformers of the four diastereoisomers of AI3-37220, four diastereoisomers of Picaridin, and DEET were optimized in a hierarchical manner using AM1 to Hartree–Fock (HF) to density functional theory (DFT) at the levels of various basis sets. The predefined hierarchy followed was: AM1 < HF/STO-3G < HF/321G < HF/6-31G < HF/6-311G < B3LYP/6-31G < B3LYP/6-311G. The objective of the hierarchical approach (Basak et al. unpublished results) was to observe the refinement in results while moving up the hierarchy and to determine the level (*basis set*) to which one should go in order to obtain reliable results. Overlay with respect to the five atoms ($C^* - N(C) - C = O$) and the optimization of the geometries using HF/321G gave the best results, and the RMSDs were found to match well with the repellent activity of the nine compounds. The most interesting outcome of the molecular overlay study was the near self-similarity of the pairs of diastereoisomers of Picaridin, namely *RR* and *RS*, and *SR* and *SS*, and this matched with the nearly same repellency observed between the compounds in the two pairs. The importance of one of the two chiral centers was also revealed because the putative pharmacophore did not contain the chiral center outside the piperidine ring. This was a very significant finding that provided important information in directing the future research and synthesis of piperidine analogs as mosquito repellents. It is to be noted that the zeroing-in on the putative pharmacophore was achieved in the absence of any information such as structure of the protein site to which the repellent molecules bind. However, this approach has one limitation: An active stereoisomer/compound must be known, and its optimized geometry should provide the template to perform the overlay. Though one can explore the 3D structure space around the active compound to screen less active and inactive compounds, it would be difficult to identify a compound more active than the standard. In the molecular overlay approach, overlay of minimum energy conformers was preferred, an idea with which many may disagree contending that the minimum energy conformer need not be the one that binds to the receptor. The flip side of this argument is that the global minimum is a single conformation and every higher energy state has multiple conformations with equal energy. Without a receptor structure, it is impossible to select one of the multiple higher energy conformations over another. These were the rationales for the use of least energy conformers in our overlay approach. The highlights of hierarchical molecular overlay are:

1. Identification of the putative pharmacophore, even in the absence of any knowledge of the receptor site

2. Formulation of an alternative scheme for 3D quantitative structure–activity relationship (3D-QSAR) modeling
3. Exploration of structures generated by different levels of molecular modeling viz., MM2, AM1, HF (STO-3G, 321G, 6-31G, 6-311G), and DFT (B3LYP/6-31G, B3LYP/6-311G), in a graduated manner whereas most available commercial software use force field calculations

6
Conclusion

In this chapter, we have developed predictive models based on two well established methods: (1) hierarchical quantitative structure–activity (HiQSAR) modeling, and (2) quantitative molecular similarity analysis (QMSA). We have reviewed published work in both of the above areas for important classes of heterocyclic compounds that have therapeutic and toxic effects. Predictive models can be developed based on experimental properties, substituent constants derived from such properties, and also theoretical descriptors which can be calculated directly from molecular structure. In view of the fact that most potential therapeutic agents and the majority of known drugs and toxicants do not have experimental data available for their evaluation, theoretical descriptors are very useful in the initial screening of compound libraries.

For the three classes of molecules for which detailed HiQSARs have been discussed here, viz., dibenzofurans, heteroaromatic and aromatic amines, and imidazole Cox-2 inhibitors, a hierarchical QSAR approach using a combination of topostructural, topochemical, 3D, and quantum chemical descriptors produced high quality models. Such models, based on purely calculated descriptors, can find application in the routine drug design process as shown in Fig. 11.

One important question arises when we look at these QSARs: Why is a set of calculated descriptors capable of predicting bioactivities of such structurally and biologically diverse chemicals? The answer lies in the principle "diversity begets diversity". What this means is that if we have a diverse pool of descriptors that characterize various aspects of molecular architecture at a fundamental level, they will have the capability of correlating diverse biochemical phenomena of structurally non-homogeneous chemicals. This is what has been achieved through the combination of topological, geometrical, and quantum chemical descriptors described in Fig. 3.

Theoretical descriptor-based QSARs can serve three purposes:

1. The parameters important for the estimation of bioactivity or toxicity can be determined and used in order to understand the molecular and submolecular basis of the property under investigation (description)

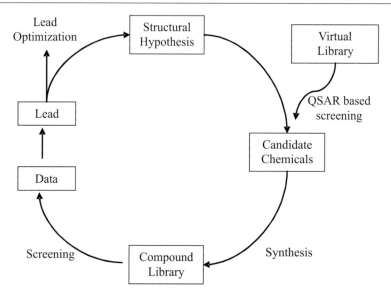

Fig. 11 QSAR in drug discovery. Reprinted with permission; source from [91]

2. Good quality QSARs based on algorithmically derived descriptors can be used for the estimation of the property of interest for any chemical, real or hypothetical (prediction)
3. A predictive QSAR for the desirable property, along with a series of QSAR for deleterious effects, can be used in the screening of real or virtual libraries in the rational design of drugs or specialty chemicals (prescription)

It is noteworthy that high quality QSARs cannot be developed if the critical factors related to bioactivity are not present in the set of descriptors chosen. This is well illustrated with the activity of diastereoisomeric insect repellents. These molecules differ only in the spatial configuration of atoms. All topological, geometrical, and quantum chemical indices will have identical (redundant) values for the various diastereomers corresponding to the same empirical formula. Such situations call for novel approaches. As evident from our results on hierarchical overlay, the comparison of good quality structures generated by quantum chemical methods is needed in such cases.

In drug design, one uses the molecular similarity concept to select analogs of interesting leads. Such selections are usually made based on user-defined methods. Therefore, such methods are always biased. We have come up with the idea of "tailored similarity", where the property of interest is used to create a structure space that is relevant to that specific property. The results indicate that such tailored methods outperform arbitrary, user-defined QMSA techniques.

It is expected from results reviewed in this chapter that both HiQSAR and tailored QMSA based on calculated molecular descriptors will be powerful tools in drug design and hazard assessment of chemicals.

Acknowledgements Research reported in this chapter was supported, in part, by Grant F49620-02-1-0138 from the United States Air Force and Cooperative Agreement Number 572112 from the Agency for Toxic Substances and Disease Registry. This chapter represents contribution number 399 from the Center for Water and the Environment of the Natural Resources Research Institute. The US Government is authorized to reproduce and distribute reprints for Governmental purposes notwithstanding any copyright notation thereon. The views and conclusions contained herein are those of the authors and should not be interpreted as necessarily representing the official policies or endorsements, either expressed or implied, of the Air Force Office of Scientific Research or the US Government.

References

1. Hammond GS (1968) Norris Award Lecture
2. Schrodinger E (1954) Nature and the Greeks. Cambridge University Press, Cambridge, UK
3. Tagore R (2005) The wakening of Siva. In: Selected poems; translated by William Radice. Penguin, London
4. Goodman AG, Wall TW, Nies AS, Taylor P (eds) (1990) Goodman and Gilman's the pharmacological basis of therapeutics. Pergamon, New York
5. 2006 Physicians' desk reference, 60th edn. (2005) Thomson Healthcare, Montvale NJ
6. Stuper AJ, Brugger WE, Jurs PC (1979) Computer-Assisted studies of chemical structure and biological function. Wiley, New York
7. Martin YC (1996) Network Science, http://www/netsci.org/Science/Combichem/feature09.html
8. Davies K (2002) BIO-IT World, http://www.bio-itworld.com/archive/071102/firstbase.html
9. Boros LG (2005) Metabolomics 1:11
10. Buchanan SG (2002) Curr Opin Drug Discov Devel 5:367
11. Molla M, Waddell M, Page D, Shavlik J (2004) AI Magazine 25:23
12. Walgren JL, Thompson DC (2004) Toxicol Lett 149:377
13. Basak SC, Magnuson VR, Niemi GJ, Regal RR (1988) Discrete Appl Math 19:17
14. Basak SC, Niemi GJ, Veith GD (1990) J Math Chem 4:185
15. Basak SC, Niemi GJ, Veith GD (1991) J Math Chem 7:243
16. Weininger SJ (1984) J Chem Educ 61:939
17. Bunge M (1973) Method, Model and Matter. D. Reidel, Dordrecht-Holland
18. Trinajstic N (1992) Chemical graph theory. CRC, Boca Raton
19. Wiener H (1947) J Am Chem Soc 69:17
20. Randic M (1975) J Am Chem Soc 97:6609
21. Kier LB, Murray WJ, Randic M, Hall LH (1976) J Pharm Sci 65:1226
22. Kier LB, Hall LH (1990) Pharm Res 8:801
23. Basak SC (1999) Information theoretic indices of neighborhood complexity and their applications. In: Devillers J, Balaban AT (eds) Topological indices and related descriptors in QSAR and QSPR. Gordon and Breach Science, The Netherlands, p 563

24. Basak SC, Gute BD, Grunwald GD (1998) Relative effectiveness of topological, geometrical, and quantum chemical parameters in estimating mutagenicity of chemicals. In: Chen F, Schuurmann G (eds) Quantitative structure–activity relationships in environmental sciences VII. SETAC, Pensacola, FL, p 245

25. Devillers J, Balaban AT (eds) (1999) Topological indices and related descriptors in QSAR and QSPR. Gordon and Breach Science, The Netherlands

26. Basak SC, Mills D, Gute BD, Grunwald GD, Balaban AT (2002) Applications of topological indices in property/bioactivity/toxicity prediction of chemicals. In: Rouvray DH, King RB (eds) Topology in chemistry: discrete mathematics of molecules. Horwood, Chichester, England, p 113

27. Bonchev D (1983) Information theoretic indices for characterization of chemical structures. Research Studies, Letchworth, UK

28. Sabljic A, Trinajstic N (1981) Acta Pharm Yugosl 31:189

29. Ray SK, Gupta S, Basak SC, Raychaudhury C, Roy AB, Ghosh JJ (1985) Indian J Chem 24B:1149

30. Seybold PG (1999) SAR QSAR Environ Res 10:101

31. Basak SC, Magnuson VR, Niemi GJ, Regal RR, Veith GD (1987) Math Model 8:300

32. Basak SC, Gute BD, Balaban AT (2004) Croat Chem Acta 77:331

33. Basak SC, Balaban AT, Grunwald GD, Gute BD (2000) J Chem Inf Comput Sci 40:891

34. Klein DJ (1986) Int J Quant Chem 30:153

35. Basak SC, Harriss DK, Magnuson VR (1988) POLLY. Copyright of the University of Minnesota

36. Filip PA, Balaban TS, Balaban AT (1987) J Math Chem 1:61

37. Molconn-Z v3.5 (2000) Hall Associates Consulting, Quincy, MA

38. Tripos Associates(1995) SYBYL v6.2. St. Louis, MO

39. Stewart JJP (1990) MOPAC v6.00, QCPE #455. Frank J Seiler Research Laboratory, US Air Force Academy, CO

40. Frisch MJ, Trucks GW, Schlegel HB, Scuseria GE, Robb MA, Cheeseman JR, Montgomery JA, Vreven T, Kudin KN, Burant JC, Millam JM, Iyengar SS, Tomasi J, Barone V, Mennucci B, Cossi M, Scalmani G, Rega N, Petersson GA, Nakatsuji M, Hada M, Ehara M, Toyota K, Fukuda J, Hasegawa J, Ishida M, Nakajima T, Honda Y, Kitao O, Nakai H, Klene M, Li X, Knox JE, Hratchian HP, Cross JB, Adamo C, Jaramillo J, Gomperts R, Stratmann RE, Yazyev O, Austin AJ, Cammi R, Pomelli C, Ochterski J, Ayala PY, Zakrzewski VG, Dapprich S, Daniels AD, Strain MC, Farkas O, Malick DK, Rabuck AD, Raghavachari K, Foresman JB, Ortiz JV, Cui Q, Baboul AG, Clifford S, Cioslowski J, Stefanov BB, Liu G, Liashenko A, Piskorz P, Komaromi I, Martin RL, Fox DJ, Keith T, Al-Laham MA, Peng CY, Nanayakkara A, Challacombe M, Gill PMW, Johnson B, Chen W, Wong MW, Gonzalez C, Pople JA (2004) Gausssian 03W, v6.0 (revision C.02)

41. Hoerl AE, Kennard RW (1970) Technometrics 12:55

42. Hoerl AE, Kennard RW (2005) Technometrics 12:69

43. Massy WF (1965) J Am Statistical Assoc 60:234

44. Wold S (1993) Technometrics 35:136

45. Rencher AC, Pun FC (1980) Technometrics 22:49

46. Frank IE, Friedman JH (1993) Technometrics 35:109

47. Hawkins DM, Basak SC, Mills D (2003) J Chem Inf Comput Sci 43:579

48. Thisted RA (1988) Elements of statistical computing: numerical computation. Chapman and Hall, New York

49. Basak SC, Gute BD, Grunwald GD (1999) SAR QSAR Environ Res 10:117

50. Basak SC, Mills DR, Balaban AT, Gute BD (2001) J Chem Inf Comput Sci 41:671

51. Basak SC, Grunwald GD, Niemi GJ (1997) Use of graph-theoretic and geometrical molecular descriptors in structure–activity relationships. In: Balaban AT (ed) From chemical topology to three-dimensional geometry. Plenum, New York, p 73
52. Basak SC, Mills D (2001) SAR QSAR Environ Res 12:481
53. Basak SC, Gute BD, Ghatak S (1999) J Chem Inf Comput Sci 39:255
54. Basak SC, Mills D (2001) J Chem Inf Comput Sci 41:692
55. Basak SC, Gute BD, Grunwald GD (1997) J Chem Inf Comput Sci 37:651
56. Basak SC, Gute BD, Grunwald GD (1996) J Chem Inf Comput Sci 36:1054
57. Gute BD, Basak SC (1997) SAR QSAR Environ Res 7:117
58. Basak SC, Mills D, El-Masri HA, Mumtaz MM, Hawkins DM (2004) Environ Toxicol Pharmacol 16:45
59. Basak SC, Mills D, Hawkins DM, El-Masri H (2003) Risk Analysis 23:1173
60. Basak SC, Mills D, Hawkins DM, El-Masri HA (2002) SAR QSAR Environ Res 13:649
61. Gute BD, Grunwald GD, Basak SC (1999) SAR QSAR Environ Res 10:1
62. Gute BD, Balasubramanian K, Geiss K, Basak SC (2004) Environ Toxicol Pharmacol 16:121
63. Basak SC, Mills D, Mumtaz MM, Balasubramanian K (2003) Indian J Chem 42A:1385
64. So SS, Karplus M (1997) J Med Chem 40:4360
65. Debnath AK, Debnath G, Shusterman AJ, Hansch C (1992) Environ Mol Mutagen 19:37
66. Basak SC, Mills D, Gute BD, Hawkins DM (2003) Predicting mutagenicity of congeneric and diverse sets of chemicals using computed molecular descriptors: a hierarchical approach. In: Benigni R (ed) Quantitative structure–activity relationship (QSAR) models of mutagens and carcinogens. CRC, Boca Raton, FL, p 207
67. Basak SC, Niemi GJ, Veith GD (1990) Recent developments in the characterization of chemical structure using graph-theoretic indices. In: Rouvray DH (ed) Computational chemical graph theory. Nova Science, New York, p 235
68. Toropov AA, Toropova AP (2001) J Mol Struct (THEOCHEM) 538:287
69. Cash GG (2001) Mutation Res Genet Toxicol Environ Mutagen 491:31
70. Maran U, Karelson M, Katritzky AR (1999) Quant Struct-Act Relat 18:3
71. Khanna IK, Weier RM, Yi Y, Xu XD, Koszyk FJ, Collins PW, Koboldt CM, Veenhuizen AW, Perkins WE, Casler JJ, Masferrer JL, Zhang YY, Gregory SA, Seibert K, Isakson PC (1997) J Med Chem 40:1634
72. Garg R, Kurup A, Mekapati SB, Hansch C (2003) Chem Rev 103:703
73. Johnson M, Maggiora GM (1990) Concepts and applications of molecular similarity. Wiley, New York
74. Starek A (2003) Int J Occup Med Environ Health 16:124
75. SAS Institute, Inc (1988) SAS/STAT user guide, release 6.03, Cary, NC
76. Basak SC, Bertelsen S, Grunwald G (1994) J Chem Inf Comput Sci 34:270
77. Basak SC, Bertelsen S, Grunwald GD (1995) Toxicology Lett 79:239
78. Basak SC, Grunwald GD, Host GE, Niemi GJ, Bradbury SP (1998) Environ Toxicol Chem 17:1056
79. Basak SC, Grunwald GD (1995) New J Chem 19:231
80. Basak SC, Grunwald GD (1994) SAR QSAR Environ Res 2:289
81. Basak SC, Grunwald GD (1995) J Chem Inf Comput Sci 35:366
82. Basak SC, Grunwald GD (1995) Chemosphere 31:2529
83. Basak SC, Grunwald GD (1995) SAR QSAR Environ Res 3:265
84. Basak SC, Gute BD, Grunwald GD (2000) Use of graph invariants in QMSA and predictive toxicology, DIMACS Series 51. In: Hansen P, Fowler P, Zheng M (eds) Discrete mathematical chemistry. American Mathematical Society, Providence Rhode Island, p 9

85. Gute BD, Grunwald GD, Mills D, Basak SC (2001) SAR QSAR Environ Res 11:363
86. Basak SC, Gute BD (1997) SAR QSAR Environ Res 7:1
87. Basak SC, Gute BD (1997) Use of graph theoretic parameters in predicting inhibition of microsomal hydroxylation of anilines by alcohols: a molecular similarity approach. In: Johnson BL, Xintaras C, Andrews JS (eds) Proceedings of the international congress on hazardous waste: impact on human and ecological health. Princeton Scientific, Princeton, NJ, p 492
88. Basak SC, Gute BD, Mills D (2002) SAR QSAR Environ Res 13:727
89. Basak SC, Gute BD, Mills D, Hawkins DM (2003) J Mol Struct (THEOCHEM) 622:127
90. Gute BD, Basak SC, Mills D, Hawkins DM (2002) Internet Electronic J Mol Design 1:374
91. Basak SC, Mills D, Gute BD (2006) Predicting Bioactivity and Toxicity of Chemicals from Mathematical Descriptors: A Chemical-cum-Biochemical Approach. In: Klein DJ, Brandas E (eds) Advances in Quantum Chemistry, Elsevier, in press
92. Basak SC, Gute BD, Grunwald GD (1998) Characterization of the molecular similarity of chemicals using topological invariants. In: Carbo-Dorca R, Mezey PG (eds) Advances in molecular similarity, vol 2. JAI, Stanford, Connecticut, p 171
93. Stinson SC (1999) Chem Eng News 77:101
94. Food and Drug Administration (2000) http://www.fda.gov/bbs/topics/NEWS/NEW00721.html
95. Takano H, Oguni Y, Kato T (1986) J Pesti Sci 11:373
96. Schultz HP, Schultz EB, Schultz TP (1995) J Chem Inf Comput Sci 35:864
97. Zabrodsky H, Avnir D (1995) J Am Chem Soc 117:462
98. Klun JA, Schmidt WF, Debboun M (2001) J Med Entomol 38:809
99. Natarajan R, Basak SC, Balaban AT, Klun JA, Schmidt WF (2005) Pest Manag Sci 61:1193

Top Heterocycl Chem (2006) 3: 81–147
DOI 10.1007/7081_027
© Springer-Verlag Berlin Heidelberg 2006
Published online: 7 April 2006

Conformational Aspects and Interaction Studies of Heterocyclic Drugs

M. N. Ponnuswamy[1] (✉) · M. Michael Gromiha[2] (✉) · S. M. Malathy Sony[1] · K. Saraboji[1]

[1]Department of Crystallography and Biophysics, University of Madras, Guindy Campus, Chennai 600 025, India
mnpsy@hotmail.com

[2]Computational Biology Research Center (CBRC), National Institute of Advanced Industrial Science and Technology (AIST), AIST Tokyo Waterfront Bio-IT Research Building, 2-42 Aomi, Koto-ku, 135-0064 Tokyo, Japan
michael-gromiha@aist.go.jp

Abstract Drug discoveries require the iterative synthesis—along with structural studies—of numerous individual analogues of biologically and medicinally active compounds. Over half of all known compounds and a large number of pharmaceutical products are heterocyclic in nature. The pharmacological activity of drugs depends mainly on interaction with their biological targets, which have a complex three-dimensional structure, and molecular recognition is guided by the nature of the intermolecular interactions. Furthermore, the drug's polymorphic nature also adversely affects its abilities. In order to address these factors, the stereochemical analysis of various piperidine and azepine derivatives, weak π-interaction analysis of isoxazole, imidazole, indole, quinoline and triazole and polymorphic analysis of two commercial drugs, valdecoxib and sildenafil citrate were carried out. Only the crystal structures were used for these analyses, of which the piperidine and azepine derivatives, valdecoxib and sildenafil citrate were solved by our group. To understand the structure-activity relationship, the results of these studies were correlated with the crystal structure of their respective drug molecules that are found in complex with the receptors. Stereochemical analysis showed that the ring conformation and orientation of the substituents correlate well with the active conformation of the drug. The π-systems prefer to form an offset stacking $\pi...\pi$ interaction geometry similar to the phenylalanine–phenylalanine interactions in proteins. In the polymorphic analysis one of the crystal conformations of valdecoxib proved to have better interaction with its receptor indicating higher activity.

Keywords Crystal structures · Conformation · Weak interaction · Polymorphism · Piperidine

1
Introduction

Considerable progress has been observed over the past few decades in molecular development and design, in particular, in drug design. A drug is any chemical agent that affects the processes of living and the creation of novel molecules with specific properties has been the cherished goal of chemists for generations. Finding new drugs is an important part of the initiatives taken by researchers in health care [1]. Drug discoveries require iterative synthesis—along with structural studies—of numerous individual analogues of biologically and medicinally active compounds and therefore structural and conformational analyses of the molecules are considered powerful paradigms of chemistry. Molecular structure holds a key to the understanding of nature's intricate design mechanism and blueprints. If these blueprints and basic materials are understood, perhaps we can begin to mimic the important products more cost effectively and with less detrimental and environmental consequences. Most biologically active compounds are heterocyclic organic compounds, which have a ring structure containing atoms such as sulfur, oxygen, or nitrogen in addition to carbon, as part of the ring. The class includes many compounds of biological importance, namely nucleic acids, vitamins, hormones, and pigments. Over half of all known compounds and a large number of pharmaceutical products are heterocyclic in nature.

The activity of any drug depends mainly on its interaction with biological targets, such as proteins, nucleic acids, and biomembranes. These targets have complex three-dimensional structures that are capable of recognizing the drug molecule in only one of its many possible arrangements in 3-D space. Structure is the complete arrangement of all the atoms of the molecule in space and includes constitution, conformation, and configuration. Molecular recognition between drugs and their receptors is guided by the nature of intermolecular interactions, such as hydrogen bonds, heteroatom interactions, $\pi - \pi$ interactions, and van der Waals forces. Hydrogen bonds, regarded as the strongest and most directional of intermolecular interactions, have been widely exploited. Depending on the specific chemical donor–acceptor combination, and the details of the contact geometry, all of these interactions influence biological activity. Therefore, knowledge of intermolecular interactions and their geometric characteristics enables one to design and manipulate molecular systems, which can be applied in the field of rational drug design.

Understanding the relationship between drug structures and biological activities forms the basis for the design of drugs. When the structural group(s) on the drug molecule that interacts with the target is known, structural modifications can be made to increase the affinity towards the desired target, decrease the affinity to an undesired target, alter the drug's ability to cross a lipid membrane, and so on. Structure-based drug design is perhaps the most elegant approach for discovering compounds exhibiting high specificity and efficacy. In reality, however, drug targets are very complex and this approach has had only limited utility. However, a number of recent successful drugs have in part or in whole emerged from a structure-based research approach.

1.1
Structural Properties

1.1.1
Influence of Chirality

Structural aspects of drugs such as shape, charge, and chirality not only influence their ability to interact with their targets but also determine the route by which the drug is administered. Chirality is very important from the point of view of drug development since most of the biological targets are enantiomeric molecules. Over 50% of all drugs are chiral and usually, one enantiomer will be more effective than its mirror structural image—suggesting a more complementary fit between the drug with its receptor-binding region. Frequently, only one stereoisomer is active and sometimes the other one is toxic (e.g., thalidomide). The (–) enantiomer of epinephrine has a more potent pressor activity than its stereoisomer, the D(–) lactoyl choline is hydrolyzed much more slowly than its L(+)-isomer and only the S-configuration of ibuprofen, an anti-inflammatory agent has pharmacological properties. Some agents used in anesthesia are typically stereospecific (eg., l-morphine and d-tubocurarine) because the enzymes that metabolize these compounds exhibit stereoselectivity. A drug upon administration undergoes a series of steps before exerting its activity and at each step the structure of the drug and hence its chirality influences the metabolism. Likewise, *cis/trans* isomers of cyclic compounds, or Z/E isomers of alkenes are expected to have variable binding potency and therefore also different biological activity [2]. In view of the above importance, a thorough understanding of the drug's chiral activity is necessary while improving its drugability.

1.1.2
Importance of Conformation

Conformation is the spatial arrangement of a molecule of a given constitution and configuration. It deals with an understanding of the floppiness of the molecules, which try to arrange themselves in the most comfortable way so as to avoid crowding and strain, and it also considers the attraction or repulsion between certain groups. The conformation is of great importance for the mode of action of drugs since it relies on the easy accessibility of the reactive groups. Knowledge of the conformation is essentially needed to explain or predict the mode of action of the drug. Saturated six-membered rings are key building blocks of most carbohydrates and also feature in compounds like inositol phosphates and cyclic 3′,5′-nucleoside phosphates. Five-membered rings are the component of nucleosides, and helps in building the double stranded DNA molecule. Therefore, it is to be noted that the conformational behavior of specific rings or molecules as a whole influence the biological activity.

In general, all rings other than three-membered and aromatic ones are non-planar. Six-membered saturated rings adopt the most favorable chair conformation, which exists in a state of rapid equilibration. The transition from chair to other forms, namely half chair, boat, and twisted boat conformations occurs through several stages depending on the energy minimum values. The boat and twist conformations have high energy and therefore are observed infrequently when the six-membered ring is fashioned with a large substituent or when it is included in a fused bicyclic structure. The conformational equilibria in a cyclopentyl system is very different from cyclohexane and extremely sensitive to the nature of substitution. The most stable conformers of cyclopentane are envelope and half-chair.

1.1.3
Effect of Polymorphism

One other property that affects the biological activity of drugs is polymorphism, which is defined as the ability of a molecule to crystallize into more than one crystal structure arrangement. It's effect on the shelf life, solubility, formulation, and processing properties can influence the different rates of uptake of a drug in the body, leading to lower or higher biological activity than desired. In extreme cases, an undesired polymorph can even be a toxic one. Chloramphenicol-3-palmitate, an antibiotic, is known to crystallize at least in three polymorphic forms. The most stable A-form is marketed since the B-form—even though it has an eightfold higher bioactivity than the A-form—creates the danger of fatal dosages when the unwanted polymorph is unwittingly administered because of alterations in process or storage conditions [3]. Aspirin can be found in many different polymorphs if experimental crystallization conditions are suitable [4]. Therefore, it is vital that the crystalline products have a polymorph that has the correct properties.

1.2
Forces that Influence the Drug-Receptor Interactions

There are three major types of chemical forces/bonds that are involved in drug-receptor interactions.

1. Covalent: they are the strongest forces and are generally "irreversible" under biological conditions. An α-adrenergic receptor antagonist, phenoxybenzamine and the receptor and DNA-alkylating chemotherapy drugs with DNA functional groups are examples for this type of interaction.
2. Electrostatic: these are weaker than covalent but are more common. Their interaction strengths vary depending on the interacting atoms. Strong interactions are found between permanently charged ionic molecules, hydrogen bonding shows weaker interactions while van der Waals forces are weaker still.
3. Hydrophobic: they are generally weak and are significant in driving lipophilic drugs into the lipid component of biological membranes. It is often found between drugs and relatively non-polar receptor regions.

An understanding of these forces will help in the development of superior drugs that have strong interactions with their target, improving dosage level and sustainability.

1.3
Techniques Used to Study the Structures

X-ray crystallographic techniques and nuclear magnetic resonance (NMR) are two of the most advanced analytical techniques currently being implemented in drug design; however, the hurdles blocking their application to the study of targets are formidable. Even though many spectroscopic techniques provide molecular structural information each has its advantages and disadvantages. IR spectroscopy can identify the chemical groups present in a small molecule, mass spectrometry shows molecular connectivity, and NMR provides some information about solution structure for both small molecules and proteins. The most information-rich technique is X-ray crystallography and it provides information about molecular structures and various possible intermolecular interactions. This information can be used either (a) a posteriori to rationalize structure–activity relationships as has been done in the design of human rhinovirus inhibitors; or (b) a priori as has been done in the production of inhibitors of human immunodeficiency viral proteases (HIV PR) that are required for viral replication.

This review mainly focuses on the structural aspects, such as configurational, conformational, and polymorphic properties, in addition to the interacting forces on many heterocyclic molecules studied in our group using X-ray crystallographic techniques.

2
Stereochemical Analysis

Configurational and conformational analysis of saturated heterocyclic rings such as piperidine, a six-membered heterocyclic ring, and azepine, a seven-membered heterocyclic ring, was carried out and is explained here in detail.

2.1
Piperidine

Piperidine, known as a basic component of the piper alkaloid—"piper nigrum" (black pepper)—is a monocyclic cyclohexane with a hetero nitrogen atom. The skeletal ring of piperidine is contained in the molecules of many synthetic and natural medicaments [5, 6]. Its various derivatives are found to possess pharmacological activity and form an essential part of the molecular structure of important drugs. For example, the piperidine ring is a characteristic feature of antihistaminic agents [7], oral anesthetics [8], narcotic analgesics [9, 10], and post-ganglionic parasympathetic agonists [11]. Clebopride, a 1,4-disubstituted piperidine, is used clinically to prevent postoperative vomiting, to speed up gastric emptying before giving anesthesia, to facilitate radiological evaluation, and to correct a variety of disturbances of gastrointestinal function [12]. Several 2,6-disubstituted piperidines are found to be useful as tranquilizers [13] and possess hypotensive activity [14], with a combination of stimulant and depressant effects on the central nervous system [15], as well as bactericidal, fungicidal, and herbicidal activities [16]. Many piperidine derivatives also form the skeleton of several alkaloids [17, 18].

4-Piperidones are the synthetic intermediates in various alkaloids and pharmaceutical products [19, 20]. These functionalized derivatives exhibit antidepressant [21, 22], antiarrhythmic [23, 24], spasmolytic [25], tranquilizing [26, 27], antithrombogenic [28, 29], and blood cholesterol-lowering activities [26, 27]. Recently, it has been shown that a series of 3,5-bis arylylidene-4-piperidones and related N-acryloyl analogues prepared as cytoxins lead to drugs that are deprived of genotoxic properties [30]. The thiosemicarbazone derivatives have a wide range of biological activities because of its reduction capability. They are very versatile tridentate ligands and can coordinate to metals as neutral molecules or, after deprotonation, as anionic ligands and can adopt a variety of different coordination modes [31–34]. Their antipathogenic and other biological activities in vitro depend on the substituent(s) at the amino group [35–37]. Heterocyclic thiosemicarbazones show important biological activities [38], for example, pyridine-2-carbaldehyde thiosemicarbazone was reported to have carcinostatic properties [39–42]. For instance, 3-ethoxy-2-oxobutyraldehyde bis-(thiosemicarbazonato) copper(II) has proved to be an efficient antitumor agent [43]. Thiosemicarbazones usually react as chelating ligands with transition metal ions by bonding through the sulfur and hydrazinic terminal nitrogen atoms. In order to understand the biological activity of the piperidine derivatives such as piperidone and thiosemicarbazone, many different derivatives were synthesized and their conformation established.

Though the piperidine derivatives are found to be pharmacologically important their N-nitroso derivatives are found to be carcinogenic in nature [44, 45]. These N-nitroso compounds are often encountered in a variety of environmental samples. Since these derivatives are fairly stable and remain in the environment for a long time, they result in the pollution of various objects [46, 47]. Also, the N-nitroso compounds are the most abundant carcinogens present in tobacco and considered to be an important risk factor for causing tobacco-related cancers [48]. Even though, the unsubstituted N-nitrosopiperidines are potential carcinogens, the carcinogenicity is found to reduce when an alkyl group is substituted at the α-position (C-2). Furthermore, if α-position (C-2 & C-6) are substituted by methyl groups, it becomes non-carcinogenic. It appears that the blocking of the α-position to the ring nitrogen atom by the methyl group in cyclic nitrosamines is responsible for reducing the carcinogenic activity [49]. When the C2 and C6 positions are substituted with a methoxyphenyl group (2,6-dimethoxyphenyl-3,5-dimethyl-N-nitrosopiperidin-4-one), activity is observed against CNS subpanels. It has

Table 1 Various piperidine derivatives used for stereochemical analysis in the present study

Structures A, B, C with ring numbering and substituents R_1, R_2, R_3, X, Y.

No.	Mol.	X	Y	R_1(2,6)	R_2	R_3	Conformation	Ref.[†]
1	A	$-\overset{O}{\underset{\parallel}{C}}-O-CH_2CH_3$	–	—⬡	–H	–H	Distorted Boat	(a)
2	A	$-\overset{O}{\underset{\parallel}{C}}-O-CH_2CH_3$	–	—⬡	–CH$_3$	–H	Distorted Boat	(b)
3	A	$-\overset{O}{\underset{\parallel}{C}}-$⬡	–	—⬡	–CH$_3$	–H	Distorted Boat	(b)
4	B	$-\overset{O}{\underset{\parallel}{C}}-O-CH_2CH_3$	–	—⬡	$<^{CH_3}_{CH_3}$	–H	Distorted Boat	(a)
5	B	$-\overset{O}{\underset{\parallel}{C}}-O-CH_2CH_3$	–	—⬡	–CH$_3$	–CH$_3$	Distorted Boat	(c)
6	B	$-\overset{O}{\underset{\parallel}{C}}-CH_2Cl$	–	—⬡	–CH$_3$	–CH$_3$	Twist Boat	(d)
7	B	$-\overset{O}{\underset{\parallel}{C}}-CH_2Cl$	–	—⬡	–CH$_2$CH$_3$	–H	Twist Boat	(d)
8	B	$-\overset{O}{\underset{\parallel}{C}}-CH_2-N\bigcirc O$	–	—⬡	–CH$_3$	–H	Twist Boat	(e)
9	B	$-\overset{O}{\underset{\parallel}{C}}-CH_2-N\bigcirc O$	–	—⬡	$-CH<^{CH_3}_{CH_3}$	–H	Chair	(e)

[†] (a) Ponnuswamy S, Venkatraj M, Jeyaraman R, Suresh Kumar M, Kumaran D and Ponnuswamy MN (2002) Indian J. Chem 41B:614; (b) Suresh Kumar M (2000) PhD thesis, University of Madras, India; (c) Suresh Kumar M, Ponnuswamy MN, Ponnuswamy S, Jeyaraman R, Panneerselvam K and Soriano M (1998) Acta Cryst C54:870; (d) Nallini A, Saraboji K, Ponnuswamy MN, Venkatraj M and Jeyaraman R (2003) Mol Cryst Liq Cryst 403:49; (e) Nallini A, Saraboji K, Ponnuswamy MN, Venkatraj M and Jeyaraman R (2003) Mol Cryst Liq Cryst 403:57

Table 1 continued

No.	Mol.	X	Y	$R_1(2,6)$	R_2	R_3	Con-formation	Ref.[†]
10	B	– COH	–	(2-chlorophenyl)	– CH_3	– CH_3	Boat	(f)
11	B	– N = O	–	(2-chlorophenyl)	– CH_3	– CH_3	Distorted Boat	(g)
12	B	– N = O	–	(3,4,5-trimethoxyphenyl)	– CH_3	– CH_3	Boat	(f)
13	B	– H	–	(phenyl)	$-\overset{O}{\overset{\|}{C}}-O-CH_2CH_3$	– H	Chair	(h)
14	B	– CH_3	–	(phenyl)	$-CH\overset{CH_3}{\underset{CH_3}{<}}$	– H	Chair	(i)
15	B	– H	–	(thienyl)	– CH_3	– CH_3	Chair	(j)
16	B	– H	–	(furyl)	– CH_3	– CH_3	Chair	(k)
17	B	– N = O	–	(furyl)	– CH_3	– CH_3	Distorted Boat	(k)

[†] (f) Kumaran D (1997) PhD thesis, University of Madras, India; (g) Sukumar N, Ponnuswamy MN, Thenmozhiyal JC and Jeyaraman R (1993) J Cryst Spec Res 23:871; (h) Sampath N, Ponnuswamy MN and Nethaji M (2004) Acta Cryst E60:o2105; (i) Sampath N, Aravindan S, Ponnuswamy MN and Nethaji M (2005) Anal Sci 21:x67; (j) Sukumar N, Ponnuswamy MN, Thenmozhiyal JC and Jeyaraman R (1995) J Chem Cryst 25:177; (k) Sukumar N, Ponnuswamy MN, Thenmozhiyal JC and Jeyaraman R (1994) Bull Chem Soc Japan 67:1069

Table 1 continued

Structures A, B, C (piperidine ring derivatives with substituents R₁, R₂, R₃, X, Y)

A   **B**   **C**

No.	Mol.	X	Y	$R_1(2,6)$	R_2	R_3	Conformation	Ref.[†]
18	C	$-CH_3$	$-NH-C(NH_2)=S$	phenyl	$-CH(CH_3)CH_3$	$-H$	Chair	(l)
19	C	$-H$	$-NH-C(NH_2)=S$	4-Cl-phenyl	CH_3	$-H$	Chair	(m)
20	C	$-CH_3$	$-NH-C(NH_2)=S$	phenyl	CH_3	$-H$	Chair	(n)
21	C	$-H$	$-NH-C(NH_2)=S$	phenyl	CH_3	$-H$	Chair	(o)
22	C	$-N=O$	$-OH$	phenyl	CH_3	$-CH_3$	Distorted Boat	(p)
23	C	$-C(=O)-CH_3$	$-O-C(=O)-CH_3$	phenyl	$-CH(CH_3)CH_3$	$-H$	Twist Boat	(q)

[†] (l) Sampath N, Malathy Sony SM, Ponnuswamy MN, Nethaji M (2003) Acta Cryst C59:o346; (m) Sampath N, Ponnuswamy MN and Nethaji M (2006) Mol Cryst Liq Cryst (accepted for publication); (n) Sampath N, Ponnuswamy MN, Nethaji M (2006) Cryst Res Tech (in press); (o) Sampath N (2006) PhD thesis, University of Madras, India; (p) Sukumar N, Ponnuswamy MN, Vijayalakshmi R and Jeyaraman R (1994) Zeits Fur Kristallographie 209:823; (q) Sukumar N (1994) PhD thesis, University of Madras, India

also been observed that the increase in bulkiness of the substituents at different positions of the piperidine ring leads to a decrease in carcinogenicity [50]. Most of the nitrosamines are found to possess carcinogenic and antitumor activity [44, 51]. Certain N-nitroso ureas are used as antitumor agents or antibiotics [52].

The importance of these piperidine derivatives lies in the clinical use of suitably substituted groups and the different ways available for making modifications in the anticipated structure of the pharmacologically active compound. Therefore, a series of piperidine derivatives with bulky aryl substitutions at the 2,6 positions and different substituents in different possible positions, especially the N-substitution, have been crystallized and their conformational features have been studied by crystallographic methods (Table 1) so as to correlate the biological activity with stereochemistry.

2.1.1
Conformational Flexibilities of Piperidine Derivatives

In six-membered saturated heterocyclic compounds such as piperidines, oxanes, and thianes, the piperidines are the most widely investigated group of compounds because of their wide occurrence in natural products [53]. For these six-membered rings, two non-planar models were proposed based on cyclohexane, namely, (i) a rigid or chair form; and (ii) a mobile form capable of taking up a variety of shapes, some of which resemble a boat form [54, 55] (Fig. 1). The flexible form estimated by indirect methods was found to have 21 to 25 kJ/mole higher energy than the rigid form [56]. Cyclohexane was thus shown to exist almost exclusively in the chair form at room temperature. Examination of the cyclohexane model shows that during gradual transition from one boat to another boat, it passes through certain forms in which flagpole interaction and eclipsing of adjacent hydrogen atoms are alleviated partially. These forms, known as skew-boats or twist forms were found to have about 6.7 kJ/mole lower energy than the boat forms [57] (Fig. 2). Though the stereochemical features of cyclohexane and piperidine have more or less the same characteristics, one notable aspect is the involvement of a lone-pair of electrons in nitrogen in the case of piperidines. Various conformational properties such as barriers to ring reversal, conformational preference of substituents on the heteroatom, and the shape of the piperidine ring systems are discussed in several reviews [58, 59].

Analogies should include correlations with cyclohexanones, as our area of study pertains to the 4-piperidones also. In cyclohexanones, minor ring distortions exist as compared to cyclohexanes. Because, the barrier of rotation about the sp^2–sp^3 carbon–carbon bond is intrinsically lower than a sp^3–sp^3 bond [60]. The cyclohexane ring shows increased flexibility in the part of the ring containing the carbonyl group [61]. Justification in extrapolation of conformational features of cyclohexane derivatives to the reduced heterocyclic piperidones is enhanced from the minute influence on the geometry witnessed upon replacing a carbon atom with only one hetero atom namely, a nitrogen.

The piperidine ring system offers a wide variety of conformational flexibilities such as chair, boat, distorted boat or chair, depending upon the

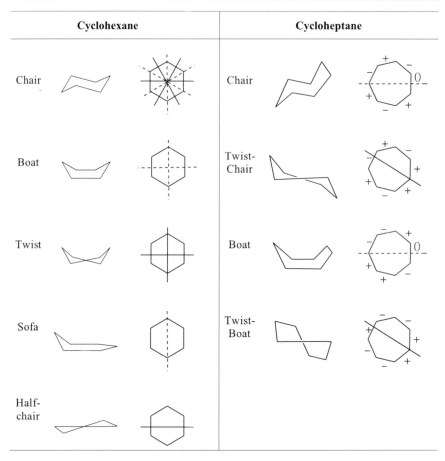

Fig. 1 Non-planar conformations adopted by cyclohexane and cycloheptane rings shown with their characteristic symmetry elements. The *dotted lines* represent the mirror plane and the *solid line* represents the 2-fold axis

position, bulkiness, and nature of the substitutions. However, either the chair or slightly distorted chair conformation has been found to be more favorable than other conformations. A complete structural analysis is required to understand the effect of increasing bulkiness of the substituents at different positions on the conformation of the piperidine ring. Ravindran and Jeyaraman [62] analyzed the stereochemistry of 2,6-diphenyl piperidines by using NMR and molecular mechanics methods. The torsional angles derived from MM2 calculation indicated that as the bulkiness of the substituents at C3 increases the flattening at this juncture also increases and deviate from equatorial orientation. When a functional group of the type $-X=Y$ is exocyclic to a system containing α-alkyl substituents, it is found to cause drastic conformational changes in the ring. Oximes and semicarbazones of cyclic

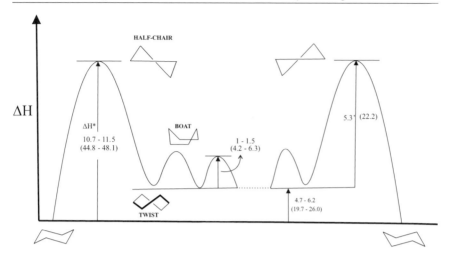

Energy profile [ΔH* kcal/mol (kJ/mol)] for cyclohexane ring reversal

Fig. 2 Energy profile for various conformations of the cyclohexane ring

ketones are examples for such systems. Similar conformational changes are observed when a nitroso or acyl group is substituted at the nitrogen atom of piperidines. Thus, the introduction of a nitroso group into conformationally rigid cyclic systems such as 2,6-diarylpiperidines, is expected to cause interesting conformational changes [63]. The presence of a partial N...N double bond character in nitrosamines caused by extensive delocalization of the lone pair of electrons on the nitrogen atom with the hetero π–electron system $(-N = O)$ was first established by NMR spectroscopy [64, 65]. In cyclic nitrosamines, such as N-nitrosopiperidines and N-nitrosopyrrolidnes, this delocalization is also observed [66]. Therefore, a stereochemical investigation of these conformationally anchored cyclic nitroso and acylamines has been undertaken for this work.

When *cis* 2,6-dimethylpiperidine, where the ring prefers a chair conformation with equatorial methyl groups, was converted to its N-nitroso derivative, the two methyl groups are found to prefer axial positions. The preference is attributed to the $A_{1,3}$ strain that was first proposed by Johnson and Malhotra [67]. In general, the $A_{1,3}$ strain can be considered as a severe non-bonded interaction between the $-X = Y$ group bearing π electrons and an alkyl group present at the allylic position of cyclic systems.

Lunzzi et al. [68], studied the stereochemistry of N-nitroso, N-imino, and N-acyl derivatives of piperidine. They considered two possible orientations of the substituents at the nitrogen atom, namely, planar and perpendicular with respect to the average plane of the ring system. It was found that the $N - X = Y$ linkage in N-nitrosopiperidines, piperidyltriazines, and piperidylsulfinylhydrazines are planar, regardless of the number of alkyl substituents at the 2 and 6 positions of the ring. Investigations of a variety of piperidine ring systems containing $N - X = Y$ functions show that the $N - X = Y$ linkage of N-acylpiperidines prefers to be coplanar with the ring regardless of the number of substituents at 2 and 6 positions.

It has been validated [69] that the axial or equatorial nature of a substituent has a bearing on its reactivity or ability to interact with its environment. Moreover, equatorial substituents are an index of greater stability and less reactivity in comparison to their axial counterparts. In this direction our group has analyzed some of the N-free, N-alkyl, and N-positions substituted by the electron withdrawing groups such as formyl, ethoxy carbonyl, and nitroso group piperidine derivatives with various substituents at different positions in a piperidine ring to understand the stereochemistry of the system (Table 1).

Chair and boat conformations are widely found for the cyclohexane ring systems. The chair form is characterized by the presence of a plane of symmetry at the atoms and a binary axis bisecting the bonds. Two planes of symmetry, one at the atom and the other bisecting the bond characterize the boat form whereas two binary axes characterize the twist form. Several ways have been proposed to denote the different conformations of the ring, including those of Hendrickson [70], Díez [71, 72], Foces-Foces [73], and Espinosa [74] that are widely used. We are following the nomenclature of Espinosa [74] to fully characterize each conformation based on the situation of the element of symmetry and on the sign of the intracyclic dihedral angles. This nomenclature is represented in Fig. 3 for all the chair and boat/twist-boat conformations of piperidine. The presence of nitrogen makes the axis or a plane of symmetry pseudo since there are differences in the atomic species and the bond lengths between $C - C$ and $C - N$ bonds.

With the above-mentioned nomenclature the 2,6-diaryl-substituted piperidine derivatives under study were characterized and are presented in Table 2. The analysis shows that the piperidine ring adopts a chair conformation if the N-position is free or substituted by methyl groups irrespective of the substituents at the other positions. If the N-position contains a $-X = Y$ group bearing π electrons it invariably adopts either a boat or twist boat conformation. In the N-substituted group if relatively more electronegative atoms are present the piperidine ring prefers a twist-boat conformation. This is in line with the reported studies [62–66].

Surprisingly, in the compounds that adopt a boat conformation the 6B1 and 6B4 conformers, wherein one of the pseudoplanes passes through the nitrogen atom, are not found. Whereas for the compounds that adopt a twist-boat conformation the 6TB3 and 6TB6 conformers, wherein one of the pseudoaxes passes through the nitrogen atom, are dominant. In the boat form, the 6B3 and 6B6 conformers, whose pseudoplane passes through the α position, dominate. The presence of the heteroatom introduces some distortions in the geometries of the boat form and the dihedral angles rarely acquire the

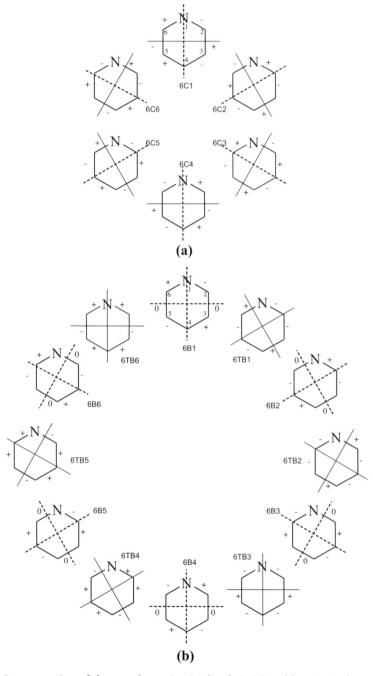

Fig. 3 Representation of the pseudorotation in the chair (**a**) and boat/twist-boat family (**b**) in piperidines. The *dashed* and *solid lines* denote the pseudoplane and pseudoaxis for each conformation, respectively

Table 2 Endocyclic torsion angles for piperidine derivatives with their conformer description

No.	Mol.	N1-C2	C2-C3	C3-C4	C4-C5	C5-C6	C6-N1	Conformation*	Conformer
1	A	13.1	− 57.8	42.8	14.6	− 56.4	41.7	Distort. Boat	6B6
2	A	14.8	− 56.6	39.6	17.3	− 57.5	41.6	Distort. Boat	6B6
3	A	26.0	− 55.8	60.3	26.7	− 64.3	57.7	Distort. Boat	6B6
4	B	− 44.6	56.0	− 21.0	− 28.6	41.5	− 3.5	Distort. Boat	6B5
5	B	9.4	− 48.2	33.2	− 20.3	− 58.0	43.7	Distort. Boat	6B6
6	B	44.4	− 59.6	21.5	32.1	− 48.2	9.7	Twist Boat	6TB6
7	B	44.2	− 57.6	18.7	35.1	− 49.1	8.1	Twist Boat	6TB6
8	B	45.7	− 58.5	18.3	36.0	− 49.3	7.0	Twist Boat	6TB6
9	B	− 54.8	48.8	− 50.1	51.9	− 51.4	55.7	Chair	6C3
10	B	4.5	45.5	− 47.8	− 1.1	49.7	− 53.5	Distort. Boat	6B3
11	B	− 1.6	45.2	− 40.7	− 8.6	49.6	− 46.6	Distort. Boat	6B3
12	B	− 49.3	52.1	− 9.3	41.1	45.6	− 0.4	Distort. Boat	6B5
13	B	63.3	− 51.0	45.7	− 47.4	53.4	− 63.7	Chair	6C4
14	B	− 55.8	46.5	− 45.3	52.1	− 58.9	61.4	Chair	6C5
15	B	− 61.2	50.6	− 49.3	51.6	− 55.9	64.5	Chair	6C1
16	B	− 63.4	55.9	− 53.7	52.8	− 53.9	62.6	Chair	6C1
17	B	2.9	40.4	− 37.2	− 9.8	50.8	− 49.6	Distort. Boat	6B3
18	C	52.5	− 49.9	52.7	− 55.9	55.0	− 54.5	Chair	6C2
19	C	− 60.5	49.2	− 45.3	47.1	− 51.9	61.0	Chair	6C1
20	C	− 59.1	54.9	− 52.0	51.8	− 52.7	57.0	Chair	6C1
21	C	61.2	− 52.0	50.6	− 52.1	55.1	− 63.1	Chair	6C4
22	C	46.1	− 57.8	19.3	34.6	− 48.0	8.0	Distort. Boat	6B2
23	C	− 38.7	63.4	− 32.9	− 21.6	47.3	− 16.2	Twist Boat	6TB3

* Distort. Boat – Distorted Boat

value of zero degrees. In the present analysis this distortion is more for compounds where N-substitution is bulkier, accordingly the dihedral angles for the compounds **1–5** deviate more (of the order of ±20°) than their lighter counterparts, compounds **10–12**, **17** and **22** (of the order of ±8°).

In the N-free and N-methyl-substituted piperidine derivatives, as expected, the bulkier aryl substitutions at the 2nd and 6th position take up an equatorial orientation in order to avoid steric hindrance. Whereas in the boat/twist-boat family they take up the opposite orientation, one axial and the other equatorial. The methyl substituted at the N-position is found in the antiperiplanar orientation whereas the − X = Y groups prefer the anticlinal orientation. In the piperidine derivatives that have a chair conformation the bulkier groups substituted at the 3rd or 5th positions are found in the equatorial position. While in the boat/twist-boat family, with one substitution in the 3rd or 5th position the equatorial orientation is preferred, but with substitutions in both positions axial and equatorial are preferred.

The substitutions at the C4 position are planar. The thiosemicarbazone moiety adopts an extended (E) conformation and the planarity arises due to the extensive electron delocalization throughout the moiety. The acetyl group of compound **23** also takes up an extended conformation. These conformational analyses provide ample information for the chemist to analyze the drugability of piperidine derivatives.

2.1.2
Stereochemical Analysis
of Piperidine Derivatives Present in the Protein Environment

In order to correlate the above stereochemical analysis of the piperidine derivatives to the biological activity, the drugs in the protein structures that contain a piperidine and piperidone moiety were extracted from Protein Data Bank (PDB)[1] [75]. A simple search of the Macromolecular Structure Database (MSD) [76] was used for this purpose. The search criteria includes only the structures solved using X-ray crystallographic techniques to a resolution less than 2.0 Å, i.e. only structures with a high resolution are taken for the analysis. Table 3 gives the PDB ID of the protein structures, the drug molecule that is present, and the conformation adopted by the piperidine ring in the drugs.

Of the 19 piperidine drugs, ten structures contain the deoxynojirimycin drug that is an α-glucosidase inhibitor with antiviral action and the derivatives have anti-HIV activity. Three structures contain free piperidine with no substitution, which is not the prime inhibitor for these proteins. Of the eight piperidin-4-one derivatives five are Factor Xa inhibitors. Except for one each in both derivatives all inhibitors adopt the chair conformation. The dominant conformer of deoxynojirimycin and the free piperidine are 6C1 and 6C4, wherein a pseudoplane passes through the nitrogen atom. The dominant conformer for piperidone is 6C6 in which the pseudoplane passes through the β position. One factor Xa inhibitor of 1qb9 structure adopts a boat conformation with the 6B2 conformer.

The environment around the protein and drugs have been analyzed; the hydrogen bonded and non-bonded contacts between the complex were calculated using the program HBPLUS [77] and the pictorial representations were drawn using the program LIGPLOT [78]. This showed that only deoxynojirimycin, its derivatives and carboxypiperidine of the piperidine class have extensive interaction with the protein residues. One of the deoxynojirimycin derivatives, XIF (piperidine-3,4-diol), which is found in protein structures 1fh8, 1vol, and 1von, has three different conformers 6C4, 6C6, and

---

[1] PDB: A database that contains the three-dimensional structures of proteins solved using X-ray crystallography, NMR, or modeling techniques.

Table 3 Conformation of drugs containing piperidine and piperidin-4-one moieties in protein structures

PDB ID	Description of the ligand	Protein to which bound	Ring conf.	Ref.‡
Piperidine				
1fh7	Derivative of deoxynojirimycin (XDN)	Xylanase	Chair (6C1)	(a)
1v0k	Derivative of deoxynojirimycin (XDN)	Xylanase	Chair (6C4)	(b)
1v0m	Derivative of deoxynojirimycin (XDN)	Xylanase	Chair (6C1)	(b)
1fh8	Derivative of deoxynojirimycin (XIF)	Xylanase	Chair (6C4)	(a)
1v0l	Derivative of deoxynojirimycin (XIF)	Xylanase	Chair (6C6)	(b)
1v0n	Derivative of deoxynojirimycin (XIF)	Xylanase	Chair (6C2)	(b)
1fh9	Derivative of deoxynojirimycin (LOX)	Xylanase	Sofa	(a)
1g6i	1-Deoxymannojirimycin (DMJ)	alpha-1,2-mannosidase	Chair (6C1)	(c)
1hxk	1-Deoxymannojirimycin (DMJ)	alpha-mannosidase II	Chair (6C4)	(d)
1i75	1-Deoxymannojirimycin (DMJ)	Cyclodextrin glucanotransferase	Chair (6C5)	(e)
1b6l	Macrocyclic peptidomimetic inhibitor 4 (PI4)	HIV-1 protease	Chair (6C6)	(f)
1fkg	Rotamase inhibitor (SB3)	FK506 binding protein	Chair (6C1)	(g)
1fkh	Rotamase inhibitor (SB3)	FK506 binding protein	Chair (6C3)	(g)
1j4r	FKBP12 ligands (001)	FK506 binding protein	Chair (6C3)	(h)

‡ (a) Notenboom V, Williams SJ, Hoos R, Withers SG, Rose DR (2000) Biochemistry 39:11553; (b) Gloster TM, Williams SJ, Roberts S, Tarling CA, Wicki J, Withers SG and Davies GJ (2004) Chem Commun 16:1794; (c) Herscovics A, Lipari F, Sleno B, Romera PA, Vallee F, Yip P and Howell PA (2002) Interdisciplinary approaches 28; (d) van den Elsen JM, Kuntz DA and Rose DR (2001) EMBO J 20:3008; (e) Kanai R, Haga K, Yamane K and Harata K (2001) J Biochem (Tokyo) 129:593; (f) Martin JL, Begun J, Schindeler A, Wickramasinghe WA, Alewood D, Alewood PF, Bergman DA, Brinkworth RI, Abbenante G, March DR, Reid RC and Fairlie DP (1999) Biochemistry 38:7978; (g) Holt DA, Luengo JI, Yamashita DS, Oh HJ, Konialian AL, Yen HK, Rozamus LW, Brandt M, Bossard MJ, Levy MA, Eggleston DS, Liang J, Schultz LW, Stout TJ and Clardy J (1993) J Am Chem Soc 115:9925; (h) Dubowchik GM, Vrudhula VM, Dasgupta B, Ditta J, Chen T, Sheriff S, Sipman K, Witmer M, Tredup J, Vyas DM, Verdoorn TA, Bollini S and Vinitsky A (2001) Org Lett 3:3987

Table 3 continued

PDB ID	Description of the ligand	Protein to which bound	Ring conf.	Ref.[‡]
1ppc	Piperidine (PIP)	Trypsin	Chair (6C1)	(i)
1pph	Piperidine (PIP)	Trypsin	Chair (6C4)	(j)
1qur	Piperidine (PIP)	Alpha-thrombin	Chair (6C4)	(k)
1qb1	Factor Xa inhibitor (974)	Trypsin	Chair (6C5)	(l)
1w3m	6-Carboxypiperidine (CPI)	Tsushimycin	Chair (6C5)	(m)
Piperidone				
1c8k	Antitumor drug–flavopiridol (CPB)	Glycogen phosphorylase	Chair (6C1)	(n)
1mts	Inhibitor specific for the blood-clotting factor Xa inhibitor (BX3)	Trypsin	Chair (6C6)	(o)
1mtu	Inhibitor specific for the blood-clotting factor Xa inhibitor (BX3)	Trypsin	Chair (6C6)	(o)
1mtv	Inhibitor specific for the blood-clotting factor Xa inhibitor (BX3)	Trypsin	Chair (6C6)	(o)
1qb9	Factor Xa inhibitor (806)	Trypsin	Boat (6B2)	(l)
1qbo	Factor Xa inhibitor (711)	Trypsin	Chair (6C6)	(l)
1tps	Inhibitor A90720A (A9A)	Trypsin	Chair (6C6)	(p)
2aid	Non-peptide inhibitor (THK)	HIV-1 Protease	Chair (6C6)	(q)

‡ (i) Bode W, Turk D, Sturzebecher J (1990) Eur J Biochem 193:175; (j) Turk D, Sturzebecher J and Bode W (1991) FEBS Lett 287:133; (k) Steinmetzer T, Renatus M, Kunzel S, Eichinger A, Bode W, Wikstrom P, Hauptmann J and Sturzebecher J (1999) Eur J Biochem 265:598; (l) Whitlow M, Arnaiz DO, Buckman BO, Davey DD, Griedel B, Guilford WJ, Koovakkat SK, Liang A, Mohan R, Phillips GB, Seto M, Shaw KJ, Xu W, Zhao Z, Light DR and Morrissey MM (1999) Acta Crystallogr D55:1395; (m) Bunkoczi G, Vertesy L and Sheldrick GM (2005) Acta Crystallogr D61:1160; (n) Oikonomakos NG, Schnier JB, Zographos SE, Skamnaki VT, Tsitsanou KE and Johnson LN (2000) J Biol Chem 275:34566; (o) Stubbs MT, Huber R and Bode W (1995) FEBS Lett 375:103; (p) Lee AY, Smitka TA, Bonjouklian R and Clardy J (1994) Chem Biol 1:113; (q) Rutenber E, Fauman EB, Keenan RJ, Fong S, Furth PS, Ortiz de Montellano PR, Meng E, Kuntz ID, DeCamp DL and Salto R (1993) J Biol Chem 268:15343

6C2, respectively, and binds to the same protein target. The study of ligand environment for these structures shows that all of them show similar interactions (Fig. 4). Since the signs of the torsion angles for these conformers are the same it can be concluded that the signs of the dihedral angles play a very important role in the interaction of drugs with its targets.

The substituents of the piperidine drugs take up an equatorial orientation in the deoxynojirimycin derivatives whereas in the deoxynojirimycin itself they take different orientations depending upon the protein targets. This confirms the influence of non-covalent interactions over the orientation of the

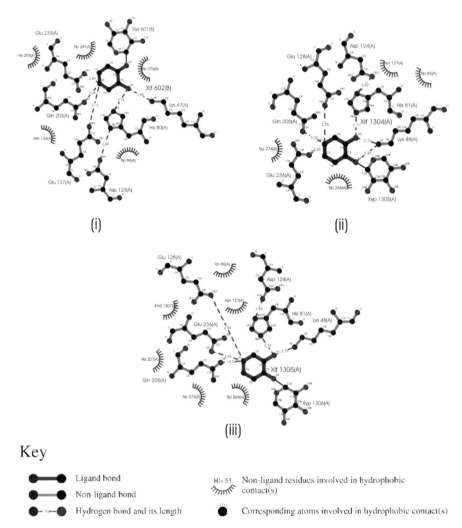

Fig. 4 LIGPLOT showing the hydrogen bonds and non-bonded contacts between the piperidine drugs and its receptor for (i) 1fh8, (ii) 1vol and (iii) 1von structures

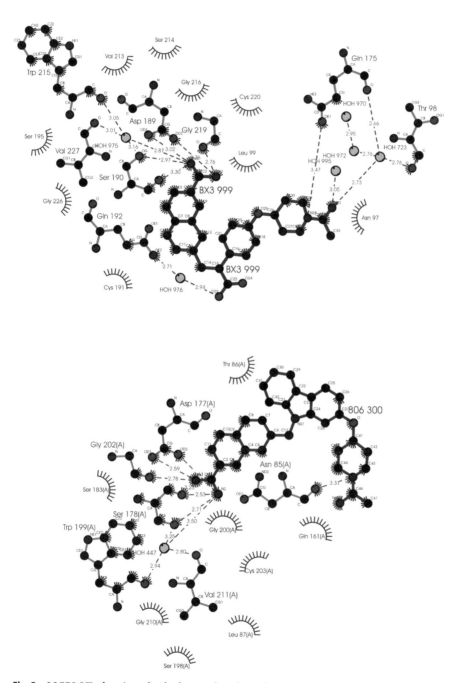

Fig. 5 LIGPLOT showing the hydrogen bonds and non-bonded contacts between the piperidone drugs and its receptor for (i) 1mts and (ii) 1qb9 structures

substituents. In all the ligands analyzed the N-substitution is oriented in the anticlinal position.

In the piperidone drugs, as in the case of piperidine, the N-substituents are anticlinally oriented. The oxygen atom in the 4th position invariably takes an axial orientation in the chair conformation, whereas it occupies an equatorial orientation in the boat conformation. The nitrogen atom present in the piperidone moiety of all the factor Xa inhibitors forms hydrogen bonds with the target protein residues implicating its importance in the drug action. Only the flavopiridol drug has free oxygen which forms a hydrogen bond with the water molecule, while all the other drugs do not have it and so no significant interaction is found for the oxygen atom of the piperidone drugs (Fig. 5).

The structure analysis of the piperidine derivatives and its drugs in the protein showed that the N-substitutions are invariably positioned in the anticlinal orientation irrespective of the different substitutions in the rings and the non-covalent interactions. The bulky groups are in a equatorial orientation in the chair conformation (piperidine ring) and assumes opposite orientations in the boat and twist-boat conformations. Since the number of drug molecules analyzed here are very small, no significant conclusion could be reached concerning the conformers.

2.2
Azepines

Azepines are a class of seven-membered monocyclic rings with nitrogen as the heteroatom. Geometrical ramifications of the azepine compounds have been looked at for a few decades. Among these, the benzo derivatives of azepine constitute a widely prescribed class of psychoactive drug. The synthesis of seven-membered ring heterocycles has focused on the 1,4-benzodiazepine structure. The 1,4-benzodiazepine class of compounds have widespread biological activities and are one of the most important classes of bioavailable therapeutic agents [79]. 1,4-Benzodiazepines such as valium have anxiolytic activity [80] and its derivatives are highly selective cholecystokinin (CCK) receptor subtype A antagonists, highly selective CCK receptor subtype B antagonists [81], k-selective opioids [82], platelet-activation factor antagonists [83], HIV Tat antagonists [84], reverse transcriptase inhibitors [85], gpIIbIIIa inhibitors [86], and ras farnesyltransferase inhibitors [87].

A wide range of diazepines have been identified as potential drugs for various diseases [88]. Most of them contain benzene rings fused to the diazepine ring at the 2,3 and 6,7 positions. They are also used as psychotropic agents, many analogous molecules have been synthesized with the aim of producing superior drugs. The heterodiazepine is effective as a hypnotic and has an extremely high affinity for the benzodiazepine binding receptor that binds stronger than the triazolobenzodiazepines [89]. Barbituric acid and its derivatives have been widely used as therapeutic agents with soporific, tranquillizing, anticonvulsant, or narcotic effectiveness. Such a wide spectrum of biological action and its duration depends on the nature of substituents at position-5 of the barbituric acid ring and on the substituents at the lactam groups. However, barbituric acid derivatives also have a high toxicity so the search is still on for new drugs that are less toxic and more differentiated in the duration of their action. Benzodiazepines are important pharmaceutical agents [90]. Tetrahydro-1H- [1, 3], oxazino [3,2-a] [1, 5], and benzodiazpin-l are examples with potential anxiolysic, anticonvulsant, and antihypnotic activities [80] and have already been synthesized and reported.

The benzodiazepines and benzothiazepines have been extensively studied for medicinal activities [91, 92]. Benzodiazepines are known for their natural occurrence in filamentous fungi and actinomycetes of the genera *Pencillium*, *Aspergillus*, and *Streptomyces* [93]. Benzodiazepines from *Aspergillus* include asperlicin, which is used for the treatment of gastrointestinal and central nervous system (CNS) disorders [93]. In recent times, benzodiazepines have replaced barbiturates that were used once for the purpose of hypnotic effects, owing to their less toxic and less severe withdrawal effects [94]. The importance of 1,5-benzodiazepines is evident from the pharmaceutical application of Globazam [95]. Globazam is a benzodiazepine psychotropic drug in which the nitrogen atoms in the heterocyclic ring are in 1,5-positions rather than in the more common 1,4-positions [96]. The other therapeutic applications of benzodiazepines [97] include antiarrhythmics [98], vasopressin antagonists [99], HIV reverse transcriptase inhibitors [100], and cholecytokinin [101].

Benzothiazepines, a class of compounds of current interest are noteworthy for specific applications. It is well known that the crucial role of L-type Ca^{2+} channels in the initiation of cardiac and smooth muscle contraction has made them major therapeutic targets for the treatment of cardiovascular diseases. These L-type channels share a common pharmacological profile, including high affinity voltage and frequency-dependent block by the phenylalkylamines, the benzothiazepines, and dihydropyridines. These drugs are thought to bind to three separate receptor sites on L-type Cl^{2+} channels that are allosterically linked. Though limited in application, the benzothiazepines are important among the list of Ca^{2+} antagonists [102–104].

The scarcity of structural studies of benzodiazepines and benzothiazepines is because of the difficulty encountered in the cyclization of these seven-membered heterocycles. In general, cyclization of linear precursors is in principle, an excellent route to heterocycles. However, for cycloheptane rings, this reaction is disfavored by entropic and enthalpic factors besides *trans* annular interactions [105, 106]. Hence, these heterocycles are usually obtained in good yield only when configurational and/or conformational constraints facilitate intramolecular cyclization. These facts kindle further interest in the study of their conformational features. Here, the structural details of a few diazepines, benzodiazepines, and benzothiazepines are discussed. Benzodiazepines involve the fusion of benzene rings with diazepine rings, whereas in benzothiazepines, the benzene rings are fused with the thiazepine rings. A comparative study of benzo derivatives of azepines always enable pharmacologists to assess the pharmacological differences from the view point of variant hetero atom substitutions in the heterocycle.

2.2.1
Stereochemical Analysis of Azepine Derivatives

In contrast to the six-membered rings, the seven-membered cyclic systems are much more flexible. The cycloheptane rings occur in complex pseudorotational equilibria that has numerous conformations of similar energy with the absence of significant pseudorotational barriers. Owing to these reasons the study of such compounds is difficult using experimental techniques and therefore condensation with a benzene ring or the introduction of a double bond freezes the pseudorotational equilibrium at a low temperature [107].

Cycloheptane [108–110] was the first compound studied that served as a model for subsequent investigations. They adopt several significant conformations with distinctly two pairs of conformers, each comprising two interconvertible forms. The chair and the twist-chair forms, as well as the boat and the twist-boat forms constitute the flexible forms of the heptagonal ring [60, 111] (Fig. 1). The chair (C) forms do not constitute real conformers as the non-bonding interactions between hydrogen atoms destabilize them and therefore the more stable conformers are the twist-chair (TC) with the C form, constituting a transition state in the pseudorotation process between two twist-chairs. Both conformations constitute the chair/twist chair family (C/TC). In the other pseudorotating family boat/twist boat (B/TB), the non-bonding interactions between hydrogen atoms turn the boat conformations into transition states of the pseudorotating process. TB conformations possess conformational energies higher than TC conformations and each TC conformation may be converted into a TB conformation by means of a pseudorotating process [107].

The conformational behavior of cycloheptane rings is changed by introducing of heteroatoms. In general, heteroatoms engender appreciable dipole moments. If there is only one heteroatom, the dipole has hardly any effect on conformation; but when there is more than one heteroatom in the ring, dipole–dipole interactions do affect conformation. Furthermore, such interactions are found to be solvent dependent. Hence, conformation of such rings may even vary with the solvent used [112]. The necessity of fusion of benzene with azepines is not unique from the medicinal point of view. Sporadic investigations [113, 114] show azepines to be less stable than azoles. The existence of only some nitrogen-substituted azepines are placed on record. However, most of them are stable only in the absence of air, or at low temperatures, or in dilute solutions. There is no enigma in realizing their benzo derivatives to be more stable, provided annulation takes place in such a way as not to destroy the aromaticity of the benzene rings. The fusion of the aromatic hydrocarbons makes the boat form of the diazepines more stable.

The non-planar arrangement of ring atoms of a seven-membered heterocycle [115] places it in the class of rings with interesting conformational properties. Chair and boat conformations are characterized by the presence of a plane of symmetry while the **TC** and **TB** conformers present a binary axis. These elements of symmetry bisect one bond of the ring and pass through the opposite carbon atom. The energy differences between conformations are small, except for the interconversions between both pseudorotating families (**TC**↔**TB** transition) [107].

This biological and conformational importance prompted us to carry out the crystallographic characterization of a few *N*-formyl-2,7-diphenyl-substituted diazepine-5-one, benzodiazepines, and benzothiazepines in order to understand the stereochemical aspects so as to correlate structure and activity (Table 4). As was explained for the cyclohexane the nomenclature of Espinosa [74] was used to fully characterize each conformation for azepine for all the boat conformations (Fig. 6).

Table 4 Various azepine derivatives used for stereochemical analysis in this study

No.	Mol.	X	R₁	R₂	R₃	R₄	Conformation
24	C	–	CH₃	H	–	–	Boat
25	C	–	CH₃	CH₃	–	–	Boat
26	C	–	—CH(CH₃)CH₃	H	–	–	Boat
27	D	N	—C(=O)—C₆H₅	—C(=O)—C₆H₅	=O	CH₃	Boat
28	D	N	H	H	=O	CH₃	Boat
29	D	S	COH	–	C₆H₅	C₆H₅	Boat
30	D	S	—C(=O)—CH₂Cl	–	CH(CH₃)CH₃	CH₃	Boat

Surprisingly, all the structures adopt a boat conformation irrespective of the different substitution and benzo fusion (Table 5). The N-formyl-2,7-diphenyl-substituted diazepine-5-one is found to prefer the 7B1 conformer that has it's pseudoplane passing through the primary nitrogen. As expected, the benzo derivative prefers the conformation whose pseudoplane bisects the fusion bond (7B6 and 7B13). Only for compound **30** of the benzothio derivative does the sign of the torsion angle change, which may be the influence of chlorine in the chloroacetyl group that is substituted at the N-position.

In cycloheptane, in addition to axial and equatorial orientations, the substituents' can be located in an isoclinal position [116]. These orientations

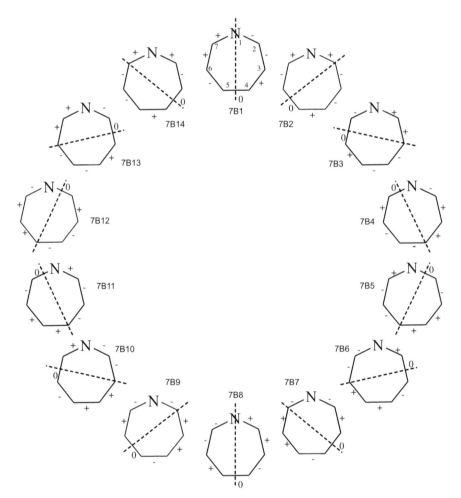

Fig. 6 Representation of the pseudorotation in the boat conformation of azepines. The *dashed* and *solid lines* denote the pseudoplane and pseudoaxis for each conformation, respectively

Table 5 Endocyclic torsion angles for azepine derivatives with their conformer description

	1–2	2–3	3–4	4–5	5–6	6–7	7–1	Confor-mation	Mirror	Con-former
A1	– 52.7	– 33.5	71.0	1.1	– 72.9	33.9	54.0	Boat	N1(N4-C5)	7B1
A2	– 54.6	– 29.2	67.8	3.1	– 76.4	38.0	50.4	Boat	N1(N4-C5)	7B1
A3	– 68.5	– 12.0	63.2	– 5.1	– 76.2	46.5	47.0	Boat	N1(N4-C5)	7B1
B4	71.3	3.7	– 46.0	– 10.9	81.5	– 45.7	– 44.5	Boat	C6(C2-C3)	7B6
B5	65.6	0.7	– 42.0	– 8.1	75.3	– 47.7	– 36.9	Boat	C6(C2-C3)	7B6
B6	67.6	5.1	– 63.8	22.5	61.2	– 68.5	– 25.0	Boat	C6(C2-C3)	7B6
B7	– 76.1	8.2	64.6	– 23.4	– 60.2	62.7	34.5	Boat	C6(C2-C3)	7B13

can be identified based on the relationship between the sign and magnitude of the endocyclic torsional angles that flank them [117]. Viewed clockwise, a β-orientated substituent is axial if the α_1 and α_2 torsion angles have the signs (+,–), (0,+), or (0,–), and is equatorial if the signs are (–,+), (+,0), or (–,0). Finally, such a substituent is isoclinal if the two torsion angles are equal in sign and magnitude [107]. The N-substitutions occupy anticlinal orientation in all the azepine derivatives. In the N-formyl-2,7-diphenyl-substituted diazepine-5-one derivatives the two phenyl rings take up isoclinal orientation, which may be due to the anomeric effect. All the other bulky substituents adopt an equatorial orientation in this derivative.

In the benzo derivative also, as is the case of the diazepine-5-one derivatives, the 5th and 7th position substituents occupy anomeric isoclinal orientation. The fusion of benzene to impute stability does not modulate the geometry of the septagonal rings in all the structures. The introduction of a conjugated system into an alicycle considerably flattens the molecule [60]. The sum of the angles around the hetero nitrogen atoms are approximately equal to 360°, which signify a flattening of the heterocyclic ring. This flattening of azepine moieties, which implies involvement of sp^2 hybridization, is a characteristic feature of this class of compounds [118, 119]. Replacement of a carbon with a nitrogen or sulfur atom has influence upon the geometry of the ring atoms adjacent to the heteroatom. Non-negligible cyclic distortions are eventually expected in comparison with their carbocyclic analogues [60]. The geometrical parameters such as unequal C – N bond distances about the hetero nitrogen atoms in the four structures under discussion are due to different environments brought by variant substituents. Furthermore, a partial double bond character due to delocalization of the nitrogen atom's lone pair of electrons enforces a shrinkage of certain N – C bond distances. The presence of another heteroatom in the benzo derivative also brings into effect the delocalization of π-electrons over the C – X – C composite, upon fusion with the benzene ring [120] (Fig. 7).

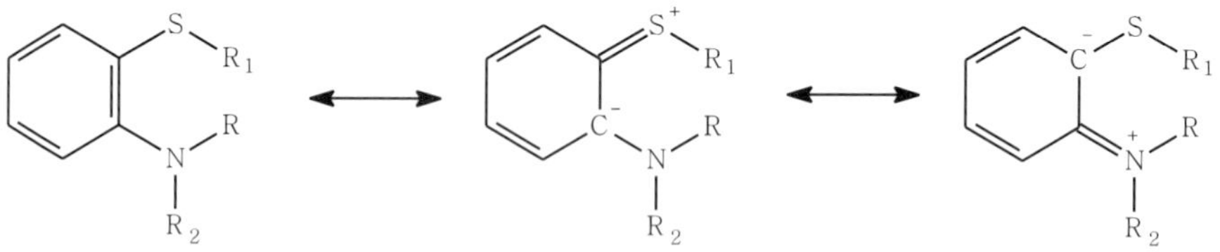

$$R = R_1 = C \qquad R_2 = \text{Formyl; Chloroacetyl}$$

Fig. 7 Partial double bond character in the azepine derivatives when an additional electro negative heteroatom is present

2.2.2
Drug Conformation in a Protein Environment

As in the case of the piperidine derivatives, azepine drugs in complex with protein were searched for in the PDB using the MSD server. Since the number of hits was very low no resolution criteria was used for this search that yielded one each in azepin and benzothiazepine and five in fused-diazepine (Table 6).

The drugs hymenialdisine and K-201 adopt a chair conformation whereas all others adopt a boat conformation. Unlike piperidine drug molecules, all the azepine drugs adopt symmetries, which correlate to the shape of the molecule. The fusion-free drug, SB203238, which is a HIV-1 protease inhibitor, takes up a 7B3 conformation such that its two linear substitutions (1st and 5th) are symmetrically placed on either side of the mirror axis. In the mono-fused drugs the pseudoplane bisects the fused-bond and in the di-fused drug (1051U91) the pseudoplane passes through the N1 atom such that the two benzene rings are on opposite sides. The only exception to this trend is the hymenialdisine drug that has a five-membered ring fused to the azepine moiety. As observed for the small molecule azepine derivatives, the N-substitutions are oriented in an anticlinal position. All other substitutions either take anticlinal or equatorial orientations.

The study of ligand interaction showed that all the azepine drugs have good hydrophobic interaction with the target protein residues and only the substituents show hydrogen bonding interactions with the protein environment. A significant hydrogen bonding interaction was found in the hymenialdisine drug in which the nitrogen atom of the azepine moiety interacts with the protein residue (Fig. 8).

The analysis showed that the N-substitution has an anticlinal orientation in both the drug as well as in the small molecules studied. The substitutions adopt equatorial orientations in both but no isoclinal orientations are seen for drugs, which may be due to the absence of substitutions at the appropriate position. Irrespective of all odds, the azepine drugs take up a conformation that correlates with its shape as seen from the synthetic derivatives that were analyzed.

Table 6 Conformation of drugs containing azepine and its benzo derivatives moiety in protein structures

PDB ID	Description of the ligand	Protein to which bound	Ring Conf.	Ref.[∈]
1dm2	Hymenialdisine – marine sponge constituent for treating neurodegenerative disorders	Human cyclin-dependent kinase 2	Chair (C7)	(a)
1hbv	SB203238	HIV-1 Protease	Boat (7B3)	(b)
1lw2	1051U91 – Inhibitor of AZT resistance HIV-RT	HIV-1 Reverse Transcriptase	Boat (7B1)	(c)
1rt3	1051U91	HIV-1 Reverse Transcriptase	Boat (7B1)	(d)
1rth	1051U91	HIV-1 Reverse Transcriptase	Boat (7B1)	(e)
2bxf	Diazepam – CNS depressant and sedative-hypnotic drugs	Human Serum Albumin	Boat (7B7)	(–)
1hak	K-201 – Calcium Channel Activity Inhibitor	Human Placental Annexin V	Chair (C4)	(f)

[∈] (a) Meijer L, Thunnissen AM, White AW, Garnier M, Nikolic M, Tsai LH, Walter J, Cleverley KE, Salinas PC, Wu YZ, Biernat J, Mandelkow EM, Kim SH and Pettit GR (2000) Chem Biol 7:51; (b) Hoog SS, Zhao B, Winborne E, Fisher S, Green DW, DesJarlais RL, Newlander KA, Callahan JF, Moore ML and Huffman WF (1995) J Med Chem 38:3246; (c) Chamberlain PP, Ren J, Nichols CE, Douglas L, Lennerstrand J, Larder BA, Stuart DI and Stammers DK (2002) J Virol 76:10015; (d) Ren J, Esnouf RM, Hopkins AL, Jones EY, Kirby I, Keeling J, Ross CK, Larder BA, Stuart DI and Stammers DK (1998) Proc Natl Acad Sci USA 95:9518; (e) Ren J, Esnouf R, Garman E, Somers D, Ross C, Kirby I, Keeling J, Darby G, Jones Y and Stuart D (1995) Nat Struct Biol 2:293; (f) Kaneko N, Ago H, Matsuda R, Inagaki E, Miyano M (1997) J Mol Biol 274:16

3
Study of Weak Interactions in a few Heterocyclic Structures

As the configurational and conformational analyses for the unsaturated heterocyclic rings are not informative owing to the adoption of planar conformation, then it is essential to study intermolecular interactions. The detailed analysis was carried out for isoxazole, imidazole, and indole molecules. Since there wasn't any significant difference in the pattern of interaction for these aromatic rings, these results can be extrapolated to all other heteroaromatic

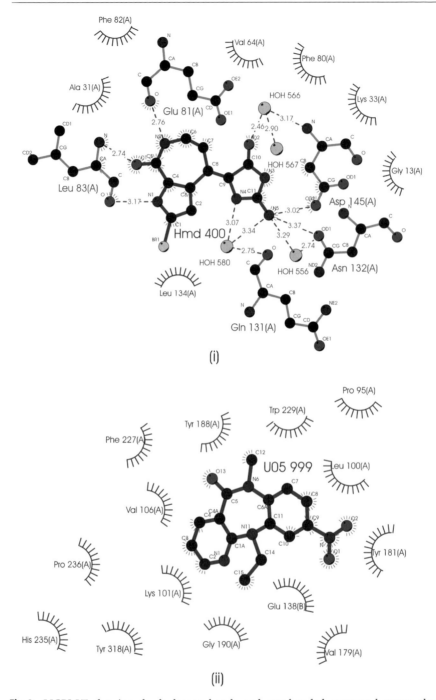

Fig. 8 LIGPLOT showing the hydrogen bonds and non-bonded contacts between the azepine drugs and its receptor for (i) 1dm2 and (ii) 1lw2 structures

rings. Similar to the saturated ring systems, the weak interaction analysis was carried out for isoxazole, imidazole, indole, quinoline, and triazole drugs.

3.1
Background

The three-dimensional crystal structure of a molecule is a free-energy minimum resulting from the optimization of attractive and repulsive intermolecular interactions with varying strengths, directional preferences, and distance–dependence properties. Intermolecular interactions in organic compounds are of two types: isotropic medium-range forces that define the shape, size, and close packing; anisotropic long-range forces that are electrostatic and include hydrogen bonds and heteroatom interactions. The observed three-dimensional architecture in the crystal is the result of interplay between the isotropic van der Waals forces and the anisotropic hydrogen-bond interactions. The distinction between hydrogen bonds and van der Waals interactions lies in their orientational and angular attributes [121].

The nature of intermolecular interactions that mediate molecular recognition for all systems are the same: strong $O-H...O$, $N-H...O$, $O-H...N$ hydrogen bonds; weak $C-H...O$, $C-H...N$ hydrogen bonds; heteroatom interactions $X...X$, $X...O$ ($X = Br$, I); $\pi-\pi$ interactions; and van der Waals forces [122]. Hydrogen bonds, regarded as the strongest and most directional of intermolecular interactions, have been widely exploited in many fields [123].

Molecular interactions are influenced by dispersion, polarization, electrostatic, charge–transfer, and exchange forces [124]. The electrostatic term is directional and of long range ($-r^{-3}$ and $-r^{-2}$) whereas polarization ($-r^{-4}$) and the charge–transfer (e^{-r}) term decreases more quickly. The dispersion ($-r^{-6}$) and exchange repulsion (as $-r^{-12}$) terms are often combined into an isotropic van der Waals contribution, which increases sharply with reducing distance. Depending on the particular chemical donor–acceptor combination, and the details of the contact geometry, all of these terms contribute with different weights [125]; but it is the non-directional van der Waals forces and the directional electrostatic interactions that contribute the most to the energy terms. At optimal geometry, van der Waals interactions contribute some tenths of a kcal/mol to the hydrogen bond energy whereas electrostatic interaction reduces with increasing distance and with reducing dipole moments or charges involved. For strong donors like $O-H$ or $N-H$, the electrostatic component is the dominant one, whereas for weakly polarized donor groups like $C-H$ the magnitude of the electrostatic component resembles to van der Waals contribution [125, 126].

Hydrogen bonding is a well-known classical structural phenomenon [127]. Knowledge of weak intermolecular interactions enables one to design and manipulate molecular systems and this can be applied in the fields of rational drug design, crystal engineering, supramolecular chemistry, and physical organic chemistry [128]. Also, these secondary interactions have ramifications in the systematic design of new materials possessing novel chemical, magnetic, optical, or electronic properties [129]. Critical assessment of the weak intermolecular interactions is a must as these may exert a substantial effect when added together. In this context, the acceptor capabilities of halogen atoms are controversial and noteworthy [130]. A second type of weak hydrogen bond established in recent years is the hydrogen bond with π acceptors [125].

The geometric characteristics of these interactions are essential since they form the basis for theoretical calculations of molecular electronic structure, molecular mechanics and dynamics, modeling studies, protein structure determination, and the derivation of new potential energy functions [131]. This approach derived from small molecular crystal structures is used as a model in structure-based drug design, pharmacophore mapping, crystal engineering, the design of supramolecular materials, nanotechnology, ligand–protein binding, and crystal structure prediction [122].

There are three techniques for studying the geometric characteristics of hydrogen bonding: X-ray crystallography, spectroscopy, and computation [122, 132]. Computation involves statistical analysis of structural data (geometrical information obtained from a large number of crystal structures), which is very important for a correct interpretation of several chemical phenomena [133], as well as for a better understanding of the nature of hydrogen bonds and intermolecular interactions. An advantage with this database approach is that any distortion in individual interaction geometry can be averaged out in the diverse sample of X-ray crystal structures determined for a variety of reasons [122].

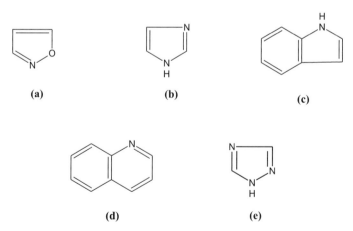

Fig. 9 Chemical structures of the heterocyclic compounds: **a** Isoxazole, **b** Imidazole, **c** Indole, **d** Quinoline, and **e** Triazole

A study of weak interactions was carried out in view of their above-mentioned importance. Statistical analysis of geometrical information obtained from a large number of crystal structures in a few selected systems was carried out. Weak interactions involving π acceptors such as isoxazole, imidazole, and indole moieties are analyzed. In order to correlate the result of this study with the activity of drug, π-interaction study was also carried out for the drug molecules that are present in the crystal structure of their respective protein targets for the aromatic heterocyclic rings isoxazole, imidazole, indole, triazole, and quinoline (Fig. 9).

3.2
Biological Importance of the Aromatic Heterocyclic Rings under Study

3.2.1
Isoxazoles

Isoxazole is a heterocyclic five-membered ring containing nitrogen and oxygen atoms adjacent to each other. They form a part of many drugs used for antidepressant therapy (propionic acid derivative [134]), act as antiviral compounds (methyltetrazole derivative [135]), as protein tyrosine phosphatase 1B inhibitors (carboxylic acid derivative [136]), and act as growth hormone secretagogue receptor (GHS-R) antagonists (carboxamide derivatives [137]). One of the important drugs that contain isoxazole is valdecoxib, which is a nonsteroidal anti-inflammatory drug (NSAID) that is used for the treatment of osteoarthritis or rheumatoid arthritis and also for the treatment of primary dysmenorrhea [138].

3.2.2
Imidazoles

Imidazole is a heterocyclic five-membered ring containing nitrogen atoms. Various imidazole derivatives are found to possess biological activities. Biotin is of interest here, it is a heterocyclic sulfur-containing B-complex vitamin with imidazole and thiophene rings fused along the $C-C$ bond. Biotin serves as a carrier of CO_2 in carboxylation reactions. Biotin-enzyme reacts with CO_2 in the presence of ATP to form a carboxybiotin-enzyme complex. This complex hands over the CO_2 to pyruvate to produce oxaloacetate [139].

Heterocyclic compounds containing two nitrogen atoms exhibit antimicrobial effects. Imidazole is a fundamental building block of many biological proteins and other biological systems. Imidazole acts as a ligand that will bind readily to a metal ion via metal ion salts in aqueous systems. Imidazole is a protonated five-membered ring which promotes chemical reactions depending on the specific physical conditions at enzyme catalytic sites [140]. Imidazole-4-acetic acid is a catabolite of histamine and is present in the brain [141–143], although its precursor in the brain is as yet unknown [144]. It is also a γ-aminobutyric acid (GABA) agonist [145, 146] and acts at the GABA receptor. Imidazoles and benzimidazoles are the components of important structures used in pharmaceuticals, agrochemicals, dyestuffs, and high-temperature polymer products. An imidazole succinic acid complex is an active non-steroidal aromatase inhibitor [147, 148]. Some imidazole derivatives serve as key intermediates for preparing the antihypertensive drug Losartan potassium [149].

3.2.3
Indoles

Indole and its derivatives are found in many natural products as plant alkaloids that have antimicrobial [150], anti-inflammatory [151], antibacterial [152], and antidepressive activities. Many sulfur-containing compounds such as sulfates, sulfones, and sulfonomides exhibit insecticidal, germicidal, antimicrobial, and antibacterial activities [153, 154] and certain phenyl sulfones show fungicidal properties [155].

3.2.4
Quinolines

Quinolines have been the interest of research for many years as a large number of natural products contain these heterocycles and they are found in numerous commercial products including pharmaceuticals, fragrances, and dyes [156]. The quinoline ring system comprises a benzene ring and a pyridine ring fused through a carbon–carbon bond. Quinoline alkaloids such as quinine, chloroquin, mefloquine, and amodia quine are used as efficient drugs for the treatment of malaria [157]. Quinolines possess interesting physiological properties and as phthalideisoquinoline alkaloids play interesting roles such as noscapine, a non-narcotic cough cure and (+)-biculine, an effective antagonist of an inhibitory neurotransmitter γ-aminobutyric acid (GABA) [158]. The 4-quinoline structure has been reported to have excellent antibiotic properties; papavarine is an opium alkaloid, which is a nonspecific smooth muscle relaxant and vasodilator [159].

The quinoline ring system is found in plants such as *Berberidaceae, Fumariaceae, Papavaraceae* [160] and is also present in the marine species [161, 162]. Quinoline derivatives, protoberbines and 8-oxoberbines, are known to possess biological properties such as antileukemic, antitumor, and anticancer activities [163]. The potent antitumor agents Dynamicin A and Virantmycin are important natural products containing the quinoline core [164].

Tetrahydroquinoline derivatives possess a wide-spectrum of biological activities. Oxaminiquine is used to eradicate blood flukes (*Schistosomo mansoni*) [165]. 2-Oxo quinoline derivatives have been found to assume inhibitory activities against blood platelet aggregation [166]. They exhibit antitumor activities [167] and also act as potent antipsychotropic agents [168], and a compound containing the tetrahydroquinoline moiety acts as an antischistosomal drug [169]. They also possess anti-inflammatory [170], antiamoebic [171], antiulcer [172], analgesic [173] and antifertility activities [174]. Synthetic tetrahydroquinoline derivatives possess high antiarrhythmic and antihypertensive activities [175, 176]. They also show inhibitory activities against acetylcholinesterase and human aldose reductase [177]. It has been reported that conformationally restrained arylquinoline derivatives may act as antagonists with enhanced selectivity towards the 5-hydroxy triptamine (5-HT) receptor [178]. Certain indoloquinoline analogues have antibacterial and antimalarial properties [179]. It has been reported that indolo[2,1-a]-isoquinolines have the structural features of dibenzopyrrocoline alkaloids, such as cryptaustoline and cryptowoline, isolated from the bark of *Cryptoocaryabowiei* [180]. Cryptolepine, a derivative of indoloquinoline alkaloids show anticryptococcal activity [181].

3.2.5
Triazoles

Generally, 1,2,4-triazole derivatives are found to be associated with diverse pharmacological activities. Recently, some new triazole derivatives have been synthesized as possible anticonvulsants, antidepressants, tranquilizers, and plant growth regulators [182–184]. Extensive studies have also been carried out on the substituted 1,2,4 triazole ligands [185–187]. It is of interest to note that some iron(II) complexes containing substituted 1,2,4-triazole ligands are spin-crossover materials that could be used as molecular-based memory devices, displays, and optical switches [188, 189]. The 1,2,4-triazole nucleus has recently been incorporated into a wide variety of therapeutically interesting drugs [190] including a histamine receptor blocker, cholinesterase active agents, CNS stimulants, antianxiety agents, and sedatives.

Substituted 1,2,4-triazoles have been actively studied as bridging ligands coordinating through their vicinal N atoms. It is of interest that some complexes containing 1,2,4-triazole ligands have particular structures and specific magnetic properties [189, 191–193]. On the other hand, some of the 1,2,4-triazole derivatives have anti-inflammatory activities [194] and some are antifungal agents [195].

Bhargava and co-workers [196] synthesized and studied the anti-inflammatory activity of some 3-(O-substituted phenyl)-4-substituted phenyl-5 alkyllalkenyl mercapto-1H-1,2,4-triazoles. Mohan and co-workers [197] synthesized and studied the antimicrobial activity of thiazolo (3,2-b)-S-triazole, S-triazolo (3,4-b) (1,3,4) thiodiazines and isomeric diazolo (2,3,-c)-S-triazoles. The compound 2,4-dihydro-1,2,4-triazole-3-thiones (disubstituted in positions 4 and 5) were prepared [198] and used as bactericides, fungicides, and pesticides. Insecticidal and ascaricidal properties of 4-methyl-3-(trifluoromethyl)-1,2,4-triazoline-5-thiones were also reported [199]. A series of 3-(2,4 dichlorophenyl)-4-aryl-5-mercapto-1,2,4-triazole compounds possess good fungitoxic properties [200]. Several 3-aryloxy methyl-1,2,4 triazoles have been synthesized and screened for their antifungal activities [201].

In view of the above-mentioned wide range of pharmacological and biological activity of the heteroaromatic rings, they were used for the π-interaction analysis.

3.3
π-Interaction Analysis for a few Heterocyclic Compounds

Molecular organization and molecular interactions are the basis of the functional properties of most molecules, and a detailed understanding of non-covalent chemistry is therefore fundamental for interpreting and predicting relationships between chemical structure and function [202] and also in understanding biological complexity. One of the less known but significant weak interactions in nature is the interaction involving π systems [203] wherein phenyl rings, various heterocycles, $C \equiv C$, $C = C$, and other π-bonded moieties are involved. Two types of interactions can be categorized with π systems: one is a hydrogen bond where the π system acts as an acceptor and the other is the interaction between the π systems ($\pi - \pi$ interactions). In this part we have analyzed both these types of π interactions for various heterocycles such as isoxazole, imidazole, and indole π-moieties (Fig. 9) and the results are presented.

3.3.1
Hydrogen Bonding Involving Heterocyclic π-Systems

Energy calculations have shown that there is significant interaction between a hydrogen bond donor and the center of a benzene ring, which is about half the strength of a normal hydrogen bond and contributes approximately 3 kcal/mol of stabilizing enthalpy. This aromatic hydrogen bond arises from small partial charges centered on the ring carbon and hydrogen atoms and there is no need to consider delocalized electrons [204]. In this hydrogen bond type, the donor group $X - H$ is placed roughly above the center of an aromatic ring, and the $X - H$ vector points at it [205].

Peptide $X - H...\pi$ interactions are functional in stabilization of helix termini, strand ends, strand edges, β-bulges, and regular turns. Side-chain $X - H...\pi$ hydrogen bonds are formed in considerable numbers in α-helices and β-sheets [205]. The conformational influence of $N - H...\pi$ and $O - H...\pi$ in organic molecules and also their role in crystal packing have been established [206, 207]. The study on $C - H...\pi$ interactions has clearly established that this weak molecular force shows a directional preference with the C – H pointing towards the center of the aromatic ring [208].

The large surface makes π-acceptors a "target that is easy to hit" and they are good reserve acceptors for hydrogen bonds, if conventional acceptors are locally lacking. X – H...π bonds are often used to fill hydrogen bond coordination shells, which allows conventional hydrogen bonds to be missed out without leading to completely unsatisfied hydrogen bond potentials [205]. The energetics and consequently the chances of the occurrence of this type of interaction are enhanced if the aromatic system contains nitrogen atoms that magnify the π-electron density. The nitrogen atom in a heterocyclic ring system is electron-rich, and any donor atom sitting on top can point its proton towards the nitrogen π electrons giving rise to the X – H...π interaction [209]. Relatively few geometrical analyses were made for heterocylic π-systems and in this study we plan to present an understanding of the influence of the nitrogen heteroatom in selected π-systems.

3.3.1.1
Methodology

Three databases were created for studying the geometries of X – H...π interactions occurring in the heterocyclic π systems isoxazole, imidazole, and indole moieties using the Cambridge Structural Database (CSD) [210], update 5.24 (November 2002). In the literature, different sets of parameters are used to describe the geometry; here, we use the parameter given in Fig. 10. Searches were carried out for donor species O – H, N – H, C – H, and S – H and the 3D constraint options within the database were used to define the center of the ring (Cg) as the centroid of all the atoms present in the heterocycle. From this point, measurements of the distances D (donor...Cg), d (H...Cg) were made and the angle θ (X – H...Cg) was recorded.

The non-bonded contact to be searched was set up as the one occurring between the ring centroid (Cg) and the interacting hydrogen and donor atoms. This was allowed to be an intermolecular interaction up to a distance of 4.0 Å for D (donor...Cg) and 3.0 Å for d (H...Cg) distances. The lower limit for interactions with Cg was set at 2.8 and 1.8 Å, respectively for D and d distances. To ensure the data analyzed had a suitable degree of accuracy, further criteria were used to include only hits with 3D coordinates, without metals and no disorders. For statistical equivalence, the crystallographic discrepancy factor was varied for each dataset (Rf $\leq$ 0.10 for isoxazole & indole and Rf $\leq$ 0.05 for imidazole). Since there were no hits for S – H...π interaction, further analyses were restricted to O – H, N – H, and C – H donor species.

Mean and standard deviation (σ) values were calculated for all the geometric parameters. All data falling outside ±4 σ cut-off of the mean values were excluded for further analysis. Angular distributions of X – H...π were plotted as histograms in 5° intervals and are corrected for the geometric error in

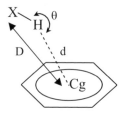

Fig. 10 Geometrical parameters used for the analysis. Benzene is used here as a model π system

the frequency of observations (cone correction), because there being a greater probability of finding the interactions on the rims of cones in increasing solid angles and the use of $N/\sin\theta$ instead of N (where N is the number of hydrogen bonds in the interval θ to $\theta + d\theta$) in the histograms effectively resolves this contradiction [122, 133, 211, 212].

Distance and directionality characteristics of hydrogen bonds can be conveniently studied in scatter plots of hydrogen bond angles versus the d and/or D distances [213]. In order to compare the different $X - H...\pi$ systems, the distances d were normalized according to the sum of the van der Waals radii:

$$R = \frac{d(H...Cg)}{\sum(r_{Cg} + r_H)}.$$

Since the centroid of the π-system is assumed to be involved in the interaction and also heteroatoms are present, the van der Waals radii for the ring is taken as the average van der Waals radii of all the atomic species. Therefore, the van der Waals radii for isoxazole, imidazole, and indole π systems are respectively 1.634, 1.64, and 1.683 Å.

Furthermore, a change of variables $x = R^3$ and $y = 1 - \cos(180 - \theta)$, was used to obtain better insight into the spatial distribution of short contacts, so that a uniform distribution of the data points in three-dimensional space transforms into a uniform distribution in the two-dimensional scattergram [128]. This normalization allows a qualitative assessment of the relative strengths of the different hydrogen bond types [212].

3.3.1.2
Results and Discussion

Mean values for the donor...Cg (D), H...Cg (d) distances and the donor angle (θ) with their respective standard deviations and the number of observations for the donor species $O - H$, $N - H$, and $C - H$ with the heterocyclic π-systems, isoxazole, imidazole, and indole are presented in Table 7. From the table it is clear that the number of observations for the $C - H$ donor species is relatively large when compared with $O - H$ and $N - H$ donor atoms. This may be due to the following two reasons; since π-interactions are weak, the

Table 7 Mean values for donor...Cg (D), H...Cg (d) and angle (θ) for X – H...π dataset. The standard errors of the mean values are given in parentheses and the number of hydrogen bonds [n] is represented in brackets. (§ na represents the absence of data for this particular screen.)

	D(Å)			d(Å)			θ(°)		
	Isoxazole	Imidazole	Indole	Isoxazole	Imidazole	Indole	Isoxazole	Imidazole	Indole
O–H	3.8(1)[16]	3.8(1)[31]	3.4(2)[5]	2.95(5)	2.88(8)	2.7(3)	163(17)	162(15)	134(21)
N–H	na§	3.8(2)[9]	3.5(1)[49]	na§	2.9(2)	2.7(2)	na§	159(18)	141(16)
C–H	3.6(1)[34]	3.6(1)[201]	3.7(1)[564]	2.9(1)	2.8(1)	2.9(1)	134(15)	140(14)	142(16)

Total no. of structures: Isoxazole – 41; Imidazole – 184; Indole – 452

strong donors like oxygen and nitrogen prefer to form stronger hydrogen-bonded interactions like N/O – H...N/O rather than preferring to form these weaker interactions [214]; and also the overall gain in energy is more when the O/N – H is engaged in typical hydrogen bonds and the C – H is simultaneously held by a π electron cloud, rather than associating the O/N – H group with the aromatic system alone [209].

The geometrical parameters for the C – H donor group are similar for all the heterocyclic π-systems depicting the invariant nature due to different π-systems, which clearly emphasize the weakness of the bond. The standard deviations of the observations for both the distances in all the categories is around 0.15 Å and this indicates approximately 95% of the data falls in the range ± 0.30 Å, which is indeed a broader distribution demonstrating once again the soft nature of the interaction. Because of this softness the donor atoms show large lateral displacements and a strong bending angle. In fact X – H may point at the aromatic center, at particular bonds, or even at an individual atoms [205]. The mean value of the H...Cg distance (d) ranges between 2.7 and 2.9 Å, which is longer when compared with the 2.5–2.7 Å range for the benzene acceptor [215].

The frequency of donor...Cg distance (D) distributions show typical strong bond characteristics for C – H whereas the H...Cg distance (d) distributions and the normalized angle distributions reveal the weaker nature of the interaction. Linear nature, as is true for any hydrogen bond, is observed in the angular distribution of O – H and N – H donor species. Comparatively the indole acceptor C – H...π is more linear than imidazole or isoxazole acceptors, this may be attributed to the increased arene size of the acceptor. The stability ranking as deduced from all the frequency distributions is given as X – H...indole > X – H...imidazole > X – H...isoxazole.

The normalized scatter plot for $R(d)^3$ vs. $1 - \cos(180 - \theta)$ is shown in Fig. 11. Generally, scatter plots require a large number of data points for any reliable conclusions. Since the dataset of the O – H and N – H donor species is very small, no reliable information could be deduced from these plots, but the linearity could be identified from the cluster around 170° [0.1 of $1 - \cos(180 - \theta)$] in all the plots. A common observation in these plots is that the π interactions occur at longer distance (greater than the van der Waals radii). The major cluster occurs around the value of 1.1 for R^3, this clearly demonstrates their electrostatic nature. The C – H...π is more diffused but still show linear behavior, the angle preference is around 130 to 180° but Allen and co-workers [214] observed a similar kind of preference for N – H...π interaction with aminophenol as the π-system. This clearly indicates that the heterocycles under study have stronger interactions than the heteroatom substitution in the phenyl ring, probably due to the larger electron-rich character. The classical image of X – H...π interactions, a T shape, with the interacting hydrogen approaching directly over the center of the aromatic ring [206, 216] can also be inferred from this study.

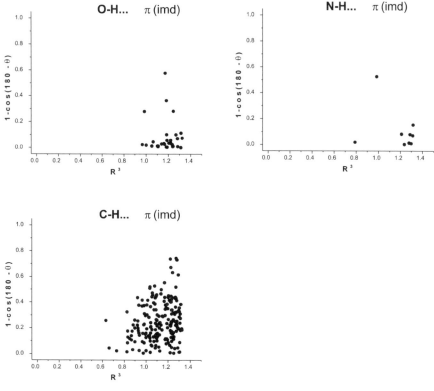

Fig. 11 Normalized scatterplots for H...acceptor distance (R^3) vs. donor angle ($1 - \cos(180 - \theta)$) for the $X - H...\pi$ dataset of imidazole (imd). (Some useful points of reference are: $R^3 = 1$ corresponds to $d(H...Cg) = \Sigma\ (r_H + r_{Cg})$; $1 - \cos(180 - \theta) = 0.0$ corresponds to $\theta = 180°$, $1 - \cos(180 - \theta) = 0.5$ corresponds to $\theta = 120°$, $1 - \cos(180 - \theta) = 0.75$ corresponds to $\theta = 104.5°$, and $1 - \cos(180 - \theta) = 1.0$ corresponds to $\theta = 90°$)

3.3.2
π-Stacking Interactions of Heterocyclic π-Systems

The three broad classes that contribute to the forces stabilizing the native structure of proteins are hydrogen bonds, electrostatic interactions, and van der Waals forces. Now that aromatic–aromatic interactions have been classified as a fourth class by Burley and Petsko [217] this clearly signifies their importance. Aromatic stacking interactions are widespread in nature [202] and are believed to provide stability to duplex DNA and thermophilic proteins, aggregation of amyloid proteins in Alzheimer's disease and are common motifs in biomolecular recognition [203, 218–221].

Hydrogen bonds are well understood; they are strong, single point interactions with a very well-defined geometry. Whereas in the π interaction there are multiple points of intermolecular contact, the geometry of interaction is variable, and a vast range of different functional groups are involved [202]. The major difference is that the surface area of intermolecular contact is large and so van der Waals interactions and desolvation are much more important. Although the electrostatic principles governing the magnitudes of H-bonds also apply to aromatic interactions, there are many more contact points where electrostatic interactions have to be considered, and so it is difficult to rationalize the behavior of aromatic interactions with straightforward rules as in the case of H-bonds [202]. Experimental estimates indicate that these interactions are energetically attractive by 1.5 kcal/mol but disfavored in solution by entropic factors because of restricted internal mobility [215].

Several geometries of the π...π interactions are attractive, and have been proposed on the basis of the electrostatic component [222], which arise due to quadrupole moments of the aromatic rings (Fig. 12). This quadrupole moment exists from the uneven distribution of charges, with greater electron-density on the face of the ring and reduced electron-density on the edge. The three most common geometries of the π...π interaction are the edge-face (T-shaped) geometry (that is a C – H...π interaction), the offset stacked (parallel displaced) orientation, and the face-to-face stacked orientation that occurs with opposite quadrupole moments [203]. Off-centered parallel displaced and T-shaped structures are the most commonly cited orientations [220] and only small energy barriers separate them [219, 223]. Small charges (< 0.153) favor parallel displaced geometries; large partial charges (> 0.3) favor T-shaped structures [220].

Hunter and Sanders [222] have devised a set of rules for non-polarized and polarized π-systems in which they argue the aromatic system is polarized either by a substituent or a heteroatom, wherein stacking interactions can be affected [202]. Heteroaromatic rings are commonly found in

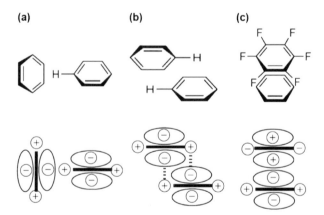

Fig. 12 Geometries of aromatic interactions. **a** Edge-face; **b** offset stacked; **c** face–to–face stacked

biomolecules and in synthetic molecules that exert specific biological effects (e.g., drugs). Non-covalent interactions involving heteroaromatic units contribute to the stability and specificity of macromolecular folding patterns and macromolecule-ligand interactions [203, 224]. A heteroatom in an aromatic ring can be electron neutral, electron rich or electron deficient and an electron deficient π-system stabilizes the interaction by decreasing the repulsion [202]. It has been reported that a nitrogen heteroatom within the ring is an electron-withdrawing perturbation which increases the tendency to stack [218]. To analyze the above fact, the following study was made.

3.3.2.1
Methodology

Nine datasets were created respectively for three heterocyclic rings, isoxazole, imidazole, and indole π-systems whose stacking interactions occur with itself (represented here as heteroaromatic–heteroaromatic), with benzene (heteroaromatic–aromatic), and benzene-benzene (aromatic–aromatic) systems. The third kind of analysis was undertaken in order to identify the influence of the heterocycles in the benzene–benzene interaction. The shape of axially symmetric aromatic rings can most naturally be represented in terms of the center of mass of the ring (the ring centroid) and the unique axis perpendicular to the ring plane (the surface normal vector) (Fig. 13). The intermolecular orientational information of one aromatic ring with respect to the other, is described by the centroid–centroid separation, R, a normal-normal angle, α, and the distance between the normal and the ring centroid, offset [220].

The CSD was searched for the non-bonded contact occurring between the ring centroid–ring centroid distance (R) of 2.8 to 5.2 Å and the dihedral angle

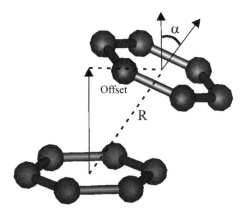

Fig. 13 Geometrical representation of various parameters used in the analysis. For simplicity benzene ring is shown as a model. The *solid arrows* represent the surface normal

(α) 0 to 90°. The secondary criteria used for choosing the data include only hits with 3D coordinates, organic structure without disorders, and the crystallographic error factor Rf $\leq$ 0.10 for isoxazole and indole and Rf $\leq$ 0.05 for imidazole (since the amount of data for this is very large). The mean and standard deviation for all the parameters and the correlation coefficient of the offset with R and dihedral angle (α) were calculated.

3.3.2.2
Results and Discussion

The mean and standard deviation values for the nine datasets for the centroid–centroid (R), the offset distances, and the dihedral angle (α) are presented in Table 8, along with the number of observations in each dataset. From the mean values and standard deviations, it is obvious that the distributions are very broad for any weaker interactions. The centroid–centroid distribution depicts the typical shape of the weak interactions and it also shows a bimodal character probably representing the two possible geometries, edge–face and face–to–face. There is no data below a 3.2 Å centroid-centroid distance in any of the datasets, which represents the minimum criteria for R. The mean values of R range between 4.4 and 4.7 Å and show a gradual increase as the arene size increases. There is no difference in the mean values of the offset distance irrespective of the incorporation of heteroatoms and/or increase in arene size, as it lies around 3 Å. The frequency distributions of the offset distance show a broader normal distribution with multiple nodes. It is clear from these distributions that the occurrence of zero offset is very rare representing that the exact face–face geometry is very uncommon even in the presence of electron-withdrawing heteroatoms [218]. Since the standard deviation for the dihedral angle is very large, the analysis of the mean values is meaningless.

The frequency distribution of the dihedral angle (α) is shown in Fig. 14. The angles between planes randomly oriented in space follow a sinusoidal distribution, with very few examples of low interplanar angles and many more at higher values [225]; but in the present case, the distribution is in contrast with this expectation, and the observed numbers in each category are very striking, and statistically significant. The distribution clearly represents the dominance of the face–face geometry. The vast majority of intermolecular plane contacts are parallel to each other, i.e. the interplanar angle is nearly zero. Yet, the two rings are parallel displaced with respect to each other which was observed in the offset distribution. A similar observation was also found for the pyridine fragments [218]. For the heteroaromatic–heteroaromatic interaction around 80% (α between 0 to 30° [225]) of the data are in the face-face geometry, whereas for heteroaromatic–aromatic and aromatic–aromatic, the percentage is around 50 and 48, respectively. This gives the stacking ranking as heteroaromatic–heteroaromatic $\gg$ heteroaromatic–aromatic >

Table 8 Mean values for centroid–centroid (R), offset distances and dihedral angle (α) for $\pi \ldots \pi$ dataset. The standard errors of the mean values are given in parentheses and the number of hydrogen bonds [n] is represented in brackets

	R(Å)			Offset(Å)			α(°)		
	Isoxazole	Imidazole	Indole	Isoxazole	Imidazole	Indole	Isoxazole	Imidazole	Indole
het–het	4.4(5)[137]	4.5(5)[605]	4.6(5)[374]	3(1)	3(1)	3(1)	15(27)	16(26)	16(27)
het–aro	4.6(5)[82]	4.5(5)[374]	4.7(5)[849]	2(1)	3(1)	2(1)	35(27)	39(29)	37(32)
aro–aro	4.7(5)[157]	4.7(4)[307]	4.7(4)[517]	3(1)	3(1)	3(1)	35(33)	37(32)	34(33)

(het – heteroaromatic; aro – aromatic; Total no. of structures: Isoxazole – 167; Imidazole – 514; Indole – 694)

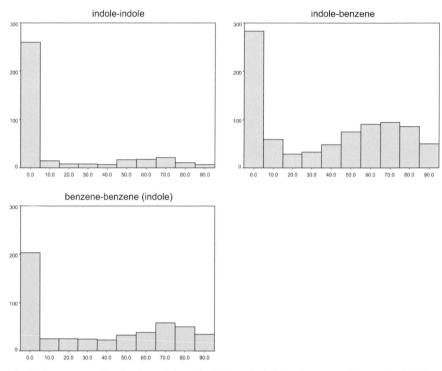

Fig. 14 Frequency distribution of the dihedral angle (α) for the $\pi\ldots\pi$ dataset for Indole

aromatic–aromatic. This observation is in line with the experimental investigations that face–face stacking of aromatic moieties shows increased stability when both partners are electron-poor [218]. This is due to the decrease in the π-electron density when electron-withdrawing atoms are added in the rings, which subsequently reduce the important π-electron repulsion.

An interesting feature observed in these analyses is that the benzene–benzene interactions also prefer face–face stacking like other interactions but normally not a preferred geometry (T-shaped geometry is the most favored one [202, 217, 226]). This may be due to the presence of the heterocyclic moiety that may induce the formation of the stacked geometry. A striking observation in this category is that, comparatively the benzene–benzene interactions for the indole moiety is more stacked (around 54%). Manual visualization of the individual structures showed that the majority of the interactions occur between the phenyl rings of the indole moiety in the face–face geometry [222] and hence the aromatic–aromatic distribution follows a similar trend to the heteroaromatic–heteroaromatic distribution.

The scatter plots for the dihedral angle with the centroid–centroid distance (R) shows a similar trend in all categories with two major clusters around 0° and 90° representing the two geometries, edge–face and face–face.

The scatterplots for the dihedral angle with the offset distance is shown in Fig. 15, which reveals the characteristic shape as seen for the Phenylalanine–Phenylalanine interaction in proteins by Hunter, Singh and Thornton [227] (Fig. 16). This indicates that heterocyclic ring systems in small molecules adopt a geometry similar to the aromatic–aromatic interaction in proteins and also follows Hunter and Sander's rules [222]. The stacked conformation is likely to be more efficient from a packing point of view, with less empty space being enclosed on folding [225] and so is widely observed. Therefore, it can be concluded that the influence of the heteroatoms in the π-systems towards a particular interaction geometry is similar for proteins and small molecules irrespective of the folding efficiency and crystal packing.

A basic element in crystal packing, molecular assembling, and in molecular recognition is the complementarity principle, which is governed by non-covalent interactions. Complementarities in shape and polarity clearly separate the hydrophilic and hydrophobic regions in crystal packing. This trend is also observed in the heteroaromatic ring systems studied and il-

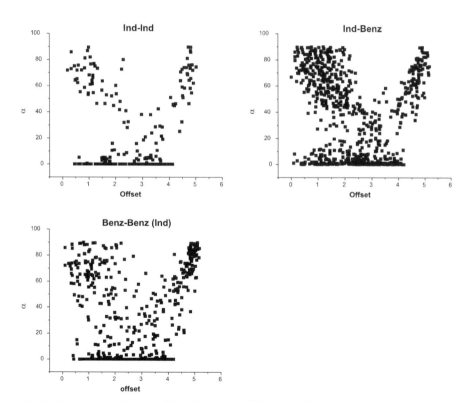

Fig. 15 Scatter plots for the offset distance vs. dihedral angle (α) for the $\pi\ldots\pi$ dataset of Indole-Indole (Ind-Ind), Indole-Benzene (Ind-Benz) and Benzene-Benzene (Benz-Benz) rings

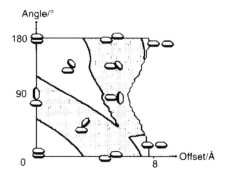

Fig. 16 The attractive orientations of Phe – Phe interactions in proteins with respect to the offset and dihedral angle (Reprinted from Hunter, Singh and Thornton [227] with permission from Elsevier)

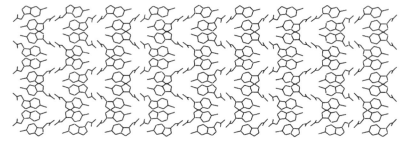

Fig. 17 Crystal packing diagram of ZUGPAS showing the hydrophilic and hydrophobic regions separately. The hydrogen atoms are removed for clarity

lustrated with an example for the indole heterocyclic ring. The compound 6-chloro-indole-3-acetic acid [228] (CSD code: ZUGPAS) has four X – H...π and 14 π...π interactions that pack the molecules into different hydrophilic and hydrophobic layers (Fig. 17).

3.4
π-Interaction Study
in the Protein Environment for Selected Aromatic Heterocycles

The wide spectrum of biological and pharmacological importance mentioned in Sect. 3.2 prompted us to study the isoxazole, indole, imidazole, triazole, and quinolines heterocycles in the protein environment so as to understand the structure activity relationship. The PDB was searched for the presence of these heterocycles as ligands using the MSD server and their various interactions with their respective protein target residues were noted (Table 9). From these only those π-interactions that involve the drug and the aromatic residues (tyrosine, histidine, tryptophan, and phenylalanine) of the protein targets were chosen for further analyses. Any interactions with other ligand

Table 9 PDB ID of the protein structures used for the π-interaction study

Imidazole (< 2.0 Å)		Indole (< 2.0 Å)	Isoxazole	Quinoline	Triazole
1aky	1ruw	185l	1d4m	1c9u	1cra
1dse	1sxu	1a53	1fsy	1cq1	1dm8
1f4t	1t8k	1a5b	1jwz	1cru	1ea1
1fhd	1u17	1beu	1nyy	1d3g	
1g0e	1u7r	1c83	1piv	1e9w	
1hs6	1ugi	1eg9	1r08	1flg	
1i12	1uux	1h5u	1ruc	1g72	
1il d	1uuy	1h69	1rud	1h4i	
1ikj	1v0m	1kfb	1rue	1h4j	
1jrl	1v0n	1kfc	1rug	1hbj	
1k6z	1v2g	1lwo	1ruh	1hxb	
1kae	1w8k	1o7n	1rui	1ida	
1keq	1weg	1oja	2r04	1ivq	
1l5n	1wkq	1qop	2r06	1jld	
1l9e	1wwj	1qoq	2r07	1k9g	
1moo	1y5e	1uuv	2rm2	1kb0	
1mun	1ywv	1wxj	2rr1	1kv9	
1muy	1yxv	2bk5	2rs1	1lrw	
1od8	1yxx	432d	2rs3	1mtb	
1odo	1z3w		2rs5	1otw	
1ouw	2izk			1w6s	
1pee	2rth			1yiq	
1phd	2rti			4aah	
1phe	2rtj				
1phf	2rtk				
1pm1	4mba				
1rky	8a3h				

molecules were also removed. Since the number of data for all categories were low no resolution criteria were given except for imidazole and indole (a resolution cutoff of 2.0 Å was given as is the case for piperidine).

It can be seen from Table 9 that triazole drugs appear only in three structures and imidazole drugs are found in 54 structures and others are found to have more than 20 structures. The analysis of percentage interaction of the drug molecules with the aromatic residues showed that the isoxazole and indole rings favor π-interaction with the tyrosine ring whereas the imidazole and the quinoline favor the tryptophan residue (Table 10). In addition, it is interesting to note that the isoxazole ring interacts only with tyrosine and phenylalanine. It is understandable that qunoline with two fused aromatic rings interacts with tryptophan, which is also a fused aromatic ring system.

Table 10 Percentage interaction of the drug molecules with the aromatic residues of protein

	Isoxazole	Imidazole	Indole	Qunoline	Triazole
Tyr	68.63	18.52	55.56	8.33	14.28
Phe	31.37	12.59	22.22	4.17	42.86
His	0.00	14.82	22.22	4.17	42.86
Trp	0.00	54.07	0.00	83.33	0.00

As explained in Sect. 3.3, the centroid–centroid distance and the angle between the plane were calculated and their distributions are shown in Fig. 18. It can be observed from the plot that except for the quinoline drugs all others are found to be spread over the whole range (interplanar angle 0 to 90°). For quinoline the major geometry is the stacked one, which is in line with the finding that as the arene size increases the tendency to stack also increases [222] and it is also seen that they follow the ideal Cg – Cg distance criteria. To some extent the isoxazole drugs follow the Cg – Cg distance criteria but the angle spread is similar to other rings.

The π-interaction analysis showed a similar trend for the intermolecular interaction in the crystal structure as well as between the drug and its receptor. Even though the dataset for the drug–receptor complex is comparatively small for the present analysis, the results obtained correlate well with the synthetic molecule, which could well be utilized in the drug discovery process.

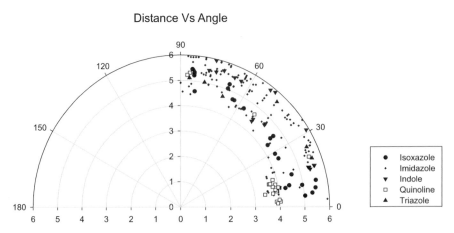

Fig. 18 Polar plot showing the distribution of the centroid–centroid distance and interplanar angle for the structure of the drug molecule with its protein targets

4
Consequences of Polymorphism

The influence of polymorphism in the drug action was explained in Sect. 1.1.3. In this section we present the stereochemical analysis of two commercially available drugs, valdecoxib and sildenafil citrate, in pseudopolymorphic conformations. The structure activity of these conformers were studied by docking them with their respective protein targets, cyclooxygenase-2 (COX-2) and phosphodiesterase-5 (PDE-5), and analyzing their drug–receptor interactions.

4.1
Valdecoxib

Valdecoxib (Fig. 19), whose brand name is Bextra, is a nonsteroidal anti-inflammatory drug (NSAID) used for the treatment of osteoarthritis or rheumatoid arthritis and also for the treatment of primary dysmenor-

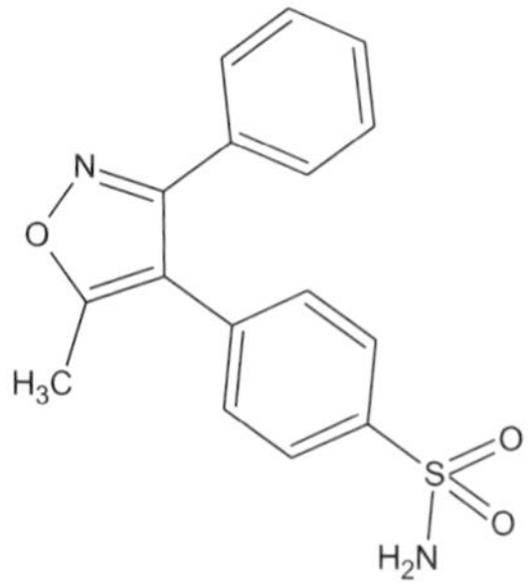

Fig. 19 Chemical structure of valdecoxib

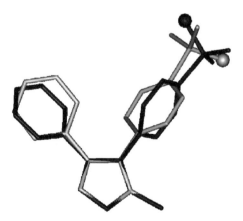

Fig. 20 Superimposition of valdecoxib with (*grey*) and without solvent (*black*)

rhea [138]. It is a diaryl-substituted isoxazole that exhibits analgesic and antipyretic properties in addition to anti-inflammatory property in animal models [229]. It is a diaryl-substituted isoxazole and one of the aryl rings is a phenylsulphonamide moiety. Two crystal structures are available for this drug, one solved by our group [230] (Vdb) and the other by Yathirajan and co-workers [231] (Vds). Both are pseudopolymorphs in the sense that Vds contains a solvent molecule, ethyl methyl ketone. Because of the inclusion of this solvent molecule in the crystal, the amino group in the sulfonamide moiety is oriented in the opposite direction (Fig. 20), which also influences the hydrogen bonding pattern.

4.2
Interaction Study of Valdecoxib with COX-2

Valdecoxib is a potent and specific inhibitor of cyclooxygenase-2 (COX-2), an isoform of cyclooxygenase. The enzyme, cyclooxygenase, is responsible for the conversion of arachidonic acid to prostaglandin H2, a key step in the generation of proinflammatory eicosanoid mediators. It is present in two forms, a constitutive (COX-1) that is widely expressed in nearly all tissues throughout the body, and the other form is inducible (COX-2), which is predominantly expressed in inflamed tissues. Therefore, selective inhibition of COX-2 should maintain the anti-inflammatory effects while reducing adverse effects (gastrointestinal bleeding, platelet effect, etc.) of traditional non-steroidal agents because of their non-selective inhibition of COX-1 [232–234]. The three-dimensional structure of the COX enzymes are strikingly similar to each other; their active site part is also similar. Both isozymes have an active site consisting of a hydrophobic channel, and amino acids in this region are nearly identical. Three amino acid differences, however, result in a larger and more accessible channel, in COX-2.

The COX-2-selective diarylheterocyclic inhibitors have been reported to be a reversible competitive inhibitor of COX-1 while demonstrating time-dependent irreversible inhibition of COX-2 which accounts for the potency and selectivity demonstrated by members of this structural class [229]. The phenylsulphonamide moiety of the diarylheterocycles associate within a side pocket present in the active site of COX-2 and this pocket is more accessible in COX-2 than in COX-1, which is the result of the substitution of a valine for an isoleucine found at position 523 in COX-1 [235].

Docking studies were performed for valdecoxib with COX-2 using the INSIGHTII package of BIOSYM on a Silicon Graphic Interface workstation [236]. The docking was performed maintaining the important hydrogen bonds between the drug molecule and its protein target and also without changing the conformation of the drug (rigid docking). The aim is to analyze the suitability of the crystal conformation of the drug molecule in the protein environment. Only manual docking was attempted since automated programs change the conformation of the ligand molecule in order to minimize the energy, which is not the goal. The hydrogen bond and non-bonded contacts between the complex were calculated using the program HBPLUS [77] and the pictorial representations are drawn using the program LIGPLOT [78].

A search of PDB revealed 23 COX-2 structures complexed with various substrates, inhibitors, and drug molecules other than valdecoxib. Analyses of all the ligand molecules in these structures indicated that two of them contain a celecoxib-like inhibitor (named SC-558; IUPAC name: 1-Phenylsulfonamide-3-trifluoromethyl-5-parabromophenylpyrazole), which is also a second generation NSAID specific for COX-2 [237] whose chemical structure resembles valdecoxib. Both these COX-2 structures that are inhibited with SC-558 (PDB ID: 1CX2 and 6COX) were solved at 3.0 Å resolution and both are essentially the same, except for their crystal symmetries [235]. The 6COX structure was chosen for the analysis, as the crystallographic parameters are comparatively better in this structure. COX-2 exists as a dimer in 6COX structure with each monomer containing one molecule of SC-558 with a similar contact environment and therefore the docking study was carried out only with one monomer unit.

The 6COX structure establishes the fact that the sulfonamide group of SC-558 shows hydrogen bonding interactions with COX-2 (Fig. 21) and therefore further studies are made giving importance to this group. The superposition of Vdb, SC-558, and Vds with respect to the sulfonamide group showed that SC-558 and Vds have similar orientations whereas Vdb shows a different orientation. The orientations of the individual rings with respect to each other in all the three structures are listed in Table 11. It can be inferred from the table that both Vdb and Vds have a similar orientation for all the rings. The difference occurs with the orientation of the central pyrazole ring of SC-558 to both the benzenesulfonamide and phenyl ring.

Rigid docking of Vdb and Vds with the COX-2 structure was performed and their interactions were calculated. This was then compared with the interactions of SC-558, which makes $N-H...O$ hydrogen bonds with Ser353, Gln192, Leu352, and His90. The Vdb molecule maintains the three hydrogen bonds (Fig. 22) out of four in SC-558 with COX-2, the one with Ser353 is converted as the weak hydrophobic interaction. The phenyl ring of Vdb makes short contacts with Val349 that could not be removed by changing the orientation of the phenyl ring of Vdb with respect to the isoxazole ring, similar to that seen in SC-558. This suggests that the torsion angle of the valine side chain is to be changed in order to accommodate this crystal conformation of Vdb. Vds maintains all the four hydrogen bonds and most of the hydrophobic contacts (Fig. 22). In addition three more hydrophobic contacts occur with Met522, Glu524, and Pro528 residues. Short contacts are noted with Val523 and Ala527 but when compared with Vdb it is minimal. Val523 is an important amino acid for the COX-2 enzyme as it is the one that differentiates it

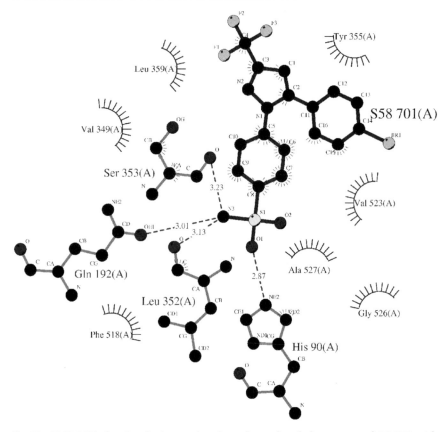

Fig. 21 LIGPLOT showing hydrogen bonds and non-bonded contacts of SC-558 with COX-2

Table 11 Dihedral angle between the various planes of valdecoxib for Vdb, SC-558, and Vds

Planes	Vdb(°)	SC-558(°)	Vds(°)
Benzenesulfonamide and phenyl	59.44	63.07	58.34
Benzenesulfonamide and heterocycle	54.34	26.35	56.18
Phenyl and heterocycle	22.24	50.48	34.56

from COX-1 by creating an active site pocket. The close contact of Vds with this crucial amino acid suggests that this conformation of valdecoxib is more specific when compared with other COX-2 specific NSAIDs. Also, in Vdb the orientation of the phenyl ring did not modify the short contacts suggesting the possibility of changing the conformation of valine and alanine.

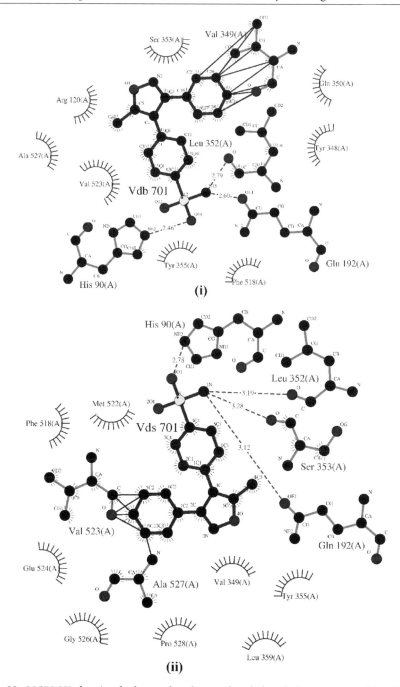

Fig. 22 LIGPLOT showing hydrogen bonds, non-bonded and short contacts (*thin lines*) between (i) valdecoxib (no solvent conformation—Vdb) and (ii) valdecoxib (solvent-induced conformation—Vds) and 6COX

These analyses suggest that solvent-induced conformation of valdecoxib (Vds) mimics the protein environment by way of changing the orientation of the amine group as for SC-558. It also shows that the orientation between various rings of valdecoxib does not significantly change the interaction patterns. The crystal conformation of Vds produces more contacts compared with SC-558 suggesting the possibility of strong binding of valdecoxib, a fact reported by Yuan and co-workers (2002). This conformation of Vds can be used as a starting model for more exhaustive docking studies.

4.3
Sildenafil Citrate

Sildenafil citrate (Fig. 23) is an anti-impotence drug, marketed as viagra, for treating male erectile dysfunction (ED) [238–241]. The drug was initially developed to treat heart disease but in trial studies, the erection-enhancing effects were noticed [242]. Sildenafil works by inhibiting the enzyme phosphodiesterase type 5, which hydrolyze cyclic guanosine monophosphate (cGMP) to inactive GMP, by occupying its active site. Sildenafil also potentiated the antivasoconstrictor activity of glyceryl trinitrate, and increased intracellular cGMP levels in response to sodium nitroprusside in intact cells [243]. The sildenafil structure was identified in a complex with the phosphodiesterase enzyme.

The crystal structure of sildenafil citrate monohydrate (Sic) was solved at the same time by our group [244] and Yathirajan and co-workers [245]. The crystal structure showed the presence of sildenafil as a cation, citrate as an anion, and a water molecule. Since this structure has citrate and water as solvent molecules, this can be considered a pseudopolymorph of the structure which was solved in a complex with its protein target.

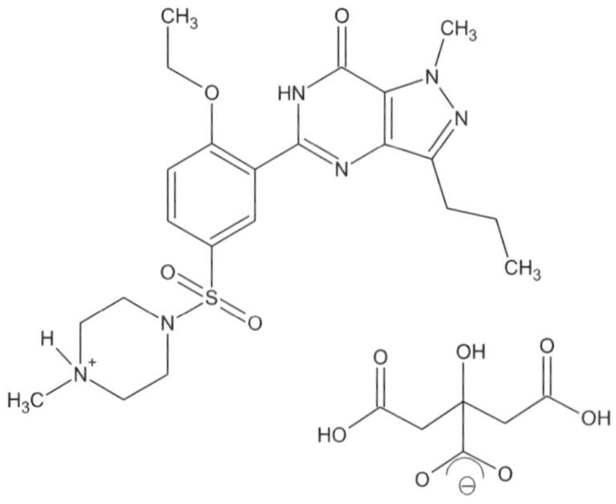

Fig. 23 Chemical diagram of sildenafil citrate

4.4
Interaction Study of Sildenafil Citrate with PDE-5

Sildenafil is a competitive inhibitor of an enzyme of the phosphodiesterase type five class (PDE-5). There are 11 types of phosphodiesterase found throughout the body and PDE-5 is present in the corpus cavernosum, platelets, skeletal muscle, and vascular and visceral muscle and it is the predominant isoform in the penis. Sexual stimulation in the non-adrenergic and non-cholinergic nerves cause the neurotransmitter nitric oxide (NO) to be released to the cavernosal smooth muscles. This NO activates the enzyme guanylate cyclase to produce an important signaling molecule cyclic guanosine monophosphate (cGMP). This molecule is a vasodilator—it relaxes the smooth muscles surrounding the blood vessels, allowing stronger blood flow leading to an erection. PDE-5 breaks down the cGMP to GMP resulting in a contraction of the smooth muscle. Anything that will potentiate the cGMP by preventing its breakdown (in effect increasing its local concentration) or by increasing its production will have a salutatory effect on erectile function and sildenafil prevents the breakdown of cGMP by blocking PDE-5 [241]. PDE-5 has 23 conserved amino acid residues in the catalytic domain [239].

Crystallographic analyses of the phosphodiesterase-5 (PDE-5) enzyme in a complex with sildenafil have been carried out; PDB identities are 1UDT [246] and 1TBF [247] at 2.3 and 1.3 Å resolution, respectively. In both structures, the sildenafil molecule forms a N – H...O type of hydrogen bond with the Gln817 residue of PDE-5 through its pyrimidine ring, even though the number of hydrophobic interacting residues differs suggesting the importance of the hydrogen bond. The difference between the two sildenafil molecules in the PDB lies in the conformation of the piperazine ring that adopts a sofa conformation in 1UDT and a chair conformation in 1TBF. When these sildenafil molecules are compared with Sic, the piperazinosulphonamide moiety in all the three molecules adopts three different orientations (Fig. 24). It has been reported that in sildenafil the piperazinosulphonamide moiety plays an important role in improving physicochemical properties by reducing lipophilicity and increasing solubility [242].

In order to verify the suitability of the Sic's conformation in the interaction with PDE-5, manual docking was carried out between PDE-5 (1UDT and 1TBF) and the sildenafil under study (Sic) as explained for valdecoxib. The rigid docking, i.e., maintaining the crucial hydrogen bond and the conformation, performed for Sic with 1UDT and 1TBF (Fig. 25) produced some short contacts between the ethoxy group of Sic and the Gln817 residue for both structures. A study of the torsion angles around the ethoxy bonds did not show any significant differences between them. An analysis of the orientation of the various rings in all the three sildenafil molecules (Table 12) showed that the planes of the sildenafil molecule in 1UDT and 1TBF assume

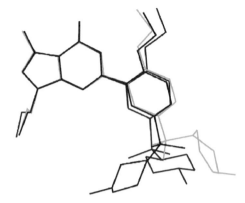

Fig. 24 Superimposition for the sildenafil molecules of Sic (*dark grey*), 1UDT (*light grey*), and 1TBF (*black*)

Table 12 Dihedral angle between the various planes of the sildenafil molecule in Sic, 1UDT and 1TBF

Planes	Sic(°)	1UDF(°)	1TBF(°)
Pyrazolo-pyrimidine and phenyl	11.9(1)	49.6(2)	47.4(6)
Pyrazolo-pyrimidine and piperazine	73.2(4)	78.7(1)	79.9(7)
Phenyl and piperazine	62.7(8)	49.3(3)	51.0(3)

similar orientations but in Sic it differs. Since the orientation of sildenafil of 1UDT and 1TBF are nearly the same, further discussion is restricted to only 1TBF owing to the high resolution data. The dihedral angles between the pyrazolo-pyrimidine and phenyl rings in Sic and 1TBF, 11.9(1) and 47.4(6)° respectively show the difference in orientation adopted by the molecule. An attempt to change the orientation to 49° for Sic removed the short contacts between Gln817 and the ethoxy group. In this conformation, Sic acquires another N – H...O hydrogen bond with the main chain atom of the MET816 residue of PDE-5 through the protonated hydrogen of the piperazine ring. This hydrogen bond interaction may not be possible, as sildenafil does not exist as an ion in the protein. In addition, this conformation also produces a few short contacts with the main chain atoms of GLY819 and PHE820. Both these residues are present in the helical region and a small conformational change upon ligand binding may produce a large energy difference. This suggested that the orientation of the piperazine ring with respect to the other rings is also very crucial. When the orientation of the piperazine ring is changed in accordance to that of the sildenafil of 1TBF, all the short contacts are removed and Sic shows similar interactions to the sildenafil of 1TBF. This analysis clearly indicates that the small molecule crystal structure of sildenafil may not be a suitable conformation for interacting with PDE-5. Assuming

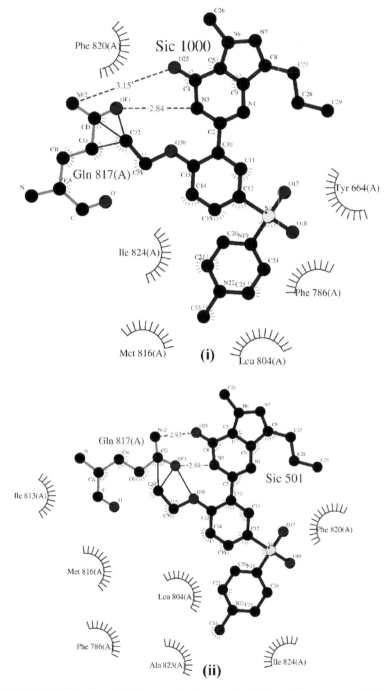

Fig. 25 LIGPLOT showing the hydrogen bonds, non-bonded and short contacts (thin lines between Gln817 and Sic) between Sic and PDE-5 of (i) 1UDT and (ii) 1TBF

this conformation of Sic to be due to the solvent molecules (citrate and water) this study predicts a different environment for the protein and the crystal containing solvent molecules, which is a different conclusion to that for valdecoxib. Furthermore, it also demonstrates, that the individual conformation of the rings of the sildenafil is not important but the orientation of the different rings is crucial as can be seen from the difference in the conformation of the piperazine ring.

The molecular conformation of the pseudopolymorphs of valdecoxib and sildenafil citrate was studied by their adaptability to their protein targets, namely cyclooxygenase-2 (COX-2) for valdecoxib and phosphodiesterase-5 (PDE-5) for sildenafil; they were analyzed through rigid manual docking. For valdecoxib, it was found that the solvent-induced crystal conformation is best suited for binding in the active site of COX-2, whereas for sildenafil this is not so, assuming the crystal conformation is due to the presence of solvents. In valdecoxib, the relative orientation of the rings does not affect the interaction pattern, whereas in sildenafil it is the relative orientation among the rings that contributes to the optimal binding in the active site. This may be due to the overall shape of the molecule. The solvent-induced crystal conformation of valdecoxib makes more contacts with COX-2 confirming the increased effectiveness of this drug compared with other NSAID's.

5
Conclusion

The importance of the stereochemical, intermolecular, and polymorphic analysis of the various heterocyclic drugs were studied. The stereochemical analysis were carried out for piperidine and azepine derivatives, weak π-interaction studies were pursued for isoxazole, imidazole, indole, quinoline, and triazole and the polymorphic analyses were done for two commercially available drugs, valdecoxib and sildenafil citrate. These analyses were correlated with the crystal structures of their respective drugs that were found in complex with the receptor. The influence of these parameters over the drug's activity were explained. Even though these analyses were based on solid-state conformations, it can provide a wealth of information for the drug design community.

Acknowledgements SMMS and KS thank the Council of Scientific and Industrial Research (CSIR), Government of India for the award of the Senior Research Fellowship and acknowledge the useful discussion with their colleague Mr. N. Sampath. MNP, SMMS, and KS thank the University Grants Commission (UGC) and the Department of Science & Technology (DST), Government of India for financial support to the Department of Crystallography and Biophysics under the UGC-SAP and DST-FIST programmes.

References

1. Luo Z, Wang R, Lai L (1996) J Chem Inf Comput Sci 36:1187
2. Williams K, Lee E (1985) Drugs 30:333
3. Bernstein J (1989) Prog Clin Biol Res 289:203
4. Payne RS, Rowe RC, Roberts RJ, Charlton MH, Docherty RJ (1999) Comput Chem 20:262
5. Melentyeva G, Antonova L (1988) Pharmaceutical Chemistry. Mir Publications, Moscow
6. Daly JW, Cordell GA (eds) (1998) Alkaloids, vol 50. Academic Press, New York
7. Cazy AF, Ison RR (1970) J Pharmacol 22:270
8. McElvain SM (1927) J Am Chem 54:4758
9. Reynolsa AK, Randall LO (1957) Morphineand Allied Drugs. University of Toronto Press, Canada
10. Lu ZY, Zaho SY, Yuav XM, Yang YL (1991) J Chem Abstract 114:815135
11. Hermons B, Van Dackle P, Van de Westeringe C, Van der Eycken C, Boey C, Docxy J, Janssen PA (1968) J Med Chem 11:797
12. Robinson OPW (1973) Postgrad Med J 49:9
13. Bochringer CF, Soehne GMBH (1961) Chem Abstr 55:24796
14. Severs WB, Kinnard WJ, Buckley JP (1965) Chem Abstr 33:10538
15. Ganellin CR, Spickett RGW (1965) J Med Chem 8:619
16. Mobo IG, Soldatenkov AT, Federov VO, Ageev EA, Sargeeva ND, Lin S, Stashenko EE, Prostakov NS, Andreeva EI, Khim Farm Zh (1990) Chem Abstr 112:7331y
17. MacConnel JM, Blun MS, Fales HM (1971) Tet Lett 27:1129
18. Hootele C, Colan B, Halin FF (1980) Tet Lett 21:5063
19. Wang CL, Wuorola MA (1992) Org Prep Proceed Int 24:585
20. Grishina GV, Gaidarova EL, Zefirov NS (1994) Chem Heterocycl Compd 30:1301
21. Champseix AA, Lefur GR (1981) Eur Pat 12:643
22. Champseix AA, Lefur GR (1981) Chem Abstr 94:15175
23. Samszuk S, Hermans HKF (1977) Ger Pat 2,642,856
24. Samszuk S, Hermans HKF (1977) Chem Abstr 87:53094
25. Abignente E, Biniecka-Picazio M (1977) Acta Pol Pharm 34:241
26. Nalanishi M, Shiraki M, Kobayakawa T, Kobayashi R (1974) Jpn Pat 74-03987
27. Nalanishi M, Shiraki M, Kobayakawa T, Kobayashi R (1974) Chem Abstr 81:12085
28. Ciba-Geigy (1981) Fr Pat 2,437,405
29. Ciba-Geigy (1981) Chem Abstr 94:83739
30. Dimmock JR, Padmanilayam MP, Puthucode RN, Nazarali AJ, Motaganahalli NL, Zello GA, Quail JW, Oloo EO, Kraatz HB, Prisciak JS, Allen TM, Santos CL, Balzarini J, Clercq ED, Manavathu EK (2001) J Med Chem 44:586
31. Campbell MJM (1975) Coord Chem Rev 15:279
32. Raper ES (1985) Coord Chem Rev 61:115
33. West DX, Carlson CS, Bouck KJ (1991) Transition Met Chem 16:271
34. Casas JS, Garcýa-Tasende MS, Sordo J (2000) Coord Chem Rev 209:49
35. Miller MC, Stineman CN, Vance JR, West DX, Hall IH (1998) Anticancer Res 18:4131
36. Miller MC, Stineman CN, Vance JR, West DX, Hall IH (1999) Appl Organomet Chem 13:9
37. Hall IH, Barnes BJ, Rowell JE, Shaffer KA, Cho SE, West DX, Stark AM (2001) Pharmazie 56:648
38. Scovill JP, Klayman DL, Franchino CF (1982) J Med Chem 25:1261
39. Agrawal KC, Sartorelli AC (1978) Prog Med Chem 15:321

40. Sartorelli AC, Agrawal KC, Tsiftsoglou AS, Moore EC (1977) Adv Enzyme Regul 15:117
41. Schenley Ind (1952) US Patent 2 719 161
42. Bymes RW, Mohan M, Antholine WE, Xu RX, Petering DH (1990) Biochemistry 29:7046
43. Kraker A, Krezoski S, Schneider J, Mingel D, Petering DH (1985) J Biol Chem 260:13710
44. Ferguson LN (1975) Chem Soc Rev 4:289
45. Magee PN, Montesano R, Preussmann R (1976) Chemical Carcinogens. American Chemical Society, Washington, DC
46. YaL K, Melamed DB (1988) Russian Chem Rev 57:88
47. Challis B, Kyrtopoulos S (1978) J Chem Soc, Perkin Trans II:1296
48. Tricker AR, Ditrich C, Preussmann R (1991) Carcinogenesis 12:257
49. Lijinsky W, Taylor HW (1975) Int J Cancer 16:318
50. Kumaran D (1997) PhD thesis, University of Madras, India
51. Loeppky RN, Outram JR (1975) N-Nitroso Compounds: Occurrence, Biological Effects, vol 41. IARC Sci Publ, France, p 459
52. Sapse AM, Allen EB, Lown JW (1988) J Am Chem Soc 110:5671
53. Blum MS, Walker JR, Callahan PS, Novak AF (1958) Science 128:306
54. Sachse H (1980) ber 23:1368
55. Sachse H (1892) Z Physik Chem 10:203
56. Johnson WS, Baver VJ, Margrave JL, Frish MA, Dreger LH, Hubbard WN (1961) J Am Chem Soc 89:3761
57. Hendrickson JB (1961) J Am Chem Soc 83:4537
58. Blackburne ID, Katritzky AR, Takeuchi Y (1975) Acc Chem Res 8:300
59. Lambert JB, Keskee RG, Carhart RE, Jovanovich AP (1967) J Am Chem Soc 89:3761
60. Testa B, Dekker M (1979) Principles of Organic Stereochemistry. Dekker, New York, p 111
61. Bucourt R (1974) The Torsion Angle Concept in Conformational Analysis. In: Eliel EL, Allinger NL (eds) Topics in Stereochemistry, vol 8. Interscience, New York
62. Ravindran T, Jeyaraman R, Murray RW, Singh M (1991) J Org Chem 56:4833
63. Stewart WE, Siddall TH (1970) III Chem Rev 70
64. Looney CE, Philips WD, Reily EL (1957) J Am Chem Soc 79:6163
65. Agostno FT, Jaffe HH (1971) J Org Chem 36:992
66. Chow YL, Colon CJ, Tam JNS (1968) Can J Chem 46:2821
67. Johnson F, Malhotra SK (1965) J Am Chem Soc 89:3761
68. Lunzzi L, Macciantelii DJ (1981) J Chem Soc, Perkin Trans II:406
69. Nogrady TH (1985) Medicinal Chemistry (A Biochemical Approach). Oxford University Press, Oxford
70. Hendrickson JB (1964) J Am Chem Soc 86:4854
71. Díez E, Esteban AL, Guilleme J, Bermejo FJ (1981) J Mol Struct 70:61
72. Esteban AL, Galiano C, Díez E, Bermejo FJ (1982) J Chem Soc, Perkin Trans II:657
73. Cano FH, Foces-Foces C (1983) J Mol Struct 94:209
74. Entrena A, Campos J, Gómez JA, Gallo MA, Espinosa A (1997) J Org Chem 62:337
75. Berman HM, Westbrook J, Feng Z, Gilliland G, Bhat TN, Weissig H, Shindyalov IN, Bourne PE (2000) Nucleic Acids Res 28:235
76. EBI Macromolecular Structure Database (MSD) available at the URL http://www.ebi.ac.uk/msd-srv/msdlite/index.html
77. McDonald IK, Thornton JM (1994) J Mol Biol 238:777
78. Wallace AC, Laskowski RA, Thornton JM (1995) Protein Eng 8:127

79. Lorin A, Thompson, Ellman JA (1996) Chem Rev 96:555
80. Sternbach LH (1979) J Med Chem 22:1
81. Bock MG, Dipardo RM, Evans BE, Rittle KE, Whitter WL, Veber DF, Anderson PS, Freidinger RM (1989) J Med Chem 32:13
82. Romer D, Buschler HH, Hill RC, Maurer R, Petcher TJ, Zeugner H, Benson W, Finner E, Milkowski W, Thies PW (1982) Nature 298:759
83. Korneki E, Erlich YH, Lenox RH (1984) Science 226:1454
84. Hsu MC, Schutt AD, Holly M, Slice LW, Sherman MI, Richman DD, Potash MJ, Volsky DJ (1991) Science 254:1799
85. Bondinell WE, Callahan JF, Huffman WF, Keenan RM, Ku TW-F, Newlander KA (1993) Pat WO 93/00095
86. Pauwels R, Andries K, Desmyter J, Schols D, Kukla MJ, Breslin HJ, Raeymaeckers A, Van Gelder J, Woestenborghs R, Heykants J, Schellekens K, Janssen MAC, Clercq ED, Jansen PAJ (1990) Nature 343:470
87. James GL, Goldstein JL, Brown MS, Rawson TE, Somers TC, McDowell RS, Crowley CW, Lucas BK, Levinson AD, Marsters JC (1993) Science 260:1937
88. Hamor HA, Martin IL (1984) In: Horn AS, De Ranter CJ (eds) X-ray Crystallography and Drug Action. Clarendon Press, Oxford
89. Hester JP, Duchamp DJ Jr, Chidester CG (1971) Tetrahedron Lett 1609
90. Bunin BA, Plunkett MJ, Ellman JA (1994) Proc Natl Acad Sci USA 91:4708
91. Krapcho J, Turk CF (1966) J Med Chem 9:191
92. Sternback LH (1971) Angew Chem Int Ed 10:34
93. Rahbaek L, Breinholt J, Frisvad JC, Christophersen C (1999) J Org Chem 64:1689
94. Gringauz A (1999) Introduction to Medicinal Chemistry. Wiley, New York
95. Butcher HJ, Hamor TA (1985) Acta Cryst C41:1081
96. Rossi S, Pirola O, Maggi R (1969) Chem Ind (Milan) 51:479
97. Lee J, Gauthier D, Rivero RA (1990) J Org Chem 64:3060
98. Selnick HG, Liverton NJ, Baldwin JJ, Butcher JW, Claremon DA, Elliotte JM, Freidinger RM, King SA, Libby BE, Mc Intyre CJ, Prisbush DA, Remy DC, Smith GR, Tebben AJ, Jurkiewicz NK, Lynch JJ, Salata JJ, Sanguinetti MC, Siegal PKS, Slaughter DE, Vyas KJ (1997) J Med Chem 40:3865
99. Albright JD, Feich MF, Santos EGD, Dusza JP, Sum FW, Venkatesan AM, Coupet J, Chan PS, Ru X, Mazandarani H, Bailey T (1998) J Med Chem 41:2442
100. Breslin HJ, Kukula MJ, Ludovici DW, Monrbacher R, Ho W, Miranda M, Rodgers JD, Hitchens TK, Leo G, Gauthier DA, Ho CY, Scott MK, De Clercq E, Pauwels R, Andries K, Janssen MAC, Janssen PA (1995) J Med Chem 38:771
101. Castro JL, Broughton HB, Russell MGN, Rathbone D, Watt AP, Ball RG, Chapman KL, Patel S, Smith AJ, Marshall GR, Matassa VG (1997) J Med Chem 40:2491
102. Hockerman GH, Peterson BZ, Johnson BD, Catterall WA (1997) Annu Rev Pharmacol Toxicol 37:361
103. Kraus RL, Hering S, Grabner M, Ostler D, Striessnig J (1998) J Biol Chem 273:27205
104. Berjukow S, Gapp F, Aczel S, Sinnergger MJ, Mitterdorfer J, Glossmann H, Hering S (1999) J Biol Chem 274:6154
105. Illuminati G, Mandolinni L (1981) Acc Chem Res 14:95
106. Gescenza A, Botta M, Corelli F, Satini A, Tafi A (1999) J Org Chem 64:3019
107. Entrena A, Campos JM, Gallo MA, Espinosa A (2005) ARKIVOC (vi):88
108. Hendrickson JB (1967) J Am Chem Soc 89:7036
109. Bocian DF, Picket MM, Rounds TC, Strauss HL (1975) J Am Chem Soc 97:687
110. Osawa E, Musso H (1984) J Comput Chem 5:307

111. Hallas G (1967) Organic Stereochemistry. McGraw-Hill Publishing Company, London
112. Eliel EL (1970) Acc Chem Res 3:1
113. Hafner A (1963) Angew Chem Int Ed Engl 75:1041
114. Maier G (1967) Edit 6:402
115. Armarego WLF (1977) In: Taylor EC, Weisberger A (eds) General Heterocycle Chemistry Series. Wiley, New York
116. Hendrickson JB (1967) J Am Chem Soc 89:7043
117. Bucourt R (1974) Top Stereochem 8:159
118. Laavanya P, Panchanatheswaram K, Venkataraj M, Jeyaraman R, Maeshall W (1999) Acta Cryst C55:1355
119. Thiruvalluvar A (1997) PhD thesis, Bharathidasan University, India
120. Baouid A, Hasnaoui A, Benharref A, Giorgi M, Pierrot M (1996) Acta Cryst C52:2281
121. Nangia A, Desiraju GR (1998) Acta Cryst A54:934
122. Nangia A (2002) Cryst Eng Comm 17:1
123. Rivas JCM, Brammer L (1998) Inorg Chem 37:4756
124. Ouvrard C, Le Questel J-Y, Berthelot M, Laurence C (2003) Acta Cryst B59:512
125. Steiner T (2002) Angew Chem Int Ed 41:48
126. Steiner T (1997) Chem Commun 727
127. Jones PG, Ahrens B (1998) Chem Commun 2307
128. Lommerse JPM, Stone AJ, Taylor R, Allen FH (1996) J Am Chem Soc 118:3108
129. Chandrasekhar V, Baskar V, Kingsley S, Nagendran S, Butcher RJ (2001) Cryst Eng Comm 17:1
130. Banerjee R, Desiraju GR, Mondal R, Howard JAK (2004) Chem Eur J 10:3373
131. Bernstein J, Davis RE, Shimoni L, Chang N-L (1995) Angew Chem Int Ed Engl 34:1555
132. Desiraju GR (2000) J Chem Soc Dalton Trans 3745
133. Ciunik Z, Desiraju GR (2001) Chem Commun 703
134. Skolnick P (2002) J Clin Psychiatry 63:19
135. Giranda VL (1995) Acta Cryst D51:496
136. Zhao H, Liu G, Xin Z, Serby MD, Pei Z, Szczepankiewicz BG, Hajduk PJ, Abad-Zapatero C, Hutchins CW, Lubben TH, Ballaron SJ, Haasch DL, Kaszubska W, Rondinone CM, Trevillyan JM, Jirousek MR (2004) Bioorg Med Chem Lett 14:5543
137. Liu B, Liu G, Xin Z, Serby MD, Zhao H, Schaefer VG, Falls HD, Kaszubska W, Collins CA, Sham HL (2004) Bioorg Med Chem Lett 14:5223
138. Scheen AJ, Malaise M (2004) Rev Med Liege 59:251
139. Satyanarayana U (1999) Biochemistry. Allied Publishers Ltd, Calcutta
140. Voet D, Voet JG (1995) Biochemistry, 2nd edn. Wiley, New York
141. Khandelwal JK, Prell GD, Morrishow AM, Green JP (1989) J Neuro Chem 52:1107
142. Prell GD, Morrishow AM (1989) J Chromatogr 472:256
143. Prell GD, Douyuon E, Sawyer W, Morrishow AM (1996) J Neurochem 66:2153
144. Prell GD, Morrishow AM (1997) J Neurochem 68:142
145. Godfraind J, Krnjevic H, Pumain R (1973) Can J Physiol Pharmacol 51:790
146. Hass H, Anderson E, Hosli L (1973) Brain Res 51:269
147. Schiavo DM, Green JD, Triana VM, Spaet R, Zaidi I (1988) Fundam Appl Toxicol 10:329
148. Bullion KA, Osawa Y, Braun DG (1990) Endocrinol Res 16:225
149. Griffiths GJ, Hauck MB, Imwinkelried R, Kohr J, Roten CA, Stucky GC (1999) J Org Chem 64:8084
150. Gadaginamath GS, Patil SA (1999) Ind J Chem 38B:1070

151. Rodriguez JG, Temprano F, Esteban-Calderon C, Martinez-Ripoll M, Garcia-Blanco S (1985) Tetrahedron 41:3813
152. Okabe N, Adachi Y (1998) Acta Cryst C54:386
153. De-Benedetti PG, Folli U, Iarossi D, Frassineti C (1985) J Chem Soc, Perkin Trans II:1527
154. Krishnaiah M, Narayana raju KV, Lu I-L, Chen Y-S, Narasinga Rao S (1995) Acta Cryst C51:2429
155. Wolf WM (1999) Acta Cryst C55:469
156. Padwa A, Brpdney MA, Liu B, Satake K, Wu T (1999) J Org Chem 64:3595
157. Robert A, Meunier B (1998) Chem Soc Rev 27:273
158. Mac Lean DB (1985) In: Brossi A (ed) The Alkaloids, vol 24. Academic Press, New York
159. Gilchrist TL (1997) Heterocyclic Chemistry, 3rd edn. Addison Wesley Longman, Essex, England
160. Blasko G, Gula DJ, Shamma M (1982) J Nat Prod 45:105
161. Ding Q, Lown JW (1999) J Org Chem 64:7965
162. Mulwad VV, Mahaddalkar BS (1999) Indian J Chem 38B:29
163. Matulenko MA, Meyers AI (1996) J Org Chem 61:573
164. Magnus P, Parry D, Illiadis T, Eisenbeis SA, Fairhuist RA (1994) J Chem Soc Chem Commun 1543
165. Rebello LH, Da Silva JR, Gutfilen B, Bernardo-Filho M (1994) J Nucl Biol Med 38:109
166. Nishi TF, Tanaka T, Shimizu T, Nakagawa K (1985) Chem Pharm Bull (Tokyo) 1140
167. Jaton JC, Roulin K, Rose K, Strotnak LM, Lewenstein A, Brunner G, Fankhauser CP, Burger U (1997) J Nat Prod 60:356
168. Norman MH, Navas F, Thomson JB, Rigdon GC (1996) J Med Chem 39:4692
169. Billings PC, Heidelberger C (1982) Cancer Res 42:2692
170. Ohnishi H, Kosuzume H, Yamaguchi K, Ohkura M, Satoh M, Uohama M, Toyonaka Y, Suzuki Y (1981) Jpn J Pharmacol 31:747
171. Bailey DM, Mount EM, Siggins J, Carlson JA, Yarinsky A, Slighter RG (1979) J Med Chem 22:599
172. Uchida M, Chihiro M, Morito S, Kanbe T, Yamashita H, Yamasaki K, Yabuuchi Y, Nakagawa K (1989) Chem Pharm Bull (Tokyo) 37:2109
173. Shaaban MA, Ghoneim KM, Khalifa M (1977) Pharmazie 32:90
174. Bell MR, Zalay AW, Oesterlin R, Schane P, Potts GO (1970) J Med Chem 13:664
175. Jones G (1977) In: Jones G (ed) Chemistry of Heterocyclic Compounds, vol 32. Wiley, New York
176. Yates FS (1984) In: Katritzky AR, Rees CW (eds) Comprehensive Heterocyclic Chemistry, vol 2. Pergamon Press, New York
177. Hua DH, Chen Y, Sin HS, Robinson PD, Meyers CY, Perchellet EM, Perchellet JP, Chaning PK, Biellmann JF (1999) Acta Cryst C55:1698
178. Anzini M, Cappelli A, Vomero S, Giorgi G, Langer T, Hamon M, Merahi N, Emerit BM, Cagnotto A, Skorupsaka M, Mennini T, Pinto JC (1995) J Med Chem 38:2692
179. Bonjean K, De Pauw-Gillet MC, Defresne MP, Colson P, Houssier C, Dassonneille L, Bailey C, Greimers R, Wright C, Quetin-Leclercq J, Tits M, Angenot L (1998) Biochemistry 31:5136
180. Orito K, Miyazawa M, Kanbayashi R, Tokuda M, Suginome H (1999) J Org Chem 64:6583
181. Ablordeppey SY, Fan P, Clark AM, Nimrod A (1999) Bio Org Med Chem 7:343
182. Bradbury RH, Rivett JE (1991) J Med Chem 34:151

183. Hirota T, Sasaki K, Yamamoto T, Nakayama T (1991) J Heterocycl Chem 28:257
184. Walser A, Flynn T, Mason C (1991) J Heterocycl Chem 28:1121
185. Cornelissen JP, van Diemen JH, Groeneveld LR, Hassnoot JG, Spek AL, Reedijk J
     (1992) Inorg Chem 31:198
186. Gupta AK, Bhargava KP (1978) Pharmazie 23:430
187. Kunkeler PJ, van Koningsbruggen PJ, Cornelissen JP, van der Horst AN, van der
     Kraan AM, Spek AL, Haasnoot JG, Reedijk J (1996) J Am Chem Soc 118:2190
188. Garcia Y, van Konningbruggen PJ, Codjovi E, Lapouyade R, Kahn O, Rabardel L
     (1997) J Mater Chem 7:857
189. Kahn O, Martinez CJ (1998) Science 279:44
190. Heindel ND, Reid JR (1980) J Heterocycl Chem 17:1087
191. Vreugdenhil W, Haasnoot JG, Reedijk J (1987) Inorg Chim Acta 129:205
192. Albada GA, van de Graaff RAG, Haasnoot JG, Reedijk J (1984) Inorg Chem 23:1404
193. Vos G, le Febre RA, de Graaff RAG, Haasnoot JG, Reedijk J (1983) J Am Chem Soc
     105:1682
194. Mazzone G, Bonina F, Panico AM, Amico-Roxas M, Caruso A, Blandino G, Vanella A
     (1987) Farmaco Ed Sci 42:525
195. Massa S, Di Santo R, Retico A, Artico M, Simonetti N, Fabrizi G, Lamba D (1992)
     Eur J Med Chem 27:495
196. Bhargava KP, Tandon M, Bhalla TN (1981) Indian J Chem 20B:1017
197. Mohan J, Anjaneyulu GSR, Verma P, Yamini KVS (1990) Indian J Chem 29B:88
198. Milcent R, Malbec F (1985) Chemical Abstracts 103:104977k
199. Cebalo T (1971) Chemical Abstracts 74:76428u
200. Goswami BN, Kataky JCS, Boruah JN (1984) Indian Chem Soc 61:530
201. Bano Q, Tiwari N, Giri S, Nizamuddin (1992) Indian J Chem 31B:714
202. Hunter CA, Lawson KR, Perkins J, Urch CJ (2001) J Chem Soc, Perkin Trans II:651
203. Waters ML (2002) Curr Opin Chem Biol 6:736
204. Levitt M, Perutz MF (1988) J Mol Biol 201:751
205. Steiner T, Koellner G (2001) J Mol Biol 305:535
206. Malone JF, Murray CM, Charlton MH, Docherty R, Lavery AJ (1997) J Chem Soc
     Faraday Trans 93:3429
207. Ciunik Z, Berski S, Latajka Z, Leszczyfiski J (1998) J Mol Struct 442:125
208. Prasanna MD, Guru Row TN (2000) Cryst Eng 3:135
209. Chakrabarti P, Samanta U (1995) J Mol Biol 251:9
210. Allen FH, Kennard O (1993) Chem Des Autom News 8:31
211. Kroon J, Kanters JA (1974) Nature 248:667
212. Brammer L, Bruton EA, Sherwood P (1999) New J Chem 23:965
213. Olovsson, Jonsson P-G (1976) The Hydrogen Bond. In: Schuster P, Zundel G, San-
     dorfy C (eds) Recent Developments in Theory and Experiments. North-Holland,
     Amsterdam
214. Allen FH, Hoy VJ, Howard JAK, Thalladi VR, Desiraju GR, Wilson CC, McIntyre GJ
     (1997) J Am Chem Soc 119:3477
215. Jennings WB, Farrell BM, Malone JF (2001) Acc Chem Res 34:885
216. Umezawa Y, Tsuboyama S, Honda K, Uzawa J, Nishio M (1998) Bull Chem Soc Japan
     71:1207
217. Burley SK, Petsko GA (1985) Science 229:23
218. Janiak C (2000) J Chem Soc, Dalton Trans 21:3885
219. Tsuzuki S, Uchimaru T, Sugawara K, Mikamia M (2002) J Chem Phys 117:11216
220. McGaughey GB, Gagné M, Rappé AK (1998) J Biol Chem 273:15458
221. Hobza P, Selzle HL, Schlag EW (1996) J Phys Chem 100:18790

222. Hunter CA, Sanders JKM (1990) J Am Chem Soc 112:5525
223. Aravinda S, Shamala N, Das C, Sriranjini A, Karle IL, Balaram P (2003) J Am Chem Soc 125:5308
224. McKay SL, Haptonstall B, Gellman SH (2001) J Am Chem Soc 123:1244
225. Mitchell JBO, Nandi CL, McDonald IK, Thornton JM, Price SL (1994) J Mol Biol 239:315
226. Burley SK, Petsko GA (1986) J Am Chem Soc 108:7995
227. Hunter CA, Singh J, Thornton JM (1991) J Mol Biol 218:837
228. Nigovic R, Kojic-Prodic B, Antolic S, Tomic S, Puntarec V, Cohen JD (1996) Acta Cryst B52:332
229. Walkeri MC, Kurumbail RG, Kiefer JR, Moreland KT, Koboldt CM, Isakson PC, Seibert K, Gierse JK (2001) Biochem J 357:709
230. Sony SMM, Charles P, Ponnuswamy MN, Yathirajan HS (2005) Acta Cryst E61:o108
231. Yathirajan HS, Narasegowda RS, Nagarajaa P, Bolte M (2005) Acta Cryst E61:o179
232. Yuan JJ, Yang D-C, Zhang JY Jr, Bible R, Karim A, Findlay JWA (2002) Drug Metabol Disposi 30:1013
233. Coats TL, Borenstein DG, Nangia NK, Brown MT (2004) Clin Ther 26:1249
234. Gierse JK, McDonald JJ, Hauser SD, Rangwala SH, Koboldt CM, Seibert K (1996) J Biol Chem 271:15810
235. Kurumbail RG, Stevens AM, Gierse JK, McDonald JJ, Stegeman RA, Pak JY, Gildehaus D, Miyashiro JM, Penning TD, Seibert K (1996) Nature (London) 384:644
236. Biosym/MSI (1995). Insight II User Guide. Biosym/MSI, San Diego
237. Carvalho WA, Carvalho RDS, Rios-Santos F (2004) Rev Bras Anestesiol 54:448
238. Chuang AT, Strauss JD, Murphy RA, Steers WD (1998) J Urol 160:257
239. Turko IV, Ballard SA, Francis SH, Corbin JD (1999) Mol Pharmacol 56:124
240. Umrani DN, Goyal RK (1999) Indian J Physiol Pharmacol 43:160
241. McCullough AR (2002) Rev Urol 4:S26
242. Campbell SF (2000) Clinical Sci 99:255
243. Moreland RB, Goldstein I, Traish A (1998) Life Sci 62:309
244. Sony SMM (2005) PhD thesis, University of Madras, India
245. Yathirajan HS, Nagaraj B, Nagaraja P, Bolte M (2005) Acta Cryst E61:o489
246. Sung J, Hwang KY, Jeon YH, Lee JI, Heo Y-S, Kim JH, Moon J, Yoon JM, Hyun Y-L, Kim E, Eum SJ, Park S-Y, Lee J-O, Lee TG, Ro S, Cho JM (2003) Nature 425:98
247. Zhang KYJ, Card GL, Suzuki Y, Artis DR, Fong D, Gillette S, Hsieh D, Neiman J, West BL, Zhang C, Milburn MV, Kim S-H, Schlessinger J, Bollag G (2004) Mol Cell 5:279

Top Heterocycl Chem (2006) 3: 149–180
DOI 10.1007/7081_036
© Springer-Verlag Berlin Heidelberg 2006
Published online: 29 April 2006

In silico Studies on PPARγ Agonistic Heterocyclic Systems

Smriti Khanna · Raman Bahal · Prasad V. Bharatam (✉)

Department of Medicinal Chemistry,
National Institute of Pharmaceutical Education and Research (NIPER), Sector 67,
160 062 S.A.S. Nagar (Mohali), India
pvbharatam@niper.ac.in

Abstract Several heterocyclic derivatives like oxazolidinediones, thiazolidinediones, tetarazoles, phenoxazines, etc. are being developed for the treatment of insulin resistance and type 2 diabetes mellitus. The heterocyclic head group in these systems binds to and activates Peroxisome Proliferator Activated Receptor γ, (PPARγ) a nuclear receptor that regulated the expression of several genes involved in the metabolism. Non-heterocyclic natural ligands like eicosonoids, prostaglandin are only weekly active against this target. Some of PPARγ agonistic agents like rosiglitazone, pioglitazone, are in daily use as anti-diabetic agents worldwide. Molecular modeling studies are being employed to study the chemistry, biology and pharmacological issues of these compounds. In this review studies based three-dimensional quantitative structure activity relationship (3D-QSAR) on the design of new chemical entities, molecular docking studies in understanding the drug receptor interactions, pharmacophore mapping studies in analyzing the important pharmacophore features and performing virtual screening, pharmacoinformatics studies in identifying new scaffolds, quantitative chemical studies in understanding the electronic structure and reactivity of these heterocyclic systems are reviewed.

Keywords 3D-QSAR · Anti-diabetic agents · Molecular docking ·
Pharmacophore mapping · Virtual screening

Abbreviations

2D	Two Dimensional
3D	Three Dimensional
ADME	Absorption, Distribution, Metabolism and Excretion
A2DVCL	Anti type 2 diabetic virtual combinatorial library
Arg	Arginine
AF	Activated Factor
AM1	Austin Model 1
AO	Atomic Orbital
BABA	Benzoyl Amine Benzoic acid
CoMFA	Comparative Molecular Field Analysis
CoMSIA	Comparative Molecular Similarity Index Analysis
Cys	Cystine
DMF	Dimethyl Formamide
DISCO	Distance Comparison
MCPBA	Meta Chloro Perbenzoic Acid
MOMCl	Methoxy methyl chloride
EC	Effective Concentration
FFA	Free Fatty Acids
G-2	Gaussian-2
Glu	Glutamine
HPLC	High Performance Liquid Chromatography
H-Bonds	Hydrogen Bonds
HF	Hartree Fock
IR	Insulin Receptor
IDDM	Insulin Dependent Diabetes Mellitus
LBD	Ligand-Binding Domain
LHO	Leave Half Out
LOO	Leave One Out
Lys	Lysine
MM	Molecular Mechanics
MP2	Moller-Plesset Second Order
NCI	National Cancer Institute
NHR	Nuclear Hormone Receptor
NIDDM	Non-Insulin Dependent Diabetes Mellitus
NPA	Natural Population Analysis
OZD	Oxazolidinedione
PDB	Protein Data Bank
PG	Prostaglandin
Phe	Phenylalanine
PLS	Partial Least Squares
PPARs	Peroxisome Proliferator Activated Receptors
QSAR	Quantitative Structure Activity Relationship
PLS	Partial Least Square
QM	Quantum Mechanical
RMS	Root Mean Square
RDF	Receptor Description File
RA	Retinoic Acid
RAR	Retinoid Activated Receptor
RMS	Root Mean Square

RXR   Retinoid Receptor X
SMILES   Simplified Molecular Input Line Entry Specification
SRC-1   Steroid Receptor Co-Activating Factor-1
TG   Triglycerides
Tyr   Tyrosine
TZDs   Thiazolidinediones
vdW   van der Waals

1
Introduction

Insulin is a hormone, which is responsible for the regulation of glucose homeostasis in liver, muscle and fat in the body. The paucity of this hormone insulin, which plays a vital role in various catabolic as well as anabolic processes of the body, leads to the metabolic disorder – diabetes mellitus. Insulin resistance is a syndrome during which period the insulin receptor refuses to recognize the presence of insulin [1–20]. Because of this, a body produces excess insulin from the pancreatic β cells, but when body cells stop responding to the excess insulin; glucose uptake gets reduced, leading to type II diabetes [3]. The period covered by insulin resistance is known as pre-diabetic stage. Insulin resistance may be observed due to several reasons, for example, defects in insulin mediated glucose uptake in muscles or adipose tissues or suppression of glucose production in liver. Insulin resistance is aggravated by obesity and physical inactivity. Treatment of insulin resistance is expected to delay the onset of frank diabetes. Several biological targets have been tried for treating insulin resistance, of which PPARγ (Peroxisome Proliferator Activated Receptors) is the most promising target. PPARγ agonists follow a path parallel to insulin in the glucose uptake and hence these systems are known as insulin mimetics or insulin sensitizers [7].

The natural ligands for PPARγ are naturally occurring fatty acids, icosanoids, prostaglandin and their metabolites, which were shown to be weak endogenous activators of PPARγ. The interaction strength between ligands and PPARγ can be improved by introducing heterocyclic rings into flexible long-chain ligands. This has been the approach in developing the insulin sensitizing anti-diabetic drugs like rosiglitazone and pioglitazone. Several novel PPARγ ligands have been proposed, all of them with heterocyclic rings [7]. An analysis of the already known lead compounds for PPARγ agonism gives us the following general description. The ligands can be divided into three main subunits with heterocyclic rings, linked by aliphatic units as shown in Fig. 1. The most important subunit is the acid head group which contains a thiazolidinedione unit in the rosiglitazone. Central subunit is either a monocyclic or bicyclic heteroaromatic ring. The lipophilic tail subunit contains one or more heterocyclic rings.

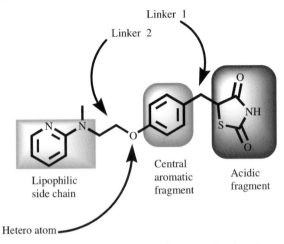

Fig. 1 A schematic diagram showing the various pharmacophoric units occupied of heterocyclic systems in the PPARγ agonists

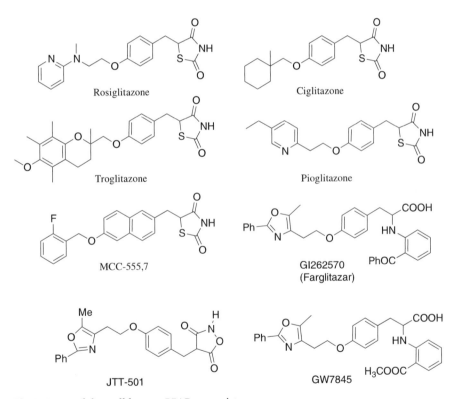

Fig. 2 Some of the well known PPARγ agonists

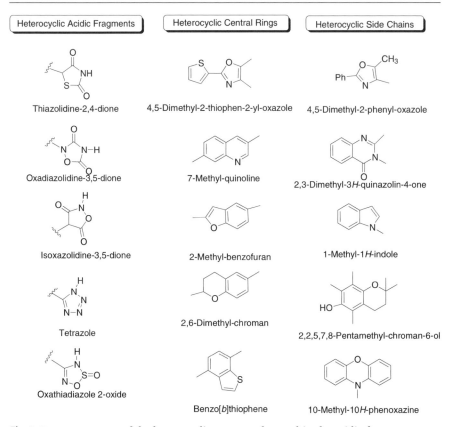

Fig. 3 Important types of the heterocyclic systems observed in the acidic fragments, central rings and the lipophilic side chain regions of the PPARγ agonists

Figure 2 shows a set of well known ligands for PPARγ agonistic activity. Figure 3 gives a selected list of five heterocyclic rings that have been employed in designing PPARγ agonists. The acidic head-group contains an active hydrogen center as the N – H unit of the thiazolidinedione [21–27]. Other acidic groups that are generally in use are oxazolidinediones [28] oxadiazolidinediones [29] tetrazoles [30] isooxazoles [31] pyrazole [32, 33] pyrazolones [32] oxathiadiazole-2-oxide [34, 35] etc. The heterocyclic variation observed in the reported central rings is quite limited. In most cases the central unit is a para-phenoxy unit, which may be replaced with heterocyclic units. The various central rings generally noticed are 7-methyl-quinoline, 2-methyl-benzofuran, 2,6-dimethyl-chroman, benzo[b]thiophene, 4,5-dimethyl-2-thiophen-2-yl-oxazole, etc. In the lipophilic tail portion of the PPARγ agonists, a wide variety of heterocyclic rings have been employed. These include the monocyclic 4,5-dimethyl-2-phenyl-oxazole,2,3-dimethyl-3H-quinazolin-4-one, 1-methyl-1H-indole, 2,2,5,7,8-pentamethyl-chroman-

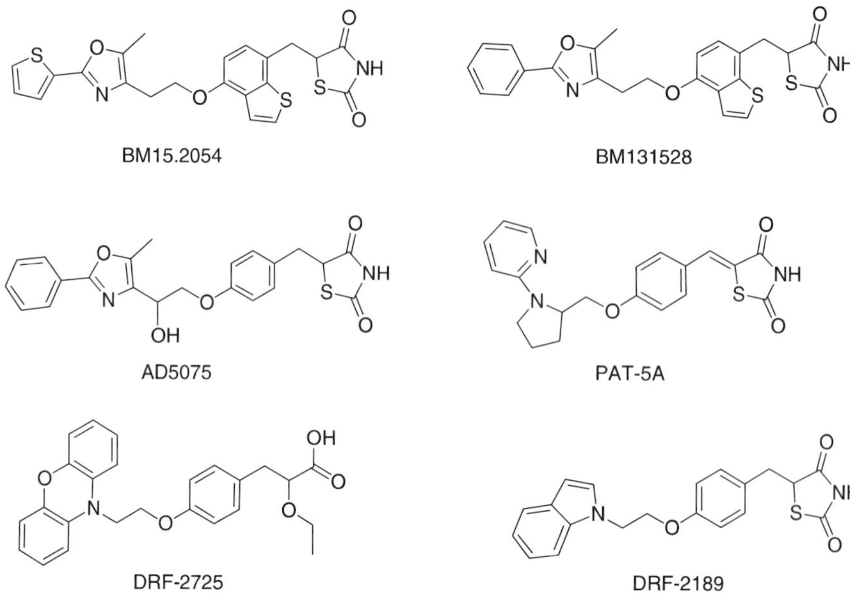

Fig. 4 A few more important leads for PPARγ agonistic activity

6-ol, 10-methyl-10H-phenoxazine, etc. several bicyclic systems like indole, quinoline, isoquinoline, benzoxazole, quinoxaline, etc. many tricyclic systems like phenoxazine, oxazine, carbazole, etc. and also two or three heterocyclic rings joined by single bonds as in BM15.2054, BM131528, AD5075, PAT-6A, DRF-2189, DRF-2725, etc. (Fig. 4).

In this review first the general synthetic methods employed in generating PPARγ agonist presented followed by discussion on the *in silico* methods which are being employed in analyzing and designing these agents including (i) QSAR studies (ii) molecular docking studies (iii) pharmacophore mapping studies (iv) electronic structure studies, etc. are presented.

2
Synthesis of PPARγ Agonistic Heterocyclic Systems

Sohda et al. [21, 27] first reported the use of thiazolidinedione-based heterocyclic systems for anti-diabetic compounds as an alternative to sulphonylureas. These active functional groups have been further employed in developing a glitazones as a series of drugs [7]. Almost all the synthetic efforts follow the general approach as shown in Scheme 1. Three representative examples of (i) marketed drug rosiglitazone (ii) Pyridine derivatives of glitazones reported from our work, and (iii) a recent example, are shown below.

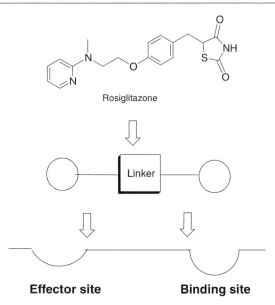

Rosiglitazone

Linker

Effector site   **Binding site**

Scheme 1 Schematic diagram showing the binding modes of PPARγ agonists in the ligand binding domain

Synthesis of Rosiglitazone [36]

Reagents: (a) Neat 160 °C, (b) 4-Fluorobenzaldehyde, NaH, DMF, (c) i) methanesulphonyl chloride, triethylamine, CH_2Cl_2 ii) 4-hydroxybenzaldehyde, K_2CO_3 DMF, 80 °C, (d) 2,4-thiazolidinedione, pyridine, C_6H_5COOH, toluene

Synthesis of Pyridine Derivative of Glitazones [37]

Reagents: (i) NaH, THF (ii) Raney Ni, HCOOH (iii) 2,4-TZD, piperidiunium acetate toluene (iv) 10% Pd/C (v) MCPBA

Synthesis of 2-methylchroman-2-carboxylic acids [38]

Reagents: (a) Cs$_2$CO$_3$, DMF, 65 °C, 68–90%, (b) 5N NaOH, i-PrOH, 70 °C, 12 h, 85–95%

3
Molecular Modeling Studies on PPARγ Agonist

Apart from synthetic methods, molecular modeling, bioinformatics, chemo-informatics methods and pharmacoinformatics etc. *in silico* methods can be used to design drugs [39–50]. The current trend is to design several leads using the *in silico* methods, find their binding potential in terms of drug re-ceptor interaction, optimize the lead using *in silico* methods, estimate the toxicological profile of the optimized lead *in silico* after obtaining all pos-sible positive signals, synthesize the promising compounds. This approach not only provides rationale for the drug development process but also helps in saving time and money. *In silico* studies can be carried out using several approaches (Scheme 2). When biological activity information about a se-ries of molecules is available one can employ statistical methods – (2D or 3D QSAR). One of the well known 3D-QSAR methods is CoMFA. When the 4–5 different classes of molecules are known to interact with a particular re-ceptor, mapping techniques can be used. When the 3D structures of leads and biomolecules are available, thermodynamical estimate of ΔG binding, ΔG docking can be made using molecular docking methods. To understand the electronic structure of molecules, quantum chemical methods can be employed. To perform a systematic search of new scaffolds for therapeutic activity, chemoinformatic methods can be adopted. To obtain the structural

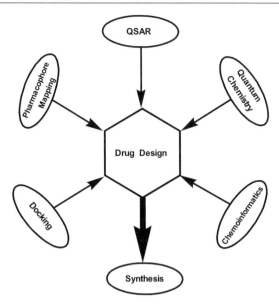

Scheme 2 A schematic diagram showing the Methods which can be employed *in silico*
Drug design

information of biomolecules, homology modeling methods can be adopted.
Integrating all the above methods for drug design purpose is known as pharmacoinformatics [51] on the heterocyclic insulin sensitizers several *in silico*
analyses have been reported. In the following sections we shall describe the
reported strategies in this direction.

3.1
3D QSAR Studies

The first reported molecular modeling study on the PPARγ agonistic anti-
diabetic agents is the 3D-QSAR study by Kulkarni et al. [52] in 1999. In 1998
the 3D structure of neither PPARγ nor any of its known ligands were available. The only option to find the pharmacophoric features was the QSAR
approach. Kulkarni et al. adopted the 3D-QSAR method Comparative Molecular Field Analysis (CoMFA) for the development of a correlation between
the observed biological activity and the 3D structure of a series of 53 glitazone molecules. Figure 5 shows the sequence of studies carried out in developing this CoMFA model. In 3D QSAR analyses, the first step is to find the
bioactive conformation of the series of molecules. Since, the bioactive conformation of any of the known drugs was not available; the authors adopted
an energy minimization approach to find a suitable template. The geometries of the compounds were modeled with the standard bond lengths and
angles using the molecular modeling package SYBYL and by using TRIPOS

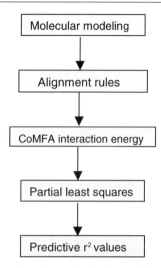

Fig. 5 Schematic diagram showing the methods adopted for the QSAR analysis

force fields. Systematic search involving the rotation of all flexible bonds by a specific angle to identify the most stable conformer was employed to perform the conformational search. The next step in 3D QSAR analysis is to perform alignment of the series of molecules into a grid such that similar functional groups and their electro-steric characters get localized into one location of a grid. Kulkarni et al. adopted two strategies for the alignment – (i) atom-based and (ii) shape-based. In atom-based alignments, atoms of the molecules were used for RMS-fitting onto the corresponding atoms of the template structure. In shape-based alignments, centroids rather than the exact superimposition of the atoms of the rings were used for RMS-fitting to the template structure. The authors generated a total of about 12 different alignments for performing CoMFA (Comparative Molecular Field Analysis) studies. The steric and electrostatic interaction energies were calculated at each lattice intersections of a regularly spaced grid box, with the help of a carbonium ion as a probe element. To identify a correlation, the statistical approach partial least square analysis (PLS) was employed. To cross validate the correlation and to estimate the robustness of the correlation, cross validation based on two methods (i) leave one out (LOO) method and (ii) leave half out method (LHO) was carried out. These procedures were repeated until all the compounds were predicted and PLS analysis gave the optimum number of components that was used to generate the final models without cross-validation. Based on the statistical analysis, CoMFA models which showed a r^2_{ncv} (0.689–0.921) and r^2_{cv} (0.624–0.764). This work leads to the initial recognition of the pharmacophoric features essential for the biological activity of this series of compounds – such as an acidic head group, a flat central aromatic ring and a lipophilic side chain. In the case of acidic fragment, the thiazolidinedione ring was replaced with various other similar or bioisosteric systems like oxazolidinedione, tetrazole, oxadiazolidinedione and alkoxy carboxylic acid derivatives. No correlation was found between the pKa and the potency of the compounds.

Based on the above features a hypothetical receptor model was proposed by the authors according to which glitazones bind to the receptor using three-point model recognition elements, (i) "cationic site or hydrophilic site", (ii) flat aromatic region and (iii) a large "hydrophobic binding site". Most of the PPARγ agonists have been designed based on this model. The derived CoMFA models were evaluated by external predictions of a set of compounds, which were not represented in the training set. Various alignment rules were derived to superimpose these ligands. All the alignments showed high internal predictivity whereas only one alignment rule showed good external predictivity. On this basis the authors suggested that the PPARγ receptor has rigid requirements in the acidic and central aromatic binding region. Poor external predictions of some alignments supported this conclusion as these alignments placed the acidic and aromatic linker fragments in the unfavorable region.

The pharmacophore modeling study through 3D (three dimensional) structure activity relationship of thiazolidinediones to understand the relationship between antihyperglycaemic and PPARγ agonistic activities was reported by Kurogi [53] also. The modeling study was carried out by conformational analysis along with Apex-3D QSAR method to identify molecular features common to a series of seven selected thiazolidinedione derivatives. The 3D structure of the selected compounds were obtained using the insight II software and were subjected to systematic conformational search with an algorithm to eliminate the high energy conformation due to steric repulsion. The conformers selected above were used to build the pharmacophore model using the Apex-3D [54] program which was used to identify molecular features common to a set of diverse chemical structures. Each identified biophore center used as the basis for molecular superimposition with the multiple conformer sets for seven compounds. The superimposition was based on the following principles: 1) biophoric centers must be superimposed for all molecules with minimal deviation, 2) the atom of the same chemical type from different molecules must be superimposed as closely as possible, 3) the closer an atom is to biophoric center, the more important is its alignment. Apex-3D uses the atomic hydrophobicity index for the QSAR analysis. The relationship between antihyperglycaemic and PPARγ agonistic activities calculated in terms of log [$1/EC_{50}$ (M)] the study by kurogi showed that the Apex-3D is a promising variation in the 3D-QSAR technology and so it has a great potential in offering design principles for new compounds.

The identification of pharmacophore and three-dimensional quantitative structure–activity studies were performed on a set of *N*-(2-Benzoylphenyl)-*l*-tyrosine for their PPARγ agonistic activity by using the logic-structural-based software Apex-3D which describes the properties and distribution of primary and secondary biophore sites in the three-dimensional space by Rathi et al. [55] in 2004. The structures of all compounds were constructed using the sketch module in Insight-II software and were converted to 3D for optimization of their geometry by selecting the force field viz. potential action, partial charge action and formal charge action. The molecular structures were minimized using the steepest descent, conjugate gradients and Newton–Raphson's algorithms in sequence followed by quasi-Newton–Raphson method. Out of a total of 23 molecules in the dataset, 20 were taken as a part of the training set and 3 in the test set. Several 3D models were generated varying in size and arrangement of features. Based on the root mean square error values from the Apex-3D analysis, two models of comparable probability were selected and a pharmacophore in terms of physicochemical properties, distances and spatial arrangements of three important biophoric sites and two secondary sites for PPARγ for binding affinity of the ligand with the receptor were identified (Fig. 6). The data were used by Apex-3D programme for automated identification of biophores (pharmacophores), superimposition of compounds and quantitative model building. Biophores represent a certain structural and electronic pattern in a bioactive molecule that is responsible for its activity through interaction with the receptor. Sites A and B were electron rich sites and were found to be involved in interactions like electrostatic, ionic and π–π stacking, with the specific receptor site and site C was involved in hydrogen bonding. The pharmacological activities of the compounds were not only dependent on the physico-chemical properties of biophoric sites A and B, but also on their biophoric distances. The 3D QSAR equations developed revealed that secondary sites are also important for the activity. The 3D-QSAR studies on N-(2-benzoylphenyl)-l-tyrosine for PPARγ agonistic activity resulted in the identification of pharmacophore in terms of physicochemical properties, distances and the spatial arrangement of the three important biophoric sites A, B and C for binding affinity of the ligand to th,e active site of PPARγ receptors. The two 3D-QSAR models describing the properties and distributions of biophoric and secondary bio-

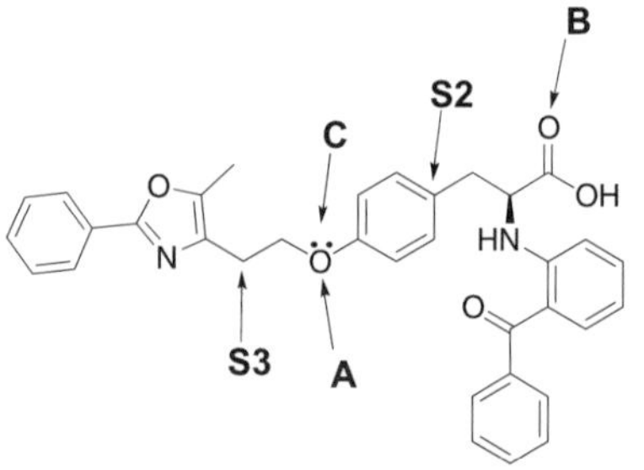

Fig. 6 Three important biophoric sites A, B and C and two secondary sites S2 and S3

phoric sites showed a good correlation between the observed and predicted activity both in training and test set and thus may be useful in designing new chemical entities as PPARγ agonists.

Liao et al. [56] have performed CoMFA and CoMSIA [57] analysis by clubbing two series of molecules i.e., thiazolidinediones and tyrosine derivatives. A training dataset including 5 thiazolidinedione derivatives and 74 tyrosine-based derivatives were used in the development of CoMFA and CoMSIA based 3D QSAR models. A set of 18 molecules were chosen for testing the models. In this case the predictive 3D-QASR models with conventional r^2 and cross validated coefficient (q^2) values up to 0.974 and 0.642 for CoMFA and 0.979 and 0.686 for CoMSIA were established using the SYBYL package. Initially, molecular mechanics-based conformational search was performed for far-glitazar to obtain a minimized structure as the template for building other molecules in the dataset. However, this leads to severe deviations from the crystal structure conformation. In order to obtain a consistent alignment, two ligands farglitazar from 1FM9 and rosiglitazone from 1FM6 were taken as the templates to carry out extraction from the crystal structures and atom-based alignment. Although there is a high degree of flexibility of compounds in the training set 3D QSAR models with good statistical significance and predictive abilities were obtained using both CoMFA and CoMSIA methodologies. The statistical results revealed that CoMSIA shows better correlations than CoMFA for this dataset. Molecular surface property (steric, electrostatic, lipophilicity and hydrogen bonding potential) mapping were integrated with CoMSIA 3D QSAR to refine what is known about the binding mode and highlight the cause of enantioselectivity. This study provided further insights to support structure based design of anti type 2 diabetic drugs to develop novel PPARγ agonists with improved activity profiles.

3.2
Homology Modeling

When glitazone series of compounds were first identified as insulin sensitizers, the target for these systems was not known. Cantello et al. [25] have identified that Peroxisome Proliferator Activated Receptor – subtype γ (PPARγ), which was known as an orphan nuclear receptor, as the target for this class of compounds in 1994. Attempts to find the structure of the ligand binding domain of this target was carried out first using homology modeling by Blaney in 1999 [58]. He undertook the construction of homology model for PPARγ using Quanta program, used the Retinoic Acid Receptor (RARγ) as a template for the homology model development. The location of binding pockets which play a vital role in the consolidation of homology model for PPARγ were identified by docking the natural prostaglandin, 15-deoxy-$\Delta^{12,14}$-PGJ2 and number of various other PPARγ agonists. The carboxylate group of the retinoic acid was found to bind to two basic residues, Lys236 and Arg278 of

RARγ which form the "mouth" of the binding pockets, supposed to share the common feature among all nuclear hormone receptors (NHRs). The charge distribution on this system was estimated using Poisson-Boltzmann calculations on the protein, which shows that an overall negative charge is present on the surface except at the mouth of the ligand-binding pocket where it is positively charged. This model also suggests that in PPAR, Lys356 forms a salt bridge with Glu341. In PPARγ the side chain of guanidine of Arg286 does extend to the same position as that of amino group of Lys356. This is one of the few residues which are very unique to the binding pocket of PPARγ, therefore, it was hypothesized that this could be responsible for the selectivity of heterocyclic thiazolidinedione compounds.

SB-219994 is a highly potent PPARγ agonist, known to be active in S configuration as compared to that of R configuration which is 1000 fold less potent on binding. However, it has been noted that the ether-side chain of the inactive R enantiomer clashed with Arg286, which reduces the strength of interaction. To test whether the arginine was indeed the key residue involved in differentiating the thiazolidinedione ligands, site-directed mutagensis was carried out on human PPARγ receptor. Mutation of Arg286 to methionine caused a complete loss of transactivation by BRL-49653. The suitable explanation for this was that arginine, being a stronger base than lysine, was able to ionize thiazolidinedione. Besides these various *Ab initio* calculations have been carried out by using 6-31G basis on 5-methylthiazolidinedione ring with Arg286 which provided ample evidence that arginine functions via selective ionization of the thiazolidinedione ring.

Ab initio electronic structure studies suggested that the transfer of hydrogen from the thiazolidinedione ring to the arginine involves about 25.9 kcal/mol for the H-transfer reaction. However, the product was found to be less stable by about 6.2 kcal/mol, thus indicating that the H-transfer reaction is not spontaneous. Further, exploration indicated that the presence of lysine in the vicinity of the ligand provides the necessary impetus to the H-transfer reaction. Because of the steric push from the lysine, the H-transfer reaction were shown to get reduced to only 18 kcal/mol. Thus the basic Arg286 as well as the sterically important Lys365 were found to be important for the therapeutic effect of the thiazolidinedione derivatives. This seminal work provided clues for further research in this area using the molecular docking analyses, quantum chemical studies, though the above suggested molecular mechanism was modified later.

3.3
Crystal Structure Analysis on PPARγ along with Ligands

Nolte et al. [59] solved the X-ray crystal structure of the human PPARγ ligand-binding domain (LBD) with rosiglitazone (PDB: 2PRG) at a resolution of 2.2 Å which revealed a large binding pocket that explained diversity of ligands found for PPARγ. PPARγ contains some general structural features of nuclear receptor family, including a central DNA binding domain and carboxy terminal domain that mediates ligand binding, dimerization and transactivation functions. PPARγ must form heterodimer with RXR to bind to DNA and activate transcription. They also described the ternary complex containing the PPARγ LBD, ligand rosiglitazone and 88 amino acids of human SRC-1 (steroid receptor co-activating factor-1).

The crystal structure revealed a large T-shaped or Y-shaped in which the ligand adopts a U-shaped conformation in the crystal structure wrapping around H3 with its central benzene ring directly behind H3 and the TZD head group and the pyridine ring wrapping around H3. The chiral center in the head group is in the *S*-configuration. Biological studies have also confirmed that it is the *S*-configuration that is more active. Rosiglitazone occupies 40%

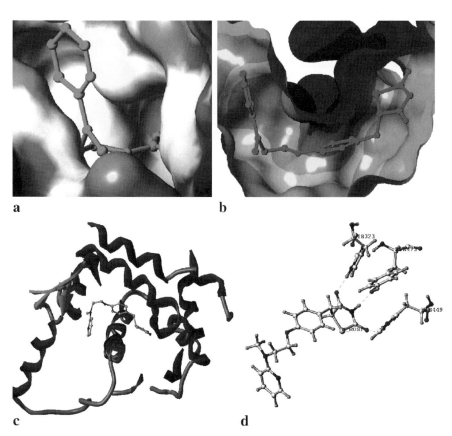

Fig. 7 The crystal structure of PPARγ in several different views; **a** Ligand visible through an opening of the receptor cavity; **b** A cross section of the molecular field with the ligand; **c** the ligand straddling helix H3; **d** ligand making three hydrogen bond with three residues His323, Tyr473 and His449

of the ligand-binding site. The TZD ring makes several interactions with amino acids in H3, H4, H10 and the AF-2 helix. A substituted carboxylic acid can act as bio-steric replacements for the thiazolidinedione head group; maintain high affinity binding and receptor activation. These carboxylic acids maintained their interaction with two histidine residues His323 and His449, which are vital for molecular recognition. These structural studies also revealed the role played by H-bonding in glitazone derivatives. The carbonyl groups of TZD form hydrogen bonds with two histidine residues, His323 and His449. Tyr473 in the AF-2 helix form secondary hydrogen bond with His323. The negatively charged nitrogen of the TZD ring is within H-bonding distance of the Tyr473 side chain hydroxyl group. The central benzene ring occupies a narrow pocket between cystine residues (Cys285) and methionine residues (Met364). The bridging oxygen atom between the benzene ring and the pyridine ring acts as a handle and provides vital geometry to the pyridine side chain. Figure 7 shows four different views of the ligand bound in the active site of PPARγ.

Many more crystal structures have been reported for PPARα and PPARγ with and without ligands which improved the understanding of the interac-

Table 1 List of all the crystal structures reported with PPAR

PDB ID	Protein	Ligand	Category	Res. (Å)	Year	Refs.
1PRG	PPARγ	—	—	2.20	1998	[59]
2PRG	PPARγ	Rosiglitazone	Agonist	2.30	1998	[59]
3PRG	PPARγ	—	—	2.90	1998	[60]
4PRG	PPARγ	GW0072	Partial agonist	2.90	1999	[22]
1GWX	PPARδ	—	—	2.5	1999	[61]
2GWX	PPARδ	—	—	2.3	1999	[61]
3GWX	PPARδ	—	—	2.4	1999	[61]
1FM6	PPARγ//RXRα	Rosiglitazone/9-cisRA	Agonist	2.10	2000	[62]
1FM9	PPARγ//RXRα	GI 262570/9-cisRA	Agonist	2.10	2000	[62]
1I7G	PPARγ	AZ-242	Agonist	2.20	2001	[63]
1I7I	PPARα	AZ-242	Agonist	2.35	2001	[63]
1K74	PPARγ//RXRα	Rosiglitazone/9-cisRA	Agonist	2.30	2001	[64]
1K7L	PPARγ//RXRα	GW 409544/9-cisRA	Agonist	2.30	2001	[64]
1KKQ	PPARα	GW6471	Antagonist	3.00	2002	[65]
1KNU	PPARγ	DRF-molecule	Agonist	2.50	2002	[66]
1NYX	PPARγ	Ragaglitazar	Agonist	2.65	2003	[67]
1WM0	PPARγ	2-BABA	Agonist	2.90	2004	[68]
1XB7	PPARγ/SHRE1	Co-factor1	—	2.50	2004	[69]
1ZGY	PPARγ/NRS	Shp	—	1.80	2004	[70]
1Y0S	PPARδ	Gw2331	Agonist	2.65	2004	[71]

4Res. — resolution, Refs. — references

tions in the active site. The list of all the crystal structures with their PDB code, ligands and resolution are tabulated in Table 1. 1PRG is a dimer of PPARγ whereas 3PRG is monomer but without ligands. 4PRG is a tetrameric protein complexed with the partial agonist GW0072. It was identified as a high-affinity ligand for PPARγ that showed weak partial agonistic activity in PPARγ transactivation studies. However, in cell culture it was a potent antagonist of adipocyte differentiation. The docking studies of GW0072 in the crystal structure of PPARγ identified several potential binding orientations different from known agonists. Its crystal structure revealed a binding mode in which the ligand occupied the region bounded by helices 3, 6, and 7. The carboxylic acid group of GW0072 was oriented between H2' and H3 and did not contact the AF-helix. The H-bonding residues His323, Tyr473, and His449, found in 2PRG, adopted conformations which were shifted from their agonist-bound positions but similar to apo-PPARγ structure.

Gampe et al. [62] have reported the crystal structures of the PPARγ with rosiglitazone and GI262570 having PDB codes 1FM6 and 1FM9 in 2000. GI262570 is a tyrosine-based compound currently known as farglitazar. The conformation of rosiglitazone in 1FM6 shows significant differences compared to the homodimer of 2PRG. The rosiglitazone side chain adopts a different gauche conformation and the pyridyl nitrogen makes a H-bond with an adjacent water molecule. Farglitazar binds to PPARγ with the binding affinity almost 50 times higher than rosiglitazone. It occurs in the same U-shaped conformation and maintains the same H-bonding interactions. The oxazole nitrogen makes a H-bond with a water molecule as a result of which the tail-end is inserted deeper into the hydrophobic pocket. At the other end the benzophenone group reaches into the large lipophilic pocket. These interactions are in addition to those observed in rosiglitazone; hence, it shows more potency.

Cronet et al. [63] have solved the crystal structure of PPARα/γ with the dual agonist AZ242 (PDB: 1I7G and 1I7I) in 2001 to provide a rationale for the ligand selectivity towards receptor subtype. The complex highlights the conserved interactions required for receptor activation. The binding mode of this is quite similar to the known ligands in the formation of hydrophobic interactions of the central ring, and the H-bond interaction between the acidic head group and Tyr464 is a conserved feature. Thus AZ242 can activate both PPARα and PPARγ. However, the observed differences in the activity can be attributed to the H-bonding head group. Tyr314 in PPARα is bulkier than its equivalent His323 in PPARγ. The carboxylate group in AZ242 is sufficiently small to form a H-bond with Tyr314 whereas the thiazolidinedione head group is sterically prevented from forming a similar hydrogen bond.

Farglitazar was again reported with PPARγ by Xu et al. [64] (PDB: 1K74) in comparison with GW409544 (PDB: 1K7L) which is modified farglitazar and is a dual activator of PPARα and PPARγ. GW409544 contains a vinylogous amide as the L-tyrosine N-substituent which contains three fewer carbon atoms in benzophenone moiety of farglitazar. Its acidic head group forms hydrogen bonds with Tyr314 and Tyr464 on the AF2-helix that stabilizes it. Superimposition of the PPARα and PPARγ crystal structures showed an overall deviation of 0.84 Å. Significant differences were observed in the positioning of farglitazar and GW409544. GW409544 occupies a position 1.5 Å deeper into the PPARα ligand binding pocket. Farglitazar cannot shift the PPARα pocket because of a steric clash with Phe273 which caps the benzophenone pocket adjacent to the AF2-helix. Thus the potency of the GW409544 as a dual PPARα/γ agonist is the result of reengineering of the ligand to accommodate the larger size of Tyr314 residue in PPARα. Point mutation of Tyr314 with His323 demonstrates that these single amino acids are responsible for subtype selectivity of fargilitazar.

Xu et al. [65] in 2002 (PDB: 1KKQ) showed the conversion of the agonistic activity of GW409544 into antagonistic activity. The H-bond interaction with GW409544 and Tyr464 that stabilizes the AF2-helix got disrupted when the acidic group was substituted with the ethylamide group. The resulting compound GW6471 showed complete inhibition of GW409544-induced activation of PPARα. Single crystal structure analysis was carried out for ragaglitazar in hPPARγ and compared with its salt form by Edrup et al. [67] in 2003 (PDB: 1NYX). By comparison, L-arginine in the single crystal it was shown that the ligand undergoes quite drastic changes to adopt to its environment. Ragaglitazar showed higher affinity in both PPARα/γ receptors and gives hope for the design of dual activators.

Ostberg et al. [68] reported the crystal structure (PDB code: 1WM0) of PPARγ with benzoylamine benzoic acid (BABA) derivatives. They reported that this class of compounds bind to PPARγ in a novel ring epitope in comparison to the other system. In general glitazones and other PPARγ agonists bind to PPARγ by the formation of 2/3 hydrogen bonds between Helix 12 of PPARγ and the acidic head group. The binding of BABA derivative does not involve the binding to Helix 12: thus, a new opportunity for PPARγ binding becomes available.

3.4
Molecular Docking Studies

The molecular modeling studies of oximes having 5-benzyl-2,4-thiazolidinesmoities (Fig. 8), ligand binding domain of PPARγ were reported by Iwata

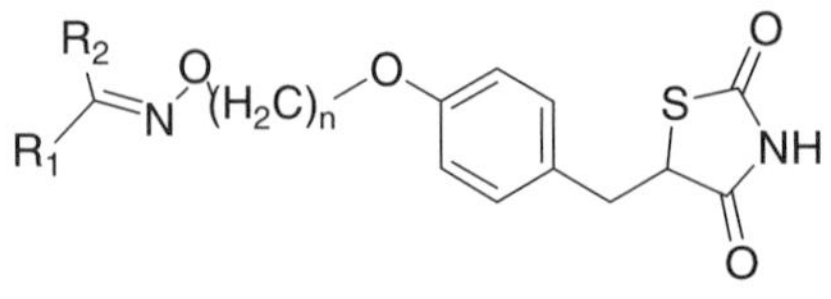

Fig. 8 Oximes containing TZD derivatives employed in the molecular docking analysis

et al. [72] in 2001. The structure of ternary complex of the PPARγ LBD rosiglitazone and SRC1 was used as the initial structure and docking studies were carried out with QUANTA/CHARMM [73] system. Extensive conformational analysis was carried with this moiety as the anchor and the torsional changes were carried out on the rest of the molecule by systematic conformational and manual modeling. During the search the structure of anchor was kept fixed in the initial conformation while the conformation of the rest of molecule was changed systematically in the cavity of receptor. The molecular mechanics energy was calculated for each conformer. Among the systematically generated conformers, only those with the lowest energies were selected for further geometry optimization and bumped structures were discarded to limit the amount of calculation to feasible values. The number of conformers generated and optimized for rosiglitazone was 324 and 30 respectively.

Both "tail-up" and "tail-down" configurations (Fig. 9) that have been observed in the binding pattern of eicosapentaenoic acid in its crystal structure complex with PPARδ were observed on the potential energy surface as the low energy conformers, however, the most potent compounds adopted a conformation similar to rosiglitazone referred by them as the "tail-down" conformation. Amongst the stable complexes generated in the docking study, the biphenyl group side chain group of the ligand was found to occupy the Y-shaped either in tail-up or tail-down conformations. Hence in the most stable structures of the active compounds the biphenyl group was located in the similar position to the pyridine ring of rosiglitazone in the crystal structure. The differences of the molecular mechanics energies between the two types were of the order of ∼ 1 kcal/mol. Although both the arms of the cavity are hydrophobic in nature, there are several hydrophilic side chains in the upward

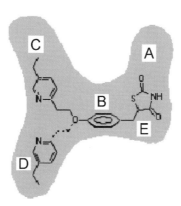

Fig. 9 μ-Shaped or H-shaped cavity: Arm A: accommodates acidic unit, arm B: adjusts the central hydrophobic unit, arm C: accommodates the tail portion of glitazones (which have been located as in 2PRG), arm D: is the empty space in both 2PRG and 1FM9 and in arm E: the benzoylphenyl unit of farglitazar in 1FM9 is located

pocket including the carboxyl group of Glu295 and Arg288 and more water molecules. Therefore the downward pocket is more hydrophobic. So the preference of tail-up and tail-own conformations of selective molecules depends on their side chains. The alkyl group and aromatic ring of the tail group of the ligands show hydrophobic interaction with receptor, and these interactions are found to be essential for the activity.

3.5
Molecular Field Analysis in the Design of Dual PPAR Activators

Each of the PPAR subtypes have been reported to be differentiated in a tissue-specific manner and play pivotal role in glucose and lipid homeostasis. Whereas PPARγ agonists enhance insulin action and promote glucose utilization in peripheral tissues. PPARα agonists improve insulin sensitivity associated with obesity and mediate their effects on lipid metabolism. PPARα/γ dual activators (Fig. 10) provide superior profile towards the control of hyperglycemia and hypertriglyceridemia. Propionic acid derivatives have been found to be dual activators for PPARα and PPARγ [36, 64, 66, 67, 74–81]. Xu et al. converted GW409544, into a dual activator of PPARα and PPARγ by suitably sprucing the side chain of farglitazar by careful analysis of the subtle differences in the active site of PPARα and PPARγ [61]. Studies from Merck showed that a series of thiazolidinedione (I) and oxazolidinedione (II) deriva-

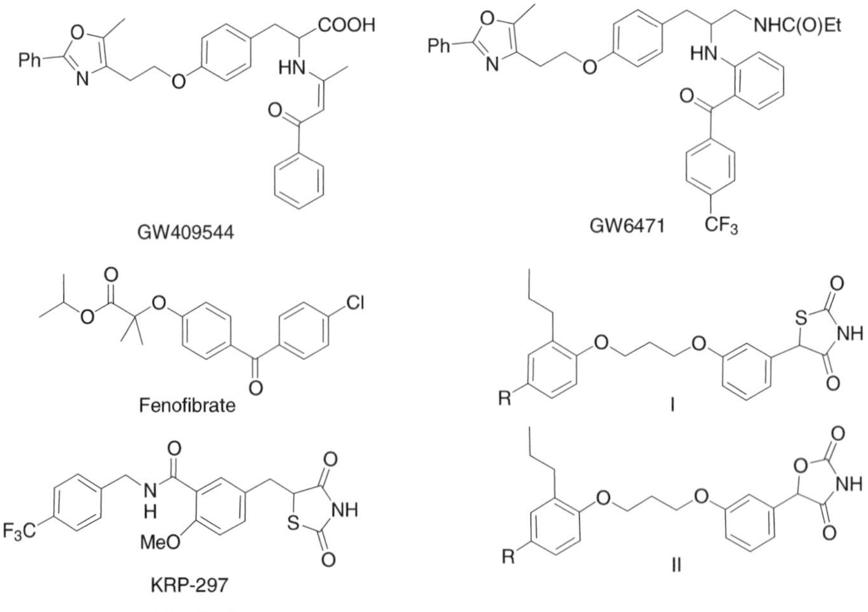

Fig. 10 Some of the dual PPARα/γ agents

tives show PPARα/γ dual activity [78, 79] this series has been derived from
the lead KRP-297. These compounds differ from standard glitazones mainly
in two ways (i) the – CH$_2$ linker between the acidic unit and phenoxy ring of
glitazones is removed and (ii) alkoxy unit and the acidic unit are meta-linked
rather than para-linked. Khanna et al. have used this set of compounds [77]
(Fig. 11) to define "additivity" of molecular fields.

A dataset consisting of a series of 5-aryl thiazolidinedione and oxazolidinedione derivatives acting as PPARα and PPARγ dual activators was chosen to
develop three CoMFA models (i) α-model (ii) γ-model and (iii) dual-model.
The dataset of 34 molecules were sorted randomly into training set and test
set comprising of 27 and 5 molecules, respectively, in the process of model
refinement for all the three CoMFA models reported herein. The biological activities were reported as the binding affinities measured as IC$_{50}$ values
in μM using radio-labeled TZD AD-5075 and recombinant PPARs by Desai
et al. [78, 79]. These were converted to pIC$_{50}$(– log IC$_{50}$) values in molar terms
and were used as the dependent variables in the CoMFA analysis. In the dual
model development, the pIC$_{50}$ values (dual) were defined as the sum of pIC$_{50}$
values for PPARα and PPARγ. Atom-based alignment method was used to
align molecules into grid. The Partial Least Squares method was used to linearly correlate the CoMFA fields to the binding affinity values. A comparison
of the contours of the dual-model with the α-model and the γ-model revealed
that the contours of the dual-model incorporates features of both the indi-

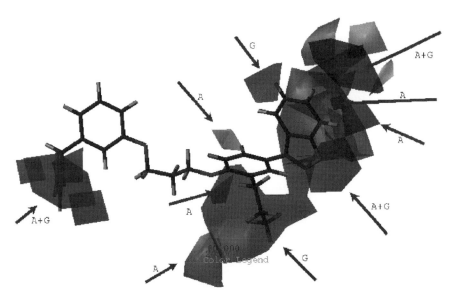

Fig. 11 Steric and electrostatic contour map for the dual-model showing the contributions
from each model. "A" depicts the contributions made by the α-model and "G" depicts the
contributions made by the γ-model

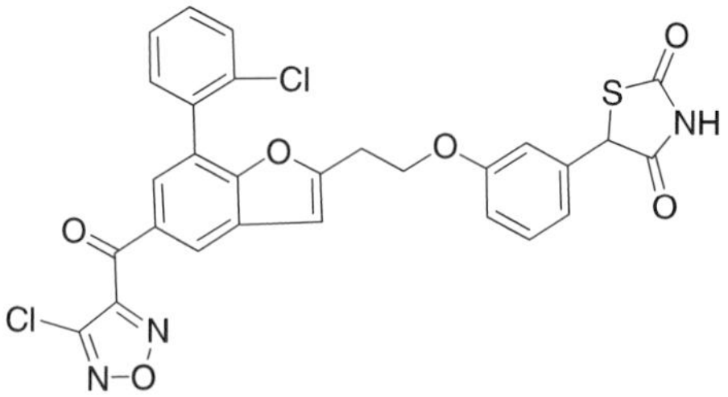

Fig. 12 One representative example showing the molecules designed using dual CoMFA models

vidual models and thus represent combination of field deviations of α and γ models. Most of the contours were found in the variable side chain region (part D) while very few were found in the acidic part (part A). The fields resulting from this CoMFA model represent "additivity fields". The additivity CoMFA model developed was validated to possess the additivity properties by (1) comparing with the individual CoMFA models for PPARα and PPARγ and (2) by carrying out docking studies. The additivity concept was further validated by comparing the molecular fit of several molecules into the contour maps of the three CoMFA models.

A set of new molecules were designed, which contain heterocyclic ring (Fig. 12). The newly designed molecules were docked into the active sites of PPARα and PPARγ using FlexX. All of them were found to dock well into the active sites of two crystal structure of 2PRG and 1FM9 (PDB codes). The FlexX and other docking scores were also showed better value than those of reference compounds.

3.6
Pharmacophore Mapping Studies

Pharmacophore mapping techniques are essential and integral components of the drug discovery program today [41] and provide important clues to discover new lead structures. We have employed the methods of pharmacophore mapping and virtual screening to identify new leads for PPARγ (Fig. 13) (Khanna S, Bharatam PV, unpublished results). In the process, different queries were generated using DISCO-based pharmacophore mapping method [82] and were validated using an inbuilt database. In this study, a two-pronged virtual screening strategy was adopted to identify new leads. The pharmacophore thus generated is being used to screen molecules from three different databases NCI [83]. Maybridge [84] (MB) and Leadquest [85] (LQ) were used for searching new lead compounds by employing the developed queries. NCI with 2 034 055 compounds, Maybridge with 55 541 molecules and Leadquest has with 41 393 molecules. Molecular docking studies were performed on two protein molecules – 2PRG and 1FM9. All the hits obtained

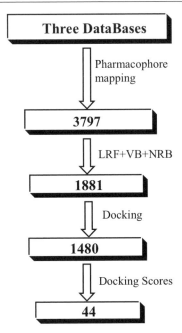

Fig. 13 Flowchart of the database search employing different filters. The final filter employing the docking methodology led to 44 hits scored better than rosiglitazone.

from the above virtual screening exercise were docked into the active sites of PPARγ as defined by the two crystal structures 2PRG and 1FM9, to further prioritize them. The final hits obtained from the FlexX [86] docking studies were analyzed in terms of their novelty in the Pharmacophoric pattern and for their diversity in terms of the various substituents in the acidic fragments, central aromatic region and side chains.

3.7
Quantum Chemical Studies on the Rapid Racemization in Thiazolidinediones

Glitazones contain a stereogenic center at C5 position of the thiazolidone-2,4-dione ring, the two enantiomers show differential activity with *S*-enantiomer being more active than *R*-enantiomer [87]. Competitive binding assay to PPARγ receptor indicated higher affinity for the *S*-enantiomer as compared to *R*-enantiomer. The finding has been further strengthened by the crystal structure analysis of the bound receptor with the *S*-enantiomer of rosiglitazone [59]. It has also been established that the higher binding affinity of *S*-enantiomer correlates with better anti-diabetic activity [88]. But attempts to use only the eutomer for therapeutic purpose turned futile when it was observed that the pure enantiomer underwent rapid racemization under physiological conditions, giving no net advantage of the tedious separation or synthesis of enantiomerically pure compounds [27]. The $t_{1/2}$ for racemization was determined to be 3 hrs at pH 7.2 [87] several non-thiazolidinedione insulin sensitizers like oxazolidinediones [28] tyrosine-based PPARγ agonists [89–91] and alkoxypropionic acid derivatives [66, 74, 75] have shown no tendency for racemization unlike glitazones and as a result show better activity of the pure compound [28, 66, 74, 75, 89–92].

It was proposed that 1,3-H shift leading to the formation of enol isomer of thiazolidinedione ring is mainly responsible for rapid racemization. The differential rates of racemization between thiazolidinediones and other insulin sensitizers were attributed to the differential rates of keto-enol tautomerism in these compounds. The π donating substituents in $CH_3 - CXO$ (X = F, NH$_2$) have been shown to reduce the probability of keto-enol tautomerism [93] and hence the tautomerisation in glitazones may not be a very favorable process.

Hulin et al. [94] have rationalized the observed rapid racemization in ciglitazone by comparing the activity of (+) and (–) isomers of thioalkyl propionic acids with that of the alkoxy propionic acids [94, 95]. They have observed thioalkyl propionic acids are less active than the corresponding alkoxy propionic acids because the former can undergo rapid racemization whereas the latter did not. This hypothesis found evidence recently from metabolizing studies of glitazones [96]. They proposed that reversible S-oxide formation in vivo [97–100] may be causing the observed differences. They also pointed out that the α-sulfinyl carboxylic acid show greater acidity by losing the chiral hydrogen. Bharatam and Khanna performed electronic structure studies to rationalize the rapid racemization [101].

Complete optimizations have been carried out on methylthiazolidinedione (1) at B3LYP/6-31 + G* and MP2(full)/6-31 + G* levels to understand the tautomerisation in glitazones using quantum chemical methods [102–109]. The calculated set geometric data (Fig. 14) were found to be comparable to that reported for the thiazolidinedione ring in various crystal structures like those of thiazolidinedione [110, 111] phenyl thiazolidinedione [112] and

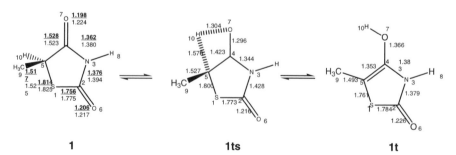

Fig. 14 Important geometrical parameters of methylthiazolidinedione, 1, its tautomer 1t and the transition state (1ts) for 1,3 hydrogen shift. The parameters are obtained using MP2/6-31 + G* level of quantum chemical calculations

troglitazone [113]. The TZD ring has been found to adopt a planar arrangement at all the computational levels, as in the case of related crystal structures. The C5 center is expected to be highly acidic because the loss of H from this center would induce sp^2 character to C5, and increase π-delocalisation.

Energy estimations have been carried out for these compounds at HF, B3LYP and MP2 levels (Table 2), the keto form is more stable than the enol form at all levels. The keto–enol energy difference at MP2(full)/6–31 + G* level in **1** (24.04 kcal/mol) is much larger than the same in acetaldehyde (16.23 kcal/mol). Higher values of ΔE and 1,3-H shift barriers suggest the enol content in **1** should be less than that of acetaldehyde. Hence, keto-enol tautomerism in glitazones is not expected to be a very favorable process.

The ionisation energy in **1** to give corresponding anion has been estimated to be 344.72 kcal/mol at MP2(full)/6–31 + G* level. This value is much less than that of acetaldehyde (363.90 kcal/mol) at the same level, indicating that the ionization of **1** by removing hydrogen at C5 is much more facile than ionization of acetaldehyde. This indicates that the higher acidity at C5 indeed can be implicated in the rapid racemization of thiazolidinediones, as suggested by Sohda et al. [27] and Gaupp and Effenberger [114]. But considering the large ΔE between the tautomers **1** and **1t**, it appears that the higher acidity at the chiral center in thiazolidinediones may not contribute to keto-enol tautomerism.

The planar structure of **1** causes the electron cloud to be π-delocalised over the whole system. This delocalisation increases the acidity of the hydrogen atom at the chiral center, as indicated by the lower values of ionizations energies calculated for **1** in comparison to acetaldehyde. Hence acid-base catalyzed ionization of glitazones is expected to play a crucial role in racemization process.

Table 2 Energy parameters (in kcal/mol) associated with tautomerisation in **1, 2** and **3** in gas phase at B3LYP/6–31 + G* and MP2 (full)/6–31 + G* and solvent phase (solvent water $\varepsilon = 78.6$) at B3LYP/6–31 + G* and MP2 (full)/6–31 + G*//B3LYP/6–31 + G*. All the values are ZPE corrected

Molecules	ΔE		E_a	
	B3LYP	MP2(full)	B3LYP	MP2(full)
Gas Phase				
1⇌1t	22.54	24.04	75.49	77.41
2⇌2t1	14.55	15.66	38.38	39.65
2⇌2t2	23.34	25.27	75.96	78.02
3⇌3t1	30.87	34.67	52.16	55.76

ΔE: Energy difference between the tautomers
E_a: Activation energy for 1, 3-H shift
IE: ionisation energy for deprotonating chiral center

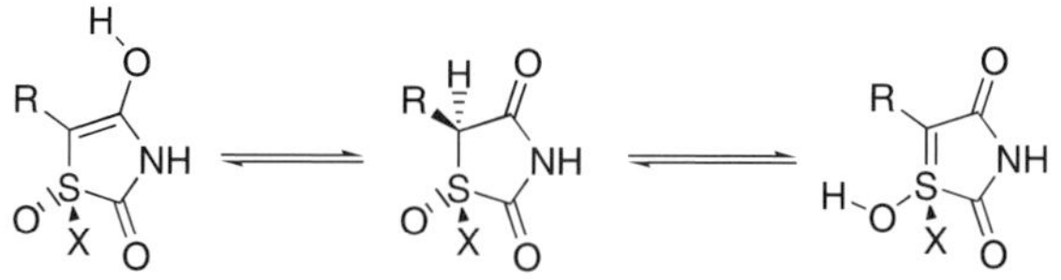

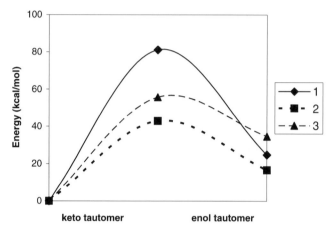

Fig. 15 Two tautomeric forms of S-oxidation products of 5-methyl thiazolidinedione; 2 (X = :) and 3 (X = O)

Fig. 16 Potential energy (PE) surface representing 1⇌1t and 2⇌2t1, 3⇌3t1 tautomeric processes at MP2(full)/6–31 + G* level

Hulin et al. proposed that a reversible S-oxidation path might be playing a role in the rapid racemization in glitazones [94, 95]. Single S-oxidation of **1** would give **2** and double oxidation would give **3**. Enolisation in the sulfoxide derivatives is possible in two ways (Fig. 15), either the sulfoxide is involved 2⇌2t1 (3⇌3t1) or the carbonyl group is involved 2⇌2t2, (3⇌3t2) in tautomerism. However, the tautomerisation involving carbonyl carbon has been found not to be effected by S-oxide, as indicated by tautomerisation energies (Table 2, Fig. 16) of (2⇌2t2) path and this path was not considered further. From the calculated ΔE and E_a (activation energies) it can be concluded that 2⇌2t1 tautomerisation is a much freely accessible path after S-oxidation.

3.8
Virtual Screening

Wei et al. [115] have reported the design of new PPARγ agonists based on virtual combinatorial library and synthesis of serine based PPARγ agonists.

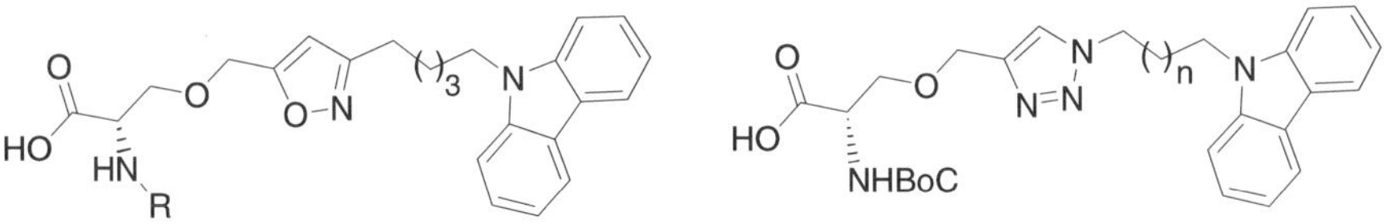

Fig. 17 Novel isoxazolyl/triazolyl – serine based PPAR agonists

A virtual combinatorial library based on the isoxazolyl-serine ligand (Fig. 17) was created and docked to the LBD of PPARγ as identified in the crystal structure of PPARγ – GW409544 complex (PDB Code: 1K7L). From the Flex-X based molecular docking studies, it was found that the isoxazolyl-serine-based ligand binds with PPARγ in a mode similar to that of in the crystal structure of PPARγ with GW409544. The phenyl moiety occupies the hydrophobic, phenylalanine rich sub pocket (Phe282, Phe360, and Phe363). Reported crystal structures have shown that the carboxyl group is important for stabilizing the *C*-terminal helix in an active conformation and that it is coordinated by two histidines (His323, His449) and a tyrosine (Tyr473) in the AF2 helix. Superposition of the PPARγ – GW409544 crystal structure with the modeled complex formed from PPARγ and the docked ligand showed that the carboxyl of these ligands is positioned in the precisely the same manner as is the carboxyl of GW409544 in the crystal structure. The isoxazolyl-serine based ligand was readily prepared by use of an intermolecular nitrile oxide cycloaddition reaction as the key step. The ligands designed using virtual combinatorial library have been found to be moderately bind to PPARγ, and they are more selective towards PPARα. Wei et al. [110] also synthesized triazole analogs, which did not show sufficient PPAR activity.

The chemoinformatics-based method has been employed by Liao et al. [116] to virtually design a library of potential PPAR modulators with large diversity. A virtual combinatorial library of 1 709 320 compounds was constructed on the basis of paradigm of PPARγ agonistic compounds using SMILES notation [117]. The method to construct virtual combinatorial library using SMILES (Simplified Molecular Input Line Entry Specification) strings was further visualized by visual basic net that can facilitate the needs of generating other types virtual combinatorial libraries. Most of fragments were collected from literature. The number of fragments obtained for effector (acidic head group); linker (central aromatic unit) and binder (lipophilic side chain) are 283, 40 and 151 respectively. The permutation of fragment of three parts of PPAR modulators gives rise to numerous molecules (1 709 320). ADME filters, based on Lipinski "Rule of 5" were employed to exclude the compounds which have poor drug like properties in this library. Besides this some other specifications were also made to enter these compounds into A2DVCL (Anti type 2 diabetic virtual combinatorial library). After filtering, the virtual library is reduced to 1 226 625 molecules. These sets of molecules were found to possess significant amounts of diversity, which is desired for such a virtual combinatorial library. This library was converted to sdf and mol2 files by the two dimensions – three-dimension (2D–3D) conversion software CONCORD4.0. All the molecules in the library were then docked to PPARγ ligand binding domain by DOCK4.0 to identify new chemical entities that may be potential drug leads against type 2 diabetes and other metabolic disease. The virtual screening by DOCK 4.0 results in identification of significant number of compounds with stronger binding potential against PPARγ than farglitazar a nanomole agonist of PPARα/γ.

4
Concluding Remarks

Molecular modeling and informatics methods are being effectively used in the design of PPARγ agonist as anti-insulin resistance agents. These studies are providing the necessary rational approach in designing new chemical entities. In future such studies are expected to increase on the heterocyclic systems with promising therapeutic potential. Several new *in silico* technologies to perform molecular dynamic analysis, predicting the ADME/T properties, estimating the metabolic pathways, etc. are emerging. Explosive growth is expected in future in the *in silico* analysis of heterocyclic systems using the fundamental theoretical chemistry techniques, molecular modeling methods and pharmacoinformatics methods.

References

1. Nolte MS, Karam JH (2000) In: Katzung BG (ed) Basic and Clinical Pharmacology, 8th ed. McGraw-Hill Publishers, USA
2. Skyler JS (2004) J Med Chem 47:4113
3. Ramarao P, Kaul CL (1999) Drugs Today 35:895
4. Matthaei S, Stumvoll M, Kellerer M, Haring HU (2000) Endo Rev 21:585
5. Goldfine AB, Type 2 Diabetes: New Drugs, New Perspectives (http://www.hosppract.com/issues/2001/09/gold.htm)
6. White MF, Kahn CR (1994) J Biol Chem 269:1
7. Lohray BB, Bhushan V (2004) Curr Med Chem 11:2467
8. Henke BR (2004) J Med Chem 47:4118
9. Nuss JM, Wagman AS (2000) Ann Rep Med Chem 35:211
10. Lebovitz HE (2002) Cleve Clinic J Med 69:809
11. Campbell IW (2000) Drugs 60:1017
12. Lebovitz HE (1999) Clin Chem 45:1339
13. Inzucchi SE (2002) J Am Med Assoc 287:360
14. Knudsen LB (2004) J Med Chem 47:4128
15. Weber AE (2004) J Med Chem 47:4135
16. Leff T, Reed EJ (2002) Curr Med Chem-Imun Endoc & Metab Agents 2:33
17. Shearer BG, Hoekstra WJ (2003) Curr Med Chem 10:267

18. Vats RK, Kumar V, Kothari A, Mittal A, Ramachandran U (2005) Curr Sci 88:241
19. Lehmann JM, Moore LB, Smith-Oliver TA, Wilkison WO, Willson TM, Kliewer SA (1995) J Biol Chem 270:12953
20. Shinkai H, Ozeki H, Motomura T, Ohta T, Furukawa N, Uchida I (1998) J Med Chem 41:5420
21. Sohda T, Mizuno K, Imamiya E, Sugiyama Y, Fujita T, Kawamatsu Y (1982) Chem Pharm Bull 30:3580
22. Oberfield JL, Collins JL, Holmes CP, Goreham DM, Cooper JP, Cobb JE, Lenhard JM, Hull-Ryde EA, Mohr CP, Blanchard SG, Parks DJ, Moore LB, Lehmann JM, Plunket K, Miller AB, Milburn MV, Kliewer SA, Willson TM (1999) Proc Natl Acad Sci 96:6102
23. Saltiel AR, Olefsky JM (1996) Diabetes 45:1661
24. Subramaniam S (1999) Clin Exper Hyp 21:121
25. Cantello BC, Cawthorne MA, Cottam GP, Duff PT, Haigh D, Hindley RM, Lister CA, Smith SA, Thurlby PL (1994) J Med Chem 37:3977
26. Parker JC (2002) Adv Drug Del Rev 54:1173
27. Sohda T, Mizuno K, Kawamatsu Y (1984) Chem Pharm Bull 32:4460
28. Dow RL, Bechle BM, Chou TT, Clark DA, Hulin B, Stevenson RW (1991) J Med Chem 34:1538
29. Malamas MS, Sredy J, Gunawan I, Mihan B, Sawicki DR, Seestaller L, Sullivan D, Flam BR (2000) J Med Chem 43:995
30. Momose Y, Maekawa T, Odaka H, Ikeda H, Sohda T (2002) Chem Pharm Bull 50:100
31. Shinkai H, Onogi S, Tanaka M, Shibata T, Iwao M, Wakitani K, Uchida I (1998) J Med Chem 41:1927
32. Kees KL, Caggiano TJ, Steiner KE, Fitzgerald JJ Jr, Kates MJ, Christos TE, Kulishoff JM Jr, Moore RD, McCaleb ML (1995) J Med Chem 38:617
33. Kees KL, Fitzgerald JJ Jr, Steiner KE, Mattes JF, Mihan B, Tosi T, Mondoro D, McCaleb ML (1996) J Med Chem 39:3920
34. Ellingboe JW, Alessi TR, Dolak TM, Nguyen TT, Tomer JD, Guzzo F, Bagli JF, McCaleb ML (1992) J Med Chem 35:1176
35. Ellingboe JW, Lombardo LJ, Alessi TR, Nguyen TT, Guzzo F, Guinosso CJ, Bullington J, Browne EN, Bagli JF, Wrenn J, Steiner K, McCaleb ML (1993) J Med Chem 36:2485
36. Lohray BB, Bhushan V, Reddy AS, Rao PB, Reddy NJ, Harikishore P, Haritha N, Vikramadityan RK, Chakrabarti R, Rajagopalan R, Katneni K (1999) J Med Chem 42:2569
37. Ramachandran U, Mital A, Bharatam PV, Khanna S, Rao PP, Srinivasan K, Kumar R, Chawala HP, Kaul CL, Raichur S, Chakrabarti R (2004) Bioorg Med Chem 12:655
38. Koyama H, Boueres JK, Miller DJ, Berger JP, MacNaul KL, Wang PR, Ippolito MC, Wright SD, Agrawal AK, Moller DE, Sahoo SP (2005) Bioorg Med Chem Lett 15:3347
39. Goodman JM (1998) Chemical Applications of Molecular Modelling. Royal Society of Chemistry, Cambridge, UK
40. Cohen NC (1996) Guidebook on Molecular Modeling in Drug Design. Academic Press Inc, California, USA
41. Guner OF (1999) Pharmacophore Perception, Development and Use in Drug Design. International University Line, La Jolla
42. Abraham DJ Ed (2003) Burger's Medicinal Chemistry & Drug Discovery, 6th ed. Wiley Interscience, New Jersey
43. Charifson PS (1997) Practical Application of Computer-Aided Drug Design. Marcel Dekker Inc, New York

44. Neamati N, Barchi JJ. New Paradigms in Drug Design and Discovery (http://www.bentham.org/sample-issues/ctmc2-3/barchi/barchi-ms.htm)
45. Gasteiger J, Engel T (2003) Chemoinformatics. Wiley, Weinheim
46. White JR, Campbell K (2000) Diabetes mellitus. In: Herfindal ET, Gourley DR (eds) The Textbook of Therapeutics: Drug and Disease Management. Amazon Publishers
47. Oprea TI (2004) Chemoinformatics in Drug Discovery. Wiley, Weinheim
48. Carloni P, Alber F. Quantum Medicinal Chemistry. Wiley, Weinheim
49. Kubinyi H (1998) In: Schleyer PvR, Allinger NL, Clark T, Gasteiger J, Kollman PA, Schaefer HF III, Schreiner PR (eds) The Encyclopedia of Computational Chemistry. Wiley, Chichester, p 448
50. Holtje HD, Sippl W, Rognan D (2003) Molecular Modeling Basic Principles and Applications, 2nd ed. Wiley, Weinheim
51. Bala S, Sharma M, Bharatam PV (2005) Curr Res Inf Pharm Sci 6:2
52. Kulkarni SS, Gediya LK, Kulkarni VM (1999) Bioorg Med Chem 7:1475
53. Kurogi Y (1999) Drug Design Disc 16:109
54. Jullian N, Brabet I, Pin JP, Acher FC (1999) J Med Chem 42:1546
55. Rathi L, Kashaw SK, Dixit A, Pandey G, Saxena AK (2004) Bioorg Med Chem 12:63
56. Liao C, Xie A, Zhou J, Shi L, Lu X (2004) J Chem Inf Comput Sci 44:230
57. Hattotuwagama CK, Doytchinova IA, Flower DR (2005) J Chem Inf Model 45:1415
58. Blaney FE (1999) Int J Quant Chem 73:97
59. Nolte RT, Wisley GB, Westin S, Cobb JE, Lambert MH, Kurokawa R, Rosenfeld MG, Willson TM, Glass CK, Milburn MV (1998) Nature 395:137
60. Uppenberg J, Svensson C, Jaki M, Bertilsson G, Jendeberg L, Berkenstam A (1998) J Biol Chem 273:31108
61. Xu HE, Lambert MH, Montana VG, Parks DJ, Blanchard SG, Brown PJ, Sternbach DD, Lehmann JM, Wisely GB, Willson TM, Kliewer SA, Milburn MV (1999) Mol Cell 3:397
62. Gampe RT Jr, Montana VG, Lambert MH, Miller AB, Bledsoe RK, Milburn MV, Kliewer SA, Willson TM, Xu HE (2000) Mol Cell 5:545
63. Cronet P, Petersen JF, Folmer R, Blomberg N, Sjoblom K, Karlsson U, Lindstedt EL, Bamberg K (2001) Structure 9:699
64. Xu HE, Lambert MH, Montana VG, Plunket KD, Moore LB, Collins JL, Oplinger JA, Kliewer SA, Gampe RT Jr, McKee DD, Moore JT, Willson TM (2001) Proc Natl Acad Sci 98:13919
65. Xu HE, Stanley TB, Montana VG, Lambert MH, Shearer BG, Cobb JE, McKee DD, Galardi CM, Plunket KD, Nolte RT, Parks DJ, Moore JT, Kliewer SA, Willson TM, Stimmel JB (2002) Nature 415:813
66. Sauerberg P, Pettersson I, Jeppesen L, Bury PS, Mogensen JP, Wassermann K, Brand CL, Sturis J, Woldike HF, Fleckner J, Andersen AST, Mortensen SB, Svensson LA, Rasmussen HB, Lehmann S, Polivka Z, Sindelar K, Panajotova V, Ynddal L, Wulff EM (2002) J Med Chem 45:789
67. Ebdrup S, Pettersson I, Rasmussen HB, Deussen HJ, Frost Jensen A, Mortensen SB, Fleckner J, Pridal L, Nygaard L, Sauerberg P (2003) J Med Chem 46:1306
68. Ostberg T, Svensson S, Selen G, Uppenberg J, Thor M, Sundborn M, Sydow-Backman M, Gustavsson AL, Jendeberg L (2004) J Biol Chem 279:41124
69. Kallen J, Schlaeppi JM, Bitsch F, Filipuzzi I, Schilb A, Riou V, Graham A, Strauss A, Geiser M, Fournier B (2004) J Biol Chem 279:49330
70. Li Y, Choi M, Suino K, Kovach A, Daugherty J, Kliewer SA, Xu HE (2005) Proc Natl Acad Sci USA 102:9505

71. Takada I, Yu RT, Xu HE, Lambert MH, Montana VG, Kliewer SA, Evans RM, Umesono K (2000) Mol Endocrinol 14:733
72. Iwata Y, Miyamoto S, Takamura M, Yanagisawa H, Kasuya A (2001) J Mol Graphics Modell 19:536
73. Fujita Y, Kasuya A, Matsushita Y, Suga M, Kizuka M, Iijima Y, Ogita T (2005) Bioorg Med Chem Lett 15:4317
74. Buckle DR, Cantello BCC, Cawthorne MA, Coyle PJ, Dean DK, Faller A, Haigh D, Hindley RM, Jefcott LJ, Lister CA, Pinto IL, Rami HK, Smith DG, Smith SA (1996) Bioorg Med Chem Lett 6:2121
75. Buckle DR, Cantello BCC, Cawthorne MA, Coyle PJ, Dean DK, Faller A, Haigh D, Hindley RM, Jefcott LJ, Lister CA, Pinto IL, Rami HK, Smith DG, Smith SA (1996) Bioorg Med Chem Lett 6:2127
76. Henke BR (2004) J Med Chem 47:4118
77. Nomura M, Tanase T, Ide T, Tsunoda M, Suzuki M, Uchiki H, Murakami K, Miyachi H (2003) J Med Chem 46:3581
78. Desai RC, Han W, Metzger EJ, Bergman JP, Gratale DF, MacNaul KL, Berger JP, Doebber TW, Leung K, Moller DE, Heck JV, Sahoo SP (2003) Bioorg Med Chem Lett 13:2795
79. Desai RC, Gratale DF, Han W, Koyama H, Metzger E, Lombardo VK, MacNaul KL, Doebber TW, Berger JP, Leung K, Franklin R, Moller DE, Heck JV, Sahoo SP (2003) Bioorg Med Chem Lett 13:3541
80. Murakami K, Tobe K, Ide T, Mochizuki T, Ohashi M, Akanuma Y, Yazaki Y, Kadowaki T (1998) Diabetes 47:1841
81. Khanna S, Bharatam PV (2005) J Med Chem 48:3015
82. Martin YC, Bures MG, Danaher EA, DeLazzer J, Lico I, Pavlik PA (1993) J Comp Aided Mol Des 7:83
83. Milne GW, Nicklaus MC, Driscoll JS, Wang S, Zaharevitz D (1994) J Chem Inf Comput Sci 34:1219
84. www.maybridge.com
85. www.leadquest.tripos.com
86. Rarey M, Kramer B, Lengauer T, Klebe G (1996) J Mol Biol 261:470
87. Parks DJ, Tomkinson NC, Villeneuve MS, Blanchard SG, Willson TM (1998) Bioorg Med Chem Lett 8:3657
88. Willson TM, Cobb JE, Cowan DJ, Wiethe RW, Correa ID, Prakash SR, Beck KD, Moore LB, Kliewer SA, Lehmann JM (1996) J Med Chem 39:665
89. Henke BR, Blanchard SG, Brackeen MF, Brown KK, Cobb JE, Collins JL, Harrington WW Jr, Hashim MA, Hull-Ryde EA, Kaldor I, Kliewer SA, Lake DH, Leesnitzer LM, Lehmann JM, Lenhard JM, Orband-Miller LA, Miller JF, Mook RA Jr, Noble SA, Oliver W Jr, Parks DJ, Plunket KD, Szewczyk JR, Willson TM (1998) J Med Chem 41:5020
90. Collins JL, Blanchard SG, Boswell GE, Charifson PS, Cobb JE, Henke BR, Hull-Ryde EA, Kazmierski WM, Lake DH, Leesnitzer LM, Lehmann J, Lenhard JM, Orband-Miller LA, Gray-Nunez Y, Parks DJ, Plunkett KD, Tong WQ (1998) J Med Chem 41:5037
91. Cobb JE, Blanchard SG, Boswell EG, Brown KK, Charifson PS, Cooper JP, Collins JL, Dezube M, Henke BR, Hull-Ryde EA, Lake DH, Lenhard JM, Oliver W Jr, Oplinger J, Pentti M, Parks DJ, Plunket KD, Tong WQ (1998) J Med Chem 41:5055
92. Zask A, Jirkovsky I, Nowicki JW, McCaleb ML (1990) J Med Chem 33:1418
93. Su C, Lin C, Wu C, Lien M (1999) J Phys Chem 103:3289

94. Hulin B, Newton LS, Lewis DM, Genereux PE, Gibbs EM, Clark DA (1996) J Med Chem 39:3897
95. Hulin B, McCarthy PA, Gibbs EM (1996) Curr Pharm Des 2:85
96. Liu DQ, Karanam BV, Doss GA, Sidler RR, Vincent SH, Hop CE (2004) Drug Metab Disp 32:1023
97. Thioethers are known to undergo S-oxidation under physiological conditions – single S-oxidation is reversible and double S-oxidation is irreversible, for example see [94, 95]
98. Damani LA (1987) Drug Metabolism-from Molecules to Man London, 581
99. Thompson HJ, Jiang C, Lu J, Mehta RG, Piazza GA, Paranka NS, Pamukcu R, Ahnen DJ (1997) Cancer Res 2:267
100. Swanson BN, Mojaverian P, Boppana VK, Dudash MR (1981) Drug Metab Dispos 6:499
101. Bharatam PV, Khanna S (2004) J Phys Chem 108:3784
102. Parr RG (1989) Density Functional Theory of Atoms and Molecules. Oxford University Press, New York
103. Frisch MJ, Trucks GW, Schlegel HB, Scuseria GE, Robb MA, Cheeseman JR, Zakrzewski VG, Montgomery JA, Stratmann RE Jr, Burant JC, Dapprich S, Millam JM, Daniels AD, Kudin KN, Strain MC, Farkas O, Tomasi J, Barone V, Cossi M, Cammi R, Mennucci B, Pomelli C, Adamo C, Clifford S, Ochterski J, Petersson GA, Ayala PY, Cui Q, Morokuma K, Malick DK, Rabuck AD, Raghavachari K, Foresman JB, Cioslowski J, Ortiz JV, Baboul AG, Stefanov BB, Liu G, Liashenko A, Piskorz P, Komaromi I, Gomperts R, Martin RL, Fox DJ, Keith T, Al-Laham MA, Peng CY, Nanayakkara A, Gonzalez C, Challacombe M, Gill PMW, Johnson B, Chen W, Wong MW, Andres L, Gonzalez C, Head-Gordon M, Replogle ES, Pople JA (1998) Gaussian 98. Gaussian Inc, Pittsburgh PA
104. Ochterski JW, Gaussian Inc. http://www.gaussian.com/g_whitepap/thermo.htm
105. Foresman JB, Frisch AE (1995) Exploring Chemistry with Electronic Structure methods, 2nd ed. Gaussian Inc, Pittsburgh
106. Hehre WJ, Radom L, Schleyer PvR, Pople JA (1985) Ab initio Molecular Orbital Theory. Wiley, New York
107. Pople JA, Beveridge DL (1970) Approximate Molecular Orbital Theory. McGraw Hill, New York
108. Cramer RD, Patterson DE, Bunce JD (1988) J Am Chem Soc 110:5959
109. Silverman BD, Platt DE (1996) J Med Chem 39:2129
110. Form GR, Raper ES, Downe TC (1975) Acta Cryst B31:2181
111. Lynch DE, McClenaghan I, Light ME (2001) Acta Cryst E57:79
112. Stankovic S, Andretti GC (1979) Acta Cryst B35:3078
113. Vyas K, Sivalakshmidevi A, Prabhakar C, Reddy GO (1999) Acta Crysta C55:411
114. Gaupp S, Effenberger F (1999) Tet Asymm 10:1777
115. Wei ZL, Petukhov PA, Bizik F, Teixeira JC, Mercola M, Volpe EA, Glazer RI, Willson TM, Kozikowski AP (2004) J Am Chem Soc 126:16714
116. Liao C, Liu B, Shi L, Zhou J, Lu XP (2005) Euro J Med Chem 40:632
117. Wininger D (1988) J Chem Inf Comput Sci 28:31

Top Heterocycl Chem (2006) 3: 181–271
DOI 10.1007/7081_038
© Springer-Verlag Berlin Heidelberg 2006
Published online: 4 May 2006

QSAR and Molecular Modeling Studies of HIV Protease Inhibitors

Rajni Garg (✉) · Barun Bhhatarai

Chemistry Department, Clarkson University, Potsdam, NY 13676-5810, USA
rgarg@clarkson.edu

Abstract Acquired immunodeficiency syndrome (AIDS) and its related disorders, caused by retrovirus human immunodeficiency virus (HIV) are a major health concern worldwide. HIV protease is one of the major viral targets for the development of new chemotherapeutics. Currently, many HIV protease inhibitors are used in combination with HIV reverse transcriptase inhibitors. However, the use of current drugs regimens has several shortcomings, such as adherence, tolerability, long-term toxicity and drug- and cross-resistance. HIV is also known to have several mutants. Therefore, the development of new inhibitors that are less toxic, more tolerable, convenient and active against drug-resistant viruses is highly desirable. Several in-silico techniques are utilized in the process of drug design and development. One such technique is quantitative structure-activity relationship (QSAR). QSAR models reveal significant correlations between the biological activity and physico-chemical parameters and molecular descriptors. QSAR model can stand alone, support other approaches or be examined in tandem with equations of a similar mechanism to truly reveal its power. This work is devoted to a discussion of QSAR and molecular modeling studies especially those pertaining to 3D-QSAR on HIV protease inhibitors. It provides an overview of new ideas and their applications as they appear in the recent literature. Examples are given of peptidic and non-peptidic protease inhibitors. The role of hydrophobic, steric and electronic interactions in the design of HIV protease inhibitors is discussed. Linking a non-peptidic moiety to a peptidic backbone may provide a way of highlighting regions of interest. Some studies on such hybrid inhibitors are also discussed. Finally, studies on mutant protease data are included as they appear to be of utmost importance.

Keywords AIDS · HIV protease inhibitors · QSAR · Molecular modeling · Heterocyclic drugs

Abbreviations

AIDS	acquired immuno deficiency syndrome
HIV	human immunodeficiency virus
PR	protease
PI	protease inhibitors
PPIs	peptidomimetic protease inhibitors
NPPIs	non-peptidic protease inhibitors
SAR	structure-activity relationship
QSAR	quantitative structure-activity relationship
CADD	computer aided drug design
CoMFA	comparative molecular field analysis
CoMSIA	comparative molecular similarity analysis
IC_{50}	minimum concentration of a compound leading to 50% inhibition of the enzyme
EC_{50} or ED_{50}	the concentration of a compound required to achieve 50% protection of MT-4 or CEM cells against the cytopathic effect of the virus
CC_{50}	concentration of a compound required to reduce the number of mock-infected MT-4 or CEM cells by 50%
K_i	enzyme inhibition constant
$\log 1/C$	logarithms of inverse of biological end-points
$C \log P$	calculated logarithm of octanol-water partition coefficient P
π	hydrophobic constant of the substituents
σ, σ^- & σ^+	Hammett electronic parameters of the substituent effects on aromatic systems
σ^*	Taft's electronic parameters for aliphatic systems
Es	Taft's steric parameter
MgVol	McGowan volume
Vw	van der Waals volume
MW	molecular weight

B1, B5 & L	Verloop's sterimol parameters for substituent width and length
CMR	calculated molar refractivity
MLR	multiple linear regression
n	number of data points
r	correlation coefficient
s	standard deviation
q^2	Cramer's coefficient for the variance in the activity
F	ratio between the variances of observed and calculated activities
$\log P_0$	optimum $C\log P$
PLS	partial least square
LOO	leave-one-out
SMF	substructure molecular fragment
THP	tetrahyropyrimidinone
US-FDA	United States Food and Drug Administration
TSI	transition state isosteres
COMBINE	comparative binding energy analysis
AHPBA	3(S)-amino-2(S)-hydroxyl-4-phenyl butanoic acid
AFMoC	reverse protein-based CoMFA

1
Introduction

Acquired immunodeficiency syndrome (AIDS) and its related disorders, caused by retrovirus human immunodeficiency virus (HIV), are a major health concern worldwide. The first report published in 1981 described this epidemic as a clinical syndrome of immune deficiency much before the identification of HIV [1]. AIDS is the end stage manifestation of prolonged infection with HIV. HIV-infected people generally die of AIDS after about 10 years of infection. Since the outbreak of the epidemic more than 60 million people have been infected worldwide [1]. There are currently 1.6 million HIV-infected people living in North America, Western and Central Europe [2]. In 2004 alone, AIDS claimed approximately 3 million human lives and close to 5 million people contracted HIV [2]. In the face of this global pandemic a major effort was mobilized in the 1980's to find the ways to combat the disease. Although a cure or treatment for global implementation is yet to be found, unprecedented progress has been made in the past two decades.

1.1
HIV: The AIDS Virus

The causative agent for AIDS was isolated in 1983 and recognized as a new retrovirus from the lymph node of a man with persistent lymphadenopathy syndrome (LAS) [3, 4]. In 1984 another human retrovirus named HTLV-III (human T-cell leukemia virus-III) was isolated from peripheral blood mononuclear cells (PBMC) of adults and pediatric AIDS patients [5]. The same year Levy et al. isolated a retrovirus from AIDS patients and called it AIDS-associated retrovirus (ARV) [6]. The three prototype viruses LAV, HTLV-III and ARV were soon recognized as antigenically indistinguishable members of the same group of retrovirus that belongs to the *lentivirdae* family. In 1986, an international committee on the taxonomy of viruses recommended giving a separate name of human immunodeficiency virus (HIV) to the AIDS virus [7].

In 1986, a second AIDS virus was discovered from AIDS patients in West Africa [8]. To distinguish these viruses, the initially discovered retrovirus associated with most of the world's HIV disease is designated as HIV-1 and the second virus detected in West Africa is designated as HIV-2. The two strains are antigenically distinct. HIV-2 differs by more than 55% from the HIV-1 strain. HIV-2 is rare in most parts of the world due to less efficient transmission. The mortality rate from HIV-2 infection is also only two-third than that of HIV-1. In common usage, "HIV" usually implies HIV-1.

Studies pursued on HIV elucidated the critical molecular events necessary for viral replication and identified several possible molecular targets for therapeutic development. The details of structural components and the life cycle of HIV have been discussed [9, 10]. Most of the research efforts for the development of new chemotherapeutic agents have been focused on two major target enzymes, reverse transcriptase (RT) and protease (PR), encoded by HIV [11, 12]. The introduction of HIV protease (HIVPR) inhibitors, in particular, has drastically decreased the mortality and morbidity associated with AIDS [13]. Several QSAR and molecular modeling studies on HIV protease inhibitors (HIVPI) have been published in the literature [14–16]. The focus of this work is to present an overview of these studies.

1.2
HIV Protease (HIVPR)

In 1985, Ratner et al. [17] reported that the second open reading frame of HIV encoded a protease, which is analogous to those found in other retroviruses. A highly characteristic amino acid triad, Asp–Thr(Ser)–Gly found within the protease sequence suggests that the enzyme is a member of the aspartic acid protease family [18]. In spite of the presence of this conserved catalytic triad, there existed significant structural difference between the classical and retroviral aspartic acid proteases. In the classical form, i.e., fungal and mammalian aspartic acid protease, the enzyme usually consists of two homologous domains, with each domain containing the key catalytic triad [19]. However, HIVPR contains about half the number of amino acid residues and only one catalytic triad [20]. On the basis of these observations it was proposed that the catalytically active form of the retroviral protease exists as a homodimer and each monomer contributes one of the two key aspartic acid residues to the catalytic site [21]. This proposal was confirmed in 1988 by site-directed mutagenesis of the active site [22], inhibition by protease inhibitors [23, 24] and finally by single crystal X-ray crystallography [25–27].

HIVPR is a C2 symmetrical homodimer [28, 29]. Each monomer has 99 residues. The C2 axis of the enzyme lies between catalytic aspartates (Asp25 and Asp25′) perpendicularly in the active site. As per the standard nomenclature the S_1 and S_1' (S_2 and S_2' etc.) subsites are structurally identical [30]. The two equivalent S1 subsites are also mostly hydrophobic except for the Asp29, Asp29′ and Asp30 and Asp30′ residues. The S_3 subsites are adjacent to the S_1 subsites and are mostly hydrophobic with the exception of Arg8 and Arg8′ residues as shown in Fig. 1 [31]. In addition to the C2 symmetry, residues 42 to 58 from each monomer form a prominent hairpin turn, or flap, which projects over the substrate-binding cleft. These flexible flaps collapse to enclose inhibitors and the substrates within the hydrophobic tunnel, which is formed by the interface of the two 99-amino acid monomers [32].

The primary sequence of an aspartic protease has two different Asp–Thr–Gly sequences and its apostructure shows these two chains running in opposite directions with a water molecule bound between two aspartates. The water molecule is believed to act as a nucleophile for the enzyme–catalyzed amide hydrolysis of the substrate [31, 33, 34]. The substrate possesses a scissile bond that is attached by the water molecule of the enzyme, in the substrate enzyme interaction (Fig. 2) [31]. Few amino acid residues of the

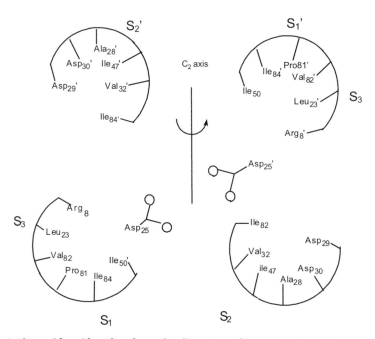

Fig. 1 Amino acid residue that forms binding sites of HIV-1 protease (Reprinted with permission from [31]. Copyright 1997 American Chemical Society)

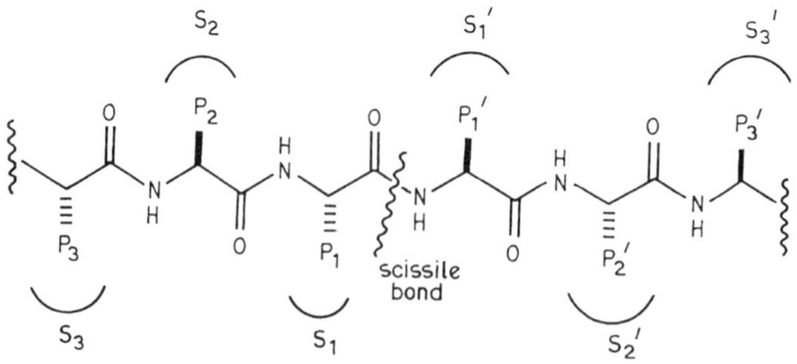

Fig. 2 Peptidic substrate of aspartic proteases. The P_1, P_2, ... P_n and P'_1, P'_2.... P'_n are amino acid residues and S_1, S_2, ... S_n and S'_1, S'_2, ... S'_n are the corresponding binding sites at the enzyme. These nomenclatures are according to the Schechter I, Burger A (1967) Biochem Biophys Res Commun 27:157 (Reprinted with permission from [31]. Copyright 1997 American Chemical Society)

substrate interact with the corresponding binding sites of the enzyme. This interaction is stabilized by several hydrogen bonds between the backbone of the substrate and the enzyme [35]. Inclusion of structural features into the inhibitors to mimic the function of this structural water provides enhanced binding energy through favorable entropic changes.

McQuade et al. [36] proposed that inhibition of HIVPR would also inhibit viral replication. Investigational experimentation proved that inactivation of the protease leads to the production of immature, non-infectious viral particles [22, 29]. Consequently, the study of the inhibition of this enzyme has drawn considerable interest of medicinal chemists for the development of anti-HIV chemotherapy [37].

1.3
HIV Protease Inhibitors (HIVPI)

The determination of the three-dimensional (3D) structure of HIVPR obtained through X-ray crystallography and the synthesis of the protease facilitated the development of HIVPI. A large number of peptidic and non-peptidic HIVPI have been synthesized and reported [38–43].

The discovery of peptide-based substrate-mimicking HIVPI was directed towards the synthesis of substrate analogs in which the scissile bond was replaced by a non-cleavable isostere with tetrahedral geometry that could mimic the tetrahedral transition-state of the proteolytic reaction. Thus, several inhibitors with hydroxyethylene or hydroxyethylamine isostere replacement were prepared to bind with the enzyme as shown in Fig. 3a [31]. However, the clinical development of peptide-derived compounds was hindered by their poor pharmacokinetics, including low oral bioavailability, rapid excretion and complex (expensive) synthesis [44, 45]. Therefore, recently more

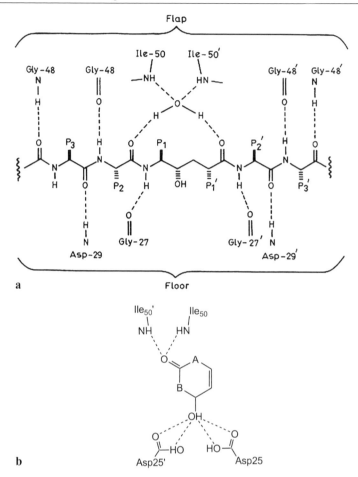

Fig. 3 a Peptidomimetic inhibitor binding at HIVPR binding site utilizing structural water (Reprinted with permission from [31]. Copyright 1997 American Chemical Society) **b** Displacement of structural water by non-peptidic moiety at the active site

focus has been on the investigation of non-peptidic inhibitors of low molecular weight that can interact with a limited number of binding sites of the enzyme critical for the inhibition.

A common feature of nearly all X-ray crystal complexes between HIVPR and peptide-derived peptidomimetic inhibitors (PPIs) is the presence of a tetracoordinated structural water molecule tightly bound between the inhibitor molecule and the flexible beta strands or flaps of the HIVPR dimer. This ubiquitous water molecule accepts two hydrogen bonds from backbone amide bonds of flap residues Ile50 and Ile50' and donates two hydrogen bonds to suitable acceptors, typically carbonyl groups, which flank the transition state isostere of the peptide mimetic inhibitor (Fig. 3a) [31]. Potent inhibition of the enzyme is attained through critical interactions between the two catalytic aspartyl residues and a hydroxyl group or similar functionality within the inhibitor molecule.

In the crystal structure of cyclic non-peptidic inhibitors (NPPIs) complexed with HIVPR, the structural water was found to be absent (Fig. 3b). The hydrogen bond accepting groups such as a carbonyl or sulfonyl group of NPPIs were found to form the requisite hydrogen bonding interactions with the flap amide nitrogen directly, without the need of a bridging water molecule. Hydroxyl or enolic hydroxyl groups present critically in the ring were found to interact with the two catalytic aspartyl residues of the HIVPR, similar to what was observed in X-ray crystal structures between linear peptide mimetic and HIVPR. The displacement of the structural water and the conformational constraints imposed by the cyclic structures of NPPIs were expected to provide a considerable energetic advantage over peptide-derived peptidomimetics [13, 31]. It is important to note that the presence of a conserved water molecule and its role in mediating hydrogen-bonding interactions between the flap of the enzyme and substrate/inhibitor is unique to all retroviral aspartyl proteases [31]. A good selectivity against mammalian aspartyl protease was anticipated with cyclic NPPIs by exploiting this novel-binding motif of NPPIs.

The United States Food and Drug Administration (US-FDA) has approved several HIVPIs for the treatment of HIV (1–7, Scheme 1) [46, 47]. These drugs have shown promising results when used in combination with HIV reverse transcriptase (HIVRT) inhibitors [48]. However, the use of these drugs has led to the development of HIV-resistant mutant strains that are less sensitive to the inhibitors [49]. Prolonged use of these drugs has also shown side effects such as nausea, diarrhea, headache, gastrointestinal problems etc. [12–14]. The emergence of a mutant virus and the occurrence of side effects suggests that the development of new HIVPIs that are less toxic, more tolerable, convenient and active against drug-resistant viruses is highly desirable.

1.4
QSAR and Molecular Modeling

Scientists in the field of drug design perceive it necessary to optimize a "lead" molecule. An essential feature of drug discovery is to synthesize analogs of lead molecules and test their biological activity in order to obtain progressively better analogs, better not only in terms of potency and efficacy but also in terms of pharmacokinetics and side effects (ADME). The principal hypothesis employed is that any change in the chemical structure produces a positive or negative change in the bioactivity. A systemic study of such cause and effect relation is called structure-activity relationship (SAR) study. The objective of SAR is to define the chemical consequences of changing the drug

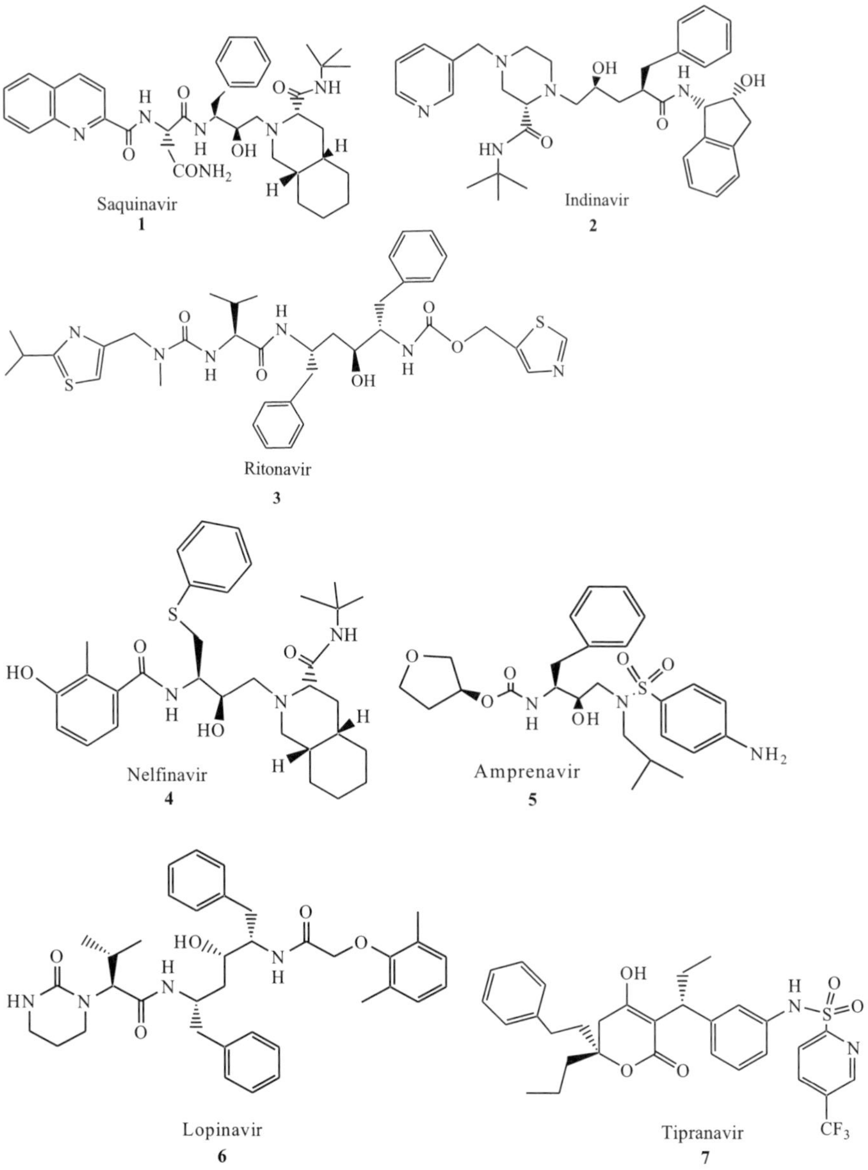

Scheme 1 FDA-approved HIV protease inhibitors [46, 47]

structure and subsequently to establish which changes in chemical structure and properties would produce better biological activities.

The researchers started analysis of SAR studies in the late 19th century. The concept of quantitatively correlating physicochemical properties of molecules with their biological activities, known as, "quantitative structure-activity relationship" (QSAR) was, however, first introduced by Corwin Hansch in the early 1960s [50, 51]. Since then, many new methods have revolutionized the field of "drug research". Several review articles describing different aspects of this field have been published [52–59]. The concept has evolved from 2D-QSAR to 3D-QSAR and lately 4D-QSAR [60].

QSAR helps in obtaining structure and affinity data on fragments that bind to protein and identifies specific groups and substituents for drug design and development. However, the 3D-structure of bioactive agents and their targets are the basis of drug discovery and development. Crystal structure determination of protein-ligand complexes has enabled computer aided drug design (CADD) and the efficient optimization of lead structure. Molecular modeling helps in building and visualizing the 3D-structure of a molecule, thus enabling exploration of the conformational space. It allows visualization of how a docked ligand is interacting with the protein-binding site. 3D-molecular modeling programs model, animate, render and export 3D-molecular graphics for optimization. A combination of QSAR and molecular modeling approach is the key for success in CADD. In the last few years several drugs have been discovered and reached the market or late stage clinical trials using these approaches [53, 54].

Many research papers have been published on the successful use of QSAR and molecular modeling studies to understand drug-receptor interactions [60–64]. Extensive applications of QSAR in the discovery of anti-HIV drugs in academia and industry have been reported [65–67]. Many molecular modeling studies on anti-HIV inhibitors using various 3D-QSAR techniques such as, comparative molecular field analysis (CoMFA), comparative molecular similarity analysis (CoMSIA), pharmacophore generation, free-energy

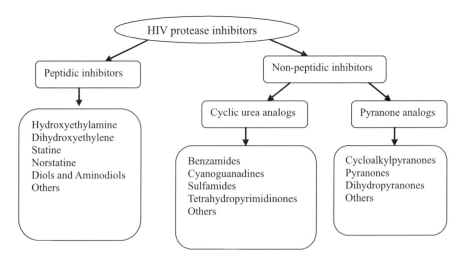

Scheme 2 Various classes of HIV protease inhibitors

binding analysis and others have been reported [16]. Several reviews have been published on the application of these QSAR and molecular modeling techniques on anti-HIV drug discovery [14–16,68]. All these publications have reviewed the studies reported on inhibitors designed to interact with different target sites of HIV.

A number of QSAR and molecular modeling studies on HIVPI were reported for biological structure-activity data gathered by various groups to determine the full potential of inhibitors belonging to various classes (Scheme 2). In this report a comprehensive overview on the applications of different QSAR and molecular modeling techniques on anti-HIV protease inhibitors published in the literature during the last decade is presented. Because of the scope of this work, mostly the molecular modeling studies as pertaining to 3D-QSAR techniques have been included. We hope that this work would be immensely helpful to those who are focusing their research efforts for the development of new HIV protease inhibitors.

2
Methods

In the QSAR reported here various biological endpoints have been used. However, in most of them the anti-HIV activity of the compounds is expressed as IC_{50} or EC_{50}. The IC_{50} refers to the minimum concentration of the compound leading to 50% inhibition of the enzyme HIVPR. EC_{50} (or ED_{50}) refers to the concentration (mol L^{-1} or mol g^{-1}) of the compound required to achieve 50% protection of MT-4 or CEM cells against the cytopathic effect of the virus. For the selectivity of the compounds, their cytotoxic effect has been measured in terms of CC_{50}, the concentration of the compound required to reduce by 50% the number of mock-infected MT-4 or CEM cells. K_i is the enzyme inhibition constant. The logarithms of inverse of these parameters refer to the biological end-points (log $1/C$) in QSAR studies. Any other biological end-points are explained in the respective section.

The correlates of these biological endpoints are various physicochemical parameters, which are hydrophobic, electronic, steric or topological in nature. Those found important in several QSAR models are briefly explained here, whereas others are discussed whenever used. For a more detail discussion of all these parameters, please see [58].

Hydrophobic parameters are mostly experimentally obtained log P or calculated log P (C log P), where P is the octanol-water partition coefficient. π is the hydrophobic constant of the substituents. The electronic parameters (Hammett constants) σ, σ^- and σ^+ applies to substituent effects on aromatic systems and Taft's σ^* applies to aliphatic systems. Steric parameters are Tafts steric parameter Es, McGowan volume MgVol, van der Waals volume Vw, molecular weight MW. Verloop's sterimol parameters B1, B5 and L are for measuring substituent width and length. B1 is a measure of the width of the first atom of a substituent, B5 is an attempt to define the overall volume and L is the substituent length. Molar refractivity MR (CMR: calculated MR) is largely a measure of volume with a small correction for polarizability. The indicator variables are often used in QSAR for special features with special effects that cannot be parameterized and has been explained wherever used. Besides, many studies have used different molecular descriptor(s), which have been discussed with respective QSAR.

The 2D-QSAR models presented in this work were mostly developed using multiple linear regression (MLR) analysis. However, the 3D-QSAR models were developed using various techniques such as CoMFA, CoMSIA, pharmacophore generation, free-energy binding analysis and many others. A note has been made along with the QSAR discussion about the technique used.

In all QSAR equations reported here, n is the number of data points, r is the correlation coefficient, s is the standard deviation, q^2 is Cramers coefficient to account for the variance in the activity [69] and the data within the parentheses are 95% confidence intervals. F is the F-ratio between the variances of observed and calculated activities.

3
Non-peptidic Protease Inhibitors (NPPI)

3.1
A Brief Description

Intense research efforts directed toward development of cyclic NPPIs for the treatment of HIV infection identified several in silico lead structures (8–11,

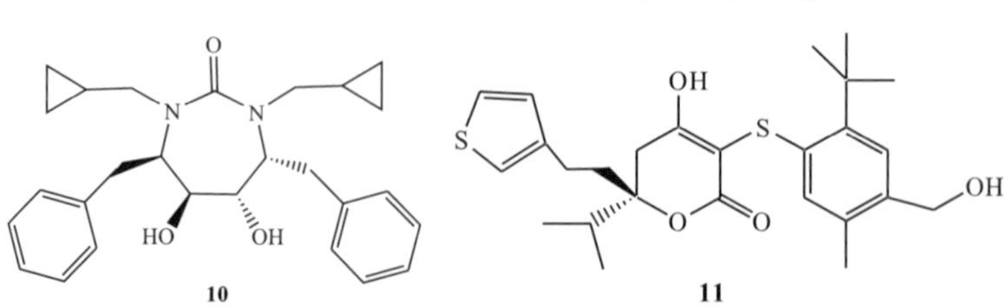

Scheme 3 Early leads on non-peptidic HIV protease inhibitors [13]

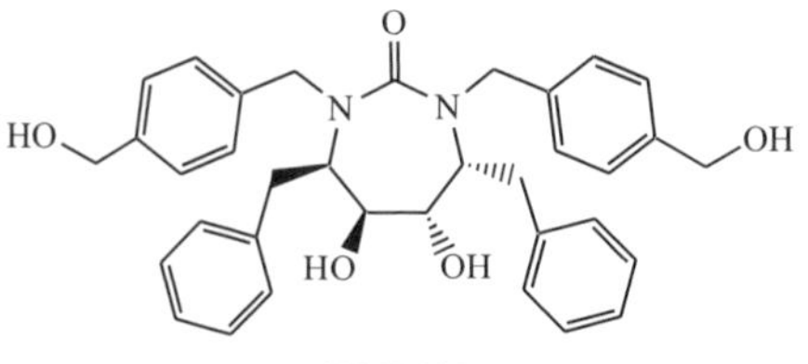

DMP 323
12

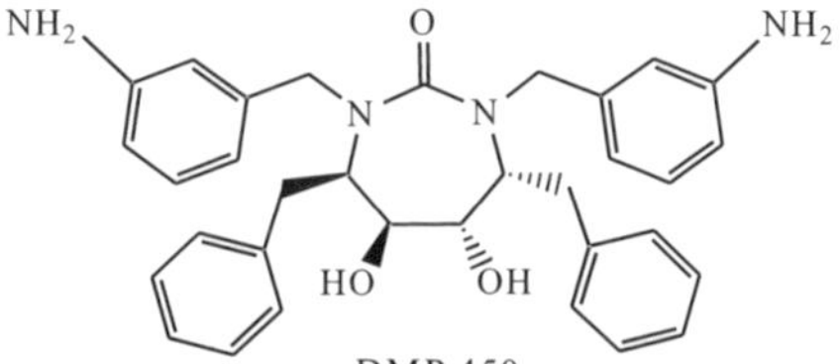

DMP 450
13

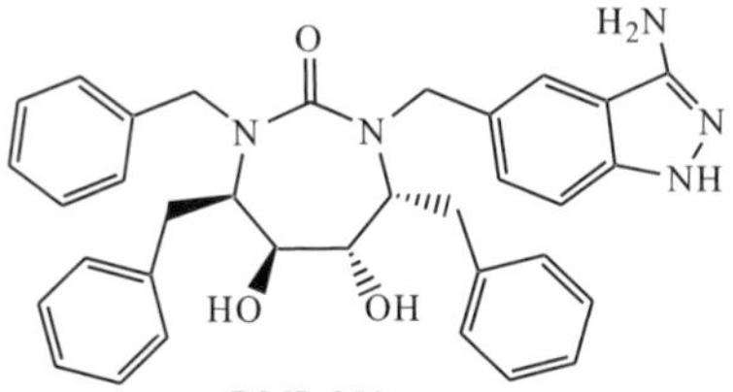

DMP 850
14

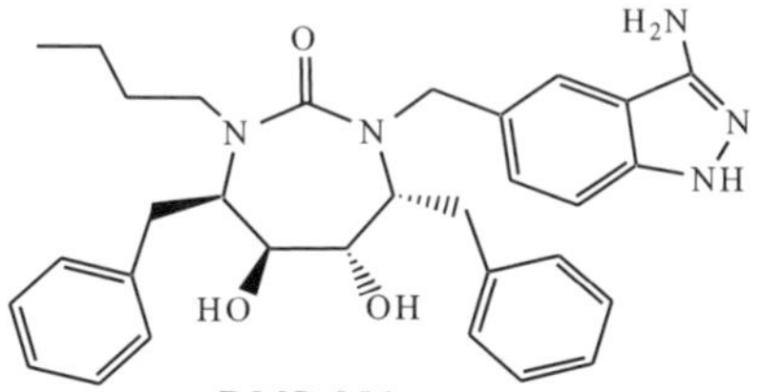

DMP 851
15

Scheme 4 Cyclic urea based on non-peptidic protease inhibitors reaching clinical trials [13]

Scheme 3) [13]. All these leads were identified using structural information gathered from extensive work on peptide leads and a computational pharmacophore analysis design approach. Much attention and focus has been placed on cyclic ureas and dihydropyranones. The research efforts on cyclic urea derivatives resulted in some of these inhibitors being selected for clinical evaluation (**12–15**, Scheme 4) [13]. The research on dihydropyranones resulted in the recent approval of Tipranavir (**7**) for HIV treatment, which is given in combination with Ritonavir (**3**) [46, 47].

3.2
Cyclic Urea Derivatives

The seven-member cyclic urea scaffold was found to be a potent inhibitor of HIVPR and has been studied extensively [70–74]. This scaffold creates an effective hydrogen bond network with the enzyme while incorporating the structural water molecule (Fig. 4) [71]. Initial SAR studies optimized the benzyl group as an ideal P_1/P_1'-substituent. Variation of nitrogen substituents at P_2/P_2' lead to clinical trials of several cyclic urea's (**12–15**, Scheme 4), which could not be pursued further due to poor bioavailability, metabolic instability, moderate potency and inadequate resistance profile as compared to other protease inhibitors. Further studies led to the identification of several other classes of cyclic urea derivatives. Subsequently QSAR and molecular modeling studies were also reported to aid in the design of an effective inhibitor belonging to this class. A discussion on these studies is presented in this section.

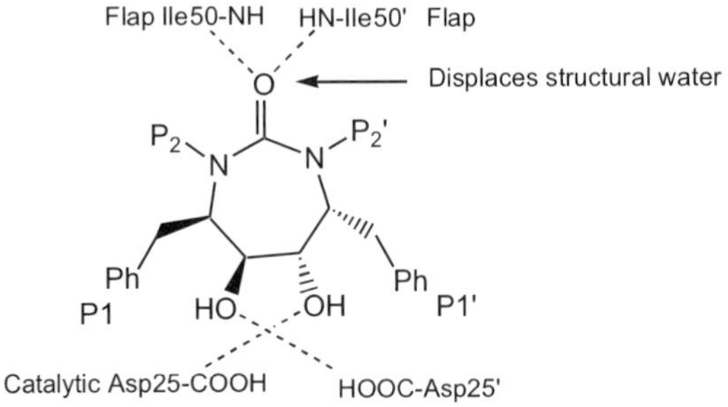

Fig. 4 Cyclic urea binding mode (Reprinted with permission from [71]. Copyright 1996 American Chemical Society)

3.2.1
QSAR Studies

3.2.1.1
Cyclic Ureas

Several research groups [70–74] studied symmetrical and non-symmetrical cyclic urea derivatives extensively. Nugiel et al. [70] reported the enzyme inhibitory activity (K_i) of symmetrical cyclic urea derivatives in which the characteristic benzyl P_1/P_1' moiety was replaced by novel P_1/P_1' substituents (**16**). P_2/P_2' substituents were either benzyl or CH_2-cyclopropyl groups. The activity was shown to be significantly correlated with the McGowan's volume of P_1/P_1'-substituent and two indicator variables [14].

$$\log 1/K_i = 4.90(\pm3.00)MgVol - 0.60(\pm0.35)(MgVol)^2 + 2.40(\pm0.57)I_a$$
$$- 1.16(\pm0.51)I_o - 3.91(\pm6.06)$$
$$n = 33, \quad r^2 = 0.858, \quad s = 0.40, \quad q^2 = 0.796,$$
$$(MgVol)_0 = 4.08(3.67-4.33) \tag{1}$$

The indicator I_a and I_o denotes a value of 1 for a P_1/P_1'-substituent containing a benzyl group and for an *ortho*-substituent on this benzyl, respectively.

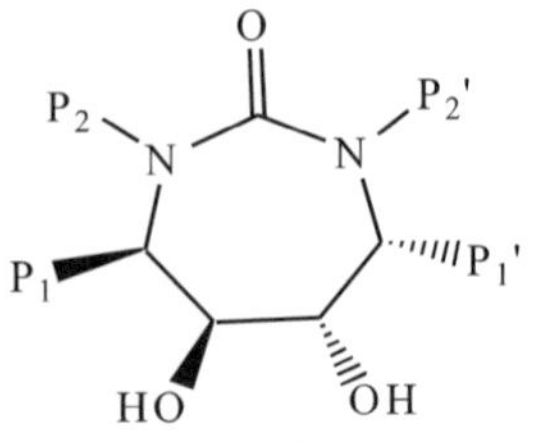

16

QSAR 1 suggests that the volume of the substituents up to an optimum of 4.08 and an aromatic moiety in them favor the activity however, an *ortho*-substituent in the aromatic moiety will be detrimental.

QSAR 2–5 were developed [14] for the structure-activity data of P_2/P_2' substituted symmetrical cyclic urea PR inhibitors reported by Lam et al. [71]. QSAR 2 and 3 were based on the SAR data reported for the derivatives of compound (16) in which P_2/P_2'-substituents were largely varied aliphatic groups. QSAR 4 & 5 represent the model using SAR data gathered for aromatic P_2/P_2'-substituents (17). P_1/P_1' groups were benzyl in both the series.

$$\log 1/K_i = 1.44(\pm0.42)C\log P - 2.13(\pm0.64)\log(\beta.10^{C\log P} + 1)$$
$$+ 0.68(\pm0.42)MR_X - 0.64(\pm2.22)$$
$$n = 21, \quad r^2 = 0.813, \quad s = 0.51, \quad q^2 = 0.727, \quad \log P_0 = 6.53(7.51-5.55),$$
$$\log \beta = -6.13 \tag{2}$$

$$\log 1/IC_{90} = 0.77(\pm0.25)C\log P - 1.24(\pm0.48)\log(\beta.10^{C\log P} + 1)$$
$$+ 1.05(\pm1.37)$$
$$n = 15, \quad r^2 = 0.813, \quad s = 0.33, \quad q^2 = 0.665, \quad \log P_0 = 6.96(7.75-6.17),$$
$$\log \beta = -6.84 \tag{3}$$

$$\log 1/K_i = -0.39(\pm0.17)C\log P - 3.82(\pm1.43)MR_{X,2}$$
$$+ 0.79(\pm0.72)MR_{X,3} + 11.73(\pm1.32)$$
$$n = 20, \quad r^2 = 0.818, \quad s = 0.52, \quad q^2 = 0.672 \tag{4}$$

$$\log 1/IC_{90} = -0.47(\pm0.12)C\log P - 1.99(\pm1.00)MR_{X,2} + 9.81(\pm0.96)$$
$$n = 20, \quad r^2 = 0.833, \quad s = 0.38, \quad q^2 = 0.728. \tag{5}$$

In these models, K_i is the enzyme inhibition constant and IC_{90} is the concentration of inhibitor resulting in 90% inhibition of viral RNA production in HIV-infected MT-2 cells. The presence of bilinear hydrophobic terms in QSAR 2 & 3, where there are aliphatic variations at the P_2/P_2' position of 16 indicate that hydrophobic interactions are important up to a rather high value (K_i 6.53, IC_{90} 6.96). Steric interaction of the X-substituent was also found to

17

be important for the inhibitory activity. Analysis of benzyl-substituted P_2/P_2' analogs (17) revealed that not only the steric effect of the *ortho*-substituent of the benzyl ring but also the overall hydrophobicity of the molecule appears to be detrimental to both enzyme inhibition and antiviral potencies (QSAR 4 & 5). However, the enzyme inhibition potency was found to be helped by the *meta*-substituent through a steric effect (QSAR 4).

Another series of derivatives of 17 was synthesized and tested by Jadhav et al. [72]. QSAR 6 & 7 reported [14] for the SAR data again showed a negative $C \log P$ term similar to QSAR 4 & 5.

$$\log 1/K_i = - 1.29(\pm 0.99)\sigma - 0.61(\pm 0.20)C \log P + 12.79(\pm 1.44)$$

$$n = 12, \quad r^2 = 0.850, \quad s = 0.57, \quad q^2 = 0.733 \tag{6}$$

$$\log 1/IC_{90} = - 1.34(\pm 0.82)\sigma - 0.56(\pm 0.17)C \log P + 10.04(\pm 1.19)$$

$$n = 12, \quad r^2 = 0.879, \quad s = 0.47, \quad q^2 = 0.786. \tag{7}$$

However, in these two QSARs the electron-donating ability of the substituents seems to be important for the activity as evident by a negative σ term. The authors [14] suggested that the rigid phenyl ring present as the P_2/P_2' substituent on 17 might not allow these substituents to interact well with the hydrophobic pocket (i.e. why a negative hydrophobic term appears in QSAR 4–7). It was also suggested that a study of flexible aliphatic substituents might be useful. Gupta et al. [73] also analyzed the data [70, 71] in detail and proposed a model of interaction of these symmetrical cyclic ureas with the receptor as shown in Fig. 5 [73].

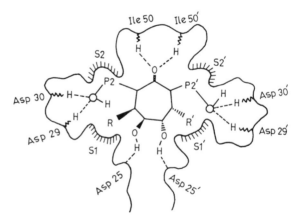

Fig. 5 Schematic representation of the interaction of cyclic urea with HIVPR (Reprinted with permission from [73]. Copyright 1998 Elsevier)

3.2.1.2
Cyclic Urea Benzamides

Wilkerson et al. [75] reported QSAR 8 and 9 for a series of N,N' di-substituted cyclic urea 3-benzamides (18) for HIV protease inhibition and antiviral activity.

$$\log 1/K_i = -1.273(\pm0.512)\text{IP} + 0.076(\pm0.021)C\log P^2$$
$$-0.866(\pm0.270)C\log P - 1.006(\pm0.330)\text{MV}/100$$
$$+0.990(\pm0.184)I + 20.770(\pm5.267)$$
$$n = 9, \quad r = 0.864, \quad s = 0.349, \quad F_{5,23} = 13.495_{(0.0001)},$$
$$\log P_0 = 5.697 \tag{8}$$

$$\log 1/\text{IC}_{90} = -0.863(\pm0.135)\log(K_i) + 0.004(\pm0.001)\text{CMR}^2$$
$$-1.062(\pm0.337)\text{HBD} - 4.226(\pm0.539)$$
$$n = 29, \quad r = 0.878, \quad s = 0.403, \quad F_{3,25} = 28.081_{(0.0001)}. \tag{9}$$

In QSAR 8 & 9, IP is ionization potential, MV/100 is a volume term and HBD was used for hydrogen bond donating characteristics of the substituent. I is an indicator variable that was used for all compounds containing nitrogen approximating that of the 2-pyridyl nitrogen. QSAR 8 developed for enzyme inhibition data suggests that a lipophilic compound up to an optimum value of 5.697 is good for activity. However, a larger volume and ionization potential will decrease the activity. 2-Pyridyl nitrogens also contribute significantly to the activity. The advantageous role of 2-pyridyl groups was attributed to its nitrogen atom that can participate in hydrogen bonding with the receptor (Fig. 6) [75]. QSAR 9 suggests that larger less hydrogen bond-donating protease inhibitors will produce a better antiviral agent. However, the quadratic term for CMR indicates that the relationship between molecular size and activity is not linear.

QSAR models on Wilkerson et al. [75] data were also developed by another group [14] in which a negative volume and IP term was found to be significant. This group [14] also reported that 2-pyridyl nitrogens contribute significantly to the activity. In contrast to the model reported by Wilkerson et al. [75] a linear or parabolic lipophilic term was not found to be important.

18

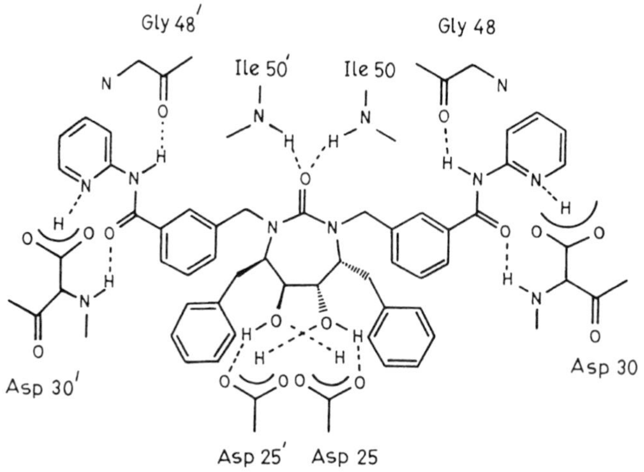

Fig. 6 Model proposed for the interaction of a 2-pyridyl containing cyclic urea benzamide derivative (Reprinted with permission from [75]. Copyright 1996 American Chemical Society)

It was also noted that there is a high mutual correlation between $C \log P$ and MgVol ($r^2 = 0.76$), thus making the role of both parameters ambiguous [14].

Wilkerson et al. [76] also synthesized and evaluated a series of unsymmetrical benzamide derivatives (**19**). They reported several QSAR models. QSAR 10 and 11 represents their best models for enzyme inhibition and antiviral activity.

$$\log 1/K_i = 0.190(\pm0.123)C \log P^2 - 2.377(\pm1.346)C \log P$$
$$- 7.398(\pm2.308)(MW/100)^2$$
$$+ 96.294(\pm30.473)(MW/100) - 304.541(\pm99.365)$$

$n = 15, \quad r = 0.852, \quad s = 0.474, \quad F = 6.599, \quad \text{prob} > F = 0.007,$

$\log P_0 = 6.40$ \hfill (10)

$$\log 1/IC_{90} = -0.114(\pm0.073)C \log P^2 + 1.581(\pm0.799)C \log P$$
$$- 0.685(\pm0.159) \log K_i \text{ pred} - 7.891(\pm2.107)$$

$n = 14, \quad r = 0.851, \quad s = 0.335, \quad F = 8.727, \quad \text{prob} > F = 0.004,$

$\log P_0 = 6.93$. \hfill (11)

A parabolic dependence on hydrophobicity and molecular weight was noticed for the enzyme inhibitory activity in QSAR 10. K_i values used for deriving QSAR 10 were measured by inhibiting recombinant single-chain dimeric HIV protease. The same authors reported another statistically better QSAR model based on $C \log P$ and the CMR term, however, $C \log P$ and CMR were highly collinear ($r^2 = 0.70$). QSAR 11 reported for antiviral activity has less predic-

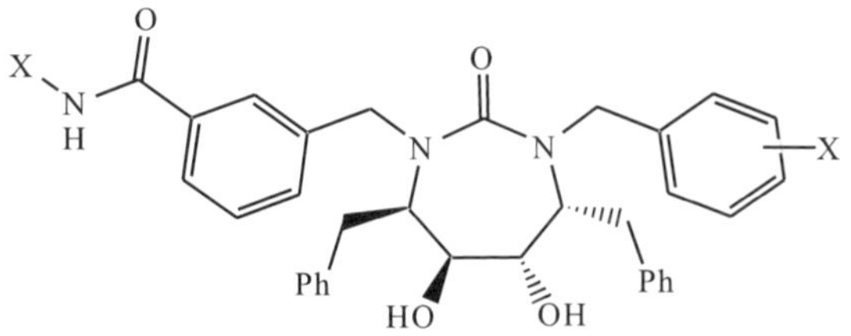

19

tive value since it relied on measured protease inhibition activity. Both these QSAR suggest that the optimum $C\log P$ for this class of compounds would lie in the range 6.4–6.9.

A very unusual negative bilinear $C\log P$ and positive linear volume term was reported significant for the enzyme inhibitory activity in another QSAR model [14]. The authors made a note that they were unable to reproduce the QSAR presented by Wilkerson et al. [76]. A re-examination of the original SAR data and methods used to calculate parameter values may shed some light on this unusual observation.

3.2.1.3
Cyclic Cyanoguanidines

Jadhav et al. [72] also observed a significant improvement in inhibitor potency for symmetrical cyclic cyanoguanidine derivatives (**20**) with the same substituents as the cyclic urea derivatives tested previously. Analysis of the given SAR data resulted in QSAR 12 & 13 [14]. It is evident that the enzyme inhibition and antiviral activities were not influenced by hydrophobicity as previously seen in QSAR models 4–7. Rather, the substitution of the ketone group of cyclic urea with a guanidine resulted in models indicating the need for smaller volume and electron-releasing substituents for good antiviral and enzyme inhibitory activity.

$$\log 1/K_i = -1.77(\pm 0.65)\text{MgVol} - 1.26(\pm 0.87)\sigma_{\text{sum}} + 16.21(\pm 3.14)$$
$$n = 11, \quad r^2 = 0.844, \quad s = 0.47, \quad q^2 = 0.747 \tag{12}$$

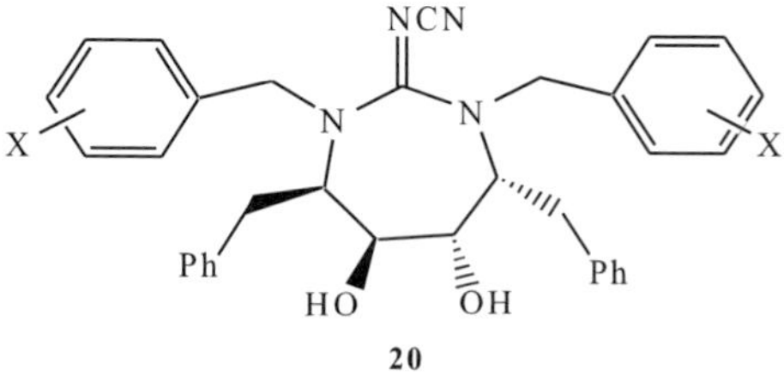

20

$$\log 1/IC_{90} = -1.53(\pm 0.51)MgVol - 1.36(\pm 0.69)\sigma_{sum} + 13.01(\pm 2.49)$$
$$n = 11, \quad r^2 = 0.877, \quad s = 0.38, \quad q^2 = 0.783 . \tag{13}$$

Han et al. [77] synthesized and analyzed another series of symmetrical cyclic cyanoguanidines (20). QSAR 14 & 15 were reported for the data [14].

$$\log 1/K_i = -1.44(\pm 0.61)I - 0.49(\pm 0.28)CMR + 19.86(\pm 5.29)$$
$$n = 9, \quad r^2 = 0.873, \quad s = 0.36, \quad q^2 = 0.651 \tag{14}$$

$$\log 1/IC_{90} = -2.18(\pm 0.77)I - 0.66(\pm 0.30)CMR + 20.85(\pm 5.93)$$
$$n = 8, \quad r^2 = 0.914, \quad s = 0.29, \quad q^2 = 0.735 . \tag{15}$$

In this series, oximes and ketones on the *meta*-position of P_2/P_2' benzyl (20) were studied. Both enzyme inhibition and antiviral activities were modeled in a similar fashion. The indicator variable, I, with a value of 1 for ketones and 0 for oximes was used in developing QSAR 14 & 15. Its negative coefficient shows that oximes are preferred over ketones for enzyme inhibition as well as antiviral activity. The negative CMR term in both the equations indicates a steric effect.

Gupta and Babu [78] also reported QSAR study on the data of Jadhav et al. [72] and proposed a model for the interaction of cyclic cyanoguanidines with HIVPR (Fig. 7) [78].

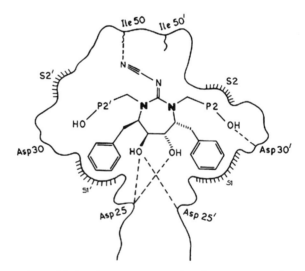

Fig. 7 A proposed model for the interaction of cyclic cyanoguanidines with HIVPR (Reprinted with permission from [78]. Copyright 1999 Elsevier)

3.2.1.4
Cyclic Urea Aminoindazole

Initially Kurup et al. [15] analyzed some series of aminoindazoles [79, 80]. Later, a comprehensive comparative QSAR study on P_2/P_2' and P_1/P_1' substituted symmetrical and non-symmetrical 3-aminoindazole cyclic urea HIV protease inhibitors (**21**) was performed [81]. The SAR data were taken from different papers [79, 80, 82–85]. Several QSAR models were developed for individual datasets. QSAR 16–18 were derived for the combined set [81].

$$\log 1/K_i = -0.34(\pm 0.07)C\log P + 12.82(\pm 0.53)$$
$$n = 28, \quad r^2 = 0.801, \quad q^2 = 0.771, \quad s = 0.195 \tag{16}$$

$$\log 1/IC_{90} = 0.46(\pm 0.11)C\log P + 1.23(\pm 0.28)I_1$$
$$- 0.19(\pm 0.10)L_4 + 3.61(\pm 0.78)$$
$$n = 27, \quad r^2 = 0.833, \quad q^2 = 0.773, \quad s = 0.234 \tag{17}$$

$$\log 1/T = 0.78(\pm 0.15)C\log P + 1.17(\pm 0.40)I_1 - 9.56(\pm 1.25)$$
$$n = 27, \quad r^2 = 0.808, \quad q^2 = 0.750, \quad s = 0.360 . \tag{18}$$

The protease inhibitory activity of these compounds was found to decrease with larger and more hydrophobic molecules, whereas the antiviral potency and translation (T) across the cell membrane increases with increase in hydrophobicity and size. QSAR 16–18 reflects that overall the hydrophobicity of these compounds is the most important parameter governing their biological activities. QSAR 17 also has a negative coefficient of parameter L_4, indicating that the length of the 4th position substituent at P_1/P_1'-benzyl is detrimental to the activities. In QSAR 17 and 18 an additional indicator parameter I_1 was used with a value of unity for non-symmetrical P_2/P_2' substituted analogs (**21**). The positive coefficient of this parameter indicates that these analogs of 3-aminoindazole cyclic ureas are more suitable for achieving good antiviral

R 1 = H/ Alkyl
R 2 = 3-aminoindazole/ alkyl/ benzyl

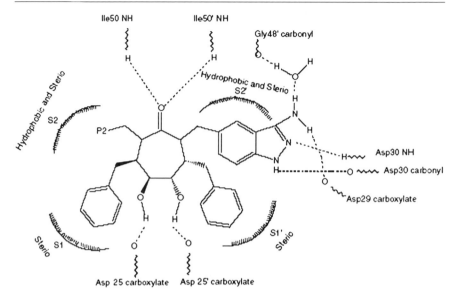

Fig. 8 The proposed model for the interaction of non-symmetrical 3-aminoindazole cyclic urea derivatives with HIVPR (Reprinted with permission from [81]. Copyright 2004 Elsevier)

activity and translation against HIV infected cells. A model for the interaction of non-symmetrical aminoindazole with the receptor was proposed as shown in Fig. 8 [81].

3.2.1.5
Tetrahydropyrimidinones (THPs)

Gayathri et al. [86] conducted QSAR studies on enzyme inhibitory and antiviral activity of P_2/P_2' tetrahydropyrimidinones (**22**) data reported by Delucca et al. [87].

$$\log 1/K_i = 0.421(\pm 0.378)^1\chi^v - 1.939(\pm 0.700)I_1$$
$$+ 1.020(\pm 0.590)I_2 + 8.293(\pm 1.266)$$
$$n = 23, \quad r = 0.895, \quad s = 0.50, \quad F_{3,19} = 25.37(5.01) \tag{19}$$

22

$$\log 1/IC_{90} = 0.348(\pm 0.258)^1\chi^v - 1.171(\pm 0.566)I_1 + 6.050(\pm 0.849)$$
$$n = 21, \quad r = 0.817, \quad s = 0.35, \quad F_{2,18} = 18.02(6.01) \tag{20}$$

$$\log 1/K_i = 0.862(\pm 0.235)^1\chi^v(P2) + 0.438(\pm 0.153)^1\chi^v(P2')$$
$$- 1.398(\pm 0.600)I_1 + 5.840(\pm 0.76)$$
$$n = 25, \quad r = 0.906, \quad s = 0.57, \quad F_{3,21} = 32.22(4.87) \tag{21}$$

$$\log 1/IC_{90} = 1.199(\pm 0.482)^1\chi^v(P2) + 0.596(\pm 0.108)^1\chi^v(P2')$$
$$- 0.942(\pm 0.578)I_1 + 2.088(\pm 1.592)$$
$$n = 14, \quad r = 0.970, \quad s = 0.25, \quad F_{3,10} = 53.40(6.55) . \tag{22}$$

The activities were found to correlate with $^1\chi^v$ (the first order valence molecular connectivity index) of P_2/P_2' substituents attached to the two nitrogens (N1 & N3) and some indicator parameters. It was suggested [86] that the less polar and more hydrophobic substituent would be beneficial for improving the activities. The authors also found that the translation of the enzyme inhibition activity to antiviral potency is more sensitive for compounds having dissimilar P2 and P2' substituents.

Later, Katritzky et al. [88] reported the CODESSA PRO-based QSAR approach to understand the importance of the charge distribution and the H-bonding interaction of both symmetrical and unsymmetrical tetrahydropyrimidinones (22). Four QSAR models (23–26) were developed [88] for the enzyme inhibition data studied by Delucca et al. [89]. The first QSAR contained a topological descriptor based on Shannons expression of information content (^{1}SIC) [90] as a term representing molecular size and shape.

$$\log 1/K_i = 2.139(\pm 0.740) + 0.167(\pm 0.018)^1\text{SIC}$$
$$r^2 = 0.646, \quad r^2_{cv} = 0.617, \quad s = 0.750, \quad F = 90 . \tag{23}$$

The second model (QSAR 24) included a term for fractional positive surface area of hydrogen donors $^{HD}FPSA^{(2)}$.

$$\log 1/K_i = 2.168(\pm 0.535) + 0.108(\pm 0.010)^2\text{SIC}$$
$$+ 12.750(\pm 1.848)^{HD}FPSA^{(2)}$$
$$r^2 = 0.816, \quad r^2_{cv} = 0.792, \quad s = 0.545, \quad F = 107 . \tag{24}$$

The third model (QSAR 25) developed using the hydrogen bond descriptor $^{HD}FCPSA^{(2)}$ (a charge-weighted analog of $^{HD}FPSA^{(2)}$ [91]), first moment of inertia (I^A) and the Balaban topological index (J, a measure of molecular centricity) has a much improved correlation.

$$\log 1/K_i = 15.901(\pm 0.917) + 261.881(\pm 24.823)^{HD}FCPSA^{(2)}$$
$$- 160.948(\pm 36.184)I^A - 5.708(\pm 0.746)J$$
$$r^2 = 0.855, \ r_{cv}^2 = 0.832, \ s = 0.489, \ F = 92. \tag{25}$$

The final model (QSAR 26) with an r^2 of 0.873 included an additional 4th parameter $^{min}E_{e\_n}$ (O – H) used for minimum electron-nuclear attraction energy for O – H bonds. Its negative coefficient indicated the importance of the H-bonding interaction within the active site of HIV protease receptor. The negative coefficient of I^A and J in QSAR 25 and 26 were proposed to indicate superiority of C2 symmetric substituents. All these models showed that symmetrical cyclic urea derivatives capable of H-bond donation and electrostatic interaction are better inhibitors of HIVPI.

$$\log 1/K_i = 42.585(\pm 8.445) + 15.601(\pm 1.776)^{HD}FPSA^{(2)}$$
$$- 141.775(\pm 34.923)I^A - 6.204(\pm 0.732)J$$
$$- 0.234(\pm 0.073)^{min}E_{e\_n}(O - H)$$
$$r^2 = 0.873, \quad r_{cv}^2 = 0.847, \quad s = 0.463, \quad F = 79. \tag{26}$$

As part of an ongoing effort in understanding the role of hydrophobicity in the design of NPPIs, a QSAR study [92] on another series of P_2/P_2' tetrahydropyrimidinone (23) was presented using enzyme inhibitory and antiviral activity data [87].

$$\log 1/K_i = - 0.738(\pm 0.416)C\log P - 1.997(\pm 1.114)MR_4 + 14.029(\pm 2.633)$$
$$n = 12, \quad r = 0.911, \quad r^2 = 0.830, \quad q^2 = 0.716, \quad s = 0.595 \tag{27}$$

$$\log 1/IC_{90} = - 0.726(\pm 0.357)C\log P - 2.371(\pm 1.281)MR_4 + 11.500(\pm 2.274)$$
$$n = 10, \quad r = 0.912, \quad r^2 = 0.832, \quad q^2 = 0.606, \quad s = 0.359. \tag{28}$$

The models suggested that the balance of hydrophobicity and a volume dependent polarizability term plays a key role in the inhibition of the viral

23

protease by these inhibitors. The size of the substituent of ligands at particular positions, which induce steric fit, was found to be crucial. It was suggested that the various substituents used to target individual pockets need to be designed cautiously to achieve better inhibition and an improved pharmacokinetic profile. It was also found that a sufficient spread in the data is required to observe the optimum value of $C\log P$ for these inhibitors.

The same authors [92] also analyzed a smaller dataset by Delucca et al. [87] to study the effect of stereochemistry on the enzyme inhibitory activity. They reported that the RRR isomer of (23) is more active than the SSS isomer. Inclusion of an electronic term for the substituents at the 3rd position of the P_2/P'_2 benzyl of (23) improved the correlation significantly and indicated that the electron-donating substituents at this position will improve the activity.

3.2.1.6
Other Classes

Han et al. [93] tested a small dataset of tricyclic urea (24) for the inhibitory action against the protease enzyme. Kurup et al. [15] derived QSAR 29, which indicates the importance of size-dependent polarizability of these compounds.

$$\log 1/K_i = 0.24(\pm 0.12)\text{CMR} + 2.73(\pm 1.45)$$
$$n = 5, \quad r^2 = 0.924, \quad q^2 = 0.876, \quad s = 0.172. \tag{29}$$

Frecer et al. [94] designed a combinatorial subset of non-symmetrical cyclic urea and predicted their biological activity using the QSAR technique. The QSAR model was derived using the SAR dataset reported by several authors [76, 80, 85, 95–97].

$$pK_i = 0.0016 \cdot \text{HB}_{score} - 0.0038 \cdot \text{VDW}_{score} + 1.8915. \tag{30}$$

A target-specific LUDI-type scoring function [98] was used to validate a set of compounds not included in the training set and to predict the inhibition constant of the generated analogs. QSAR analysis showed HB_{score} and VDW_{score} as two important parameters, that characterize the hydrogen bonding and lipophilic interactions between the ligand and the receptor. The predicted in-

24

hibition constants of the new lead were up to 2 orders of magnitude lower than the K_i's of the training set.

3.2.2
Molecular Modeling Studies

Several molecular modeling studies using various 3D-QSAR techniques have been reported in the literature on cyclic urea HIV protease inhibitors. Most of them were based on a heterogeneous dataset taken from the work of different authors. Because of the nature of the dataset studied, it was difficult to group them in distinct classes; however, an attempt has been made to organize the following discussion according to the previous subsection for comparison purposes.

3.2.2.1
Cyclic Urea Benzamides

Many QSAR and molecular modeling studies have been published on cyclic urea benzamide derivatives first reported by Wilkerson et al. [75, 76]. 3D-QSAR studies on symmetrical bis-benzamide cyclic urea HIVPI was reported by Debnath [99] using the CoMFA method. CoMFA-derived 3D-QSAR has been shown to be of great value in many publications [100–106]. The CoMFA results through the 3D contour plots of the steric and electrostatic fields, along with information on X-ray crystallographic coordinates of the ligand–HIVPR complexes help in the design of better HIVPI. The anti-HIVPR activity (K_i) data taken from Wilkerson et al. [75] was modeled for a dataset consisting of 29 training set molecules and 8 test set molecules. Sybyl 6.4 software [107] was used for conducting all the modeling studies. The systematic search routine within Sybyl was used for all conformational searches. Maximin2 minimizer was used to minimize the structures. The biopolymer module within Sybyl was used to retrieve and analyze all protein structures. The CoMFA feature of the QSAR module within Sybyl was used to develop all QSAR models. The crystal structure of the cyclic urea derivative (XV638) complexed with HIVPR was used as a template to construct all the training and test set molecules. CoMFA default settings were used to calculate the steric and electrostatic field energies using sp^3 carbon probe atom with a + 1 charge. The effect of grid spacing on this study with 2.0, 1.5 and 1.0 Å was analyzed. The best results were obtained with a grid spacing of 1.5 Å. The PLS method with leave-one-out (LOO) cross-validation was used for the analysis. The best model reported has a significant cross-validated q^2 of 0.724, r^2 of 0.971 and a low standard error of estimate(s) of 0.119. The CoMFA contour maps revealed several important contributions of electrostatic and steric fields and conform well to the interactions observed with the template protease complex. This analysis provided valuable information in designing a more active analog of the bis-benzamide cyclic urea.

QSAR 8 mentioned earlier was derived for the same dataset [75]. Ionization potential (IP), $C \log P$, molar volume (as MV/100) and an indicator variable (I) were used to model the activity. A parabolic model with $C \log P$ unfolded the ideal $C \log P$ value ($C \log P_0 = 5.697$) for maximum anti-protease activity. It is of note that the activities of the test set compounds (total eight) were predicted well both by the 2D-QSAR model ($r^2 = 0.746$) of Wilkerson et al. [75] and the CoMFA model ($r^2 = 0.971$) of Debnath [99].

Debnath [108] conducted another 3D-QSAR study employing CoMFA on a large series of cyclic urea derivatives. HIVPR inhibitory activity (K_i) data from Nugiel et al. [70] and Wilkerson et al. [75, 76] was used for deriving 3D-QSAR models. Since the dataset has a large number of compounds, instead of one model the author decided to derive three different models from 93 different sets of compounds and left the rest (25) as test sets. Although there were overlaps in compounds in the training sets no overlap was allowed in the test set to ensure rigorous validation of the predictive models. The starting geometry of all the molecules was based on the crystal structures of two known inhibitors in complex with HIVPR, XV638 and DMP450. Sybyl 6.4 [107] was used for all the modeling. The methods used were the same as discussed earlier. The PLS method along with the LOO cross-validation technique was used to develop highly predictive models (q^2 from 0.699 to 0.727; r^2 from 0.965 to 0.973 and standard error of estimates (S) from 0.239 to 0.265). These models revealed approximately 45% contributions from steric effects and 55% from electrostatic effects. The HINT method [109] was used to verify whether there was any additional effect from hydrophobicity. It revealed a significant contribution of hydrophobicity towards HIVPR inhibitory activity. This study helped in understanding the HIVPR-inhibitor interactions and provided valuable information for designing more effective inhibitors of the cyclic urea class.

Tervo et al. [110] conducted a CoMFA and CoMSIA study of 113 flexible cyclic ureas and compared their quality and productiveness by manual (M) and automated (A) alignment methods. They also used the same dataset as used by Debnath [71, 76, 108]. A reference cyclic-urea derivative (XV638) complexed with the HIVPR enzyme was used as template molecule and all the inhibitor structures were built based on this template. The inhibitor was then docked using the GOLD program [111] and scores were calculated. The inhibitors were minimized inside the active site with the Steepest Descent method using Tripos force fields [112], Gasteiger-Huckel charges for inhibitors and Kollman all-atom charges [113] for HIVPR. Steric and electrostatic CoMFA descriptors were calculated in a 3D cubic box extending 4 Å beyond the aligned inhibitors and with 1.5 Å grid spacing. For CoMSIA two different sets of models were developed using (1) steric and electrostatic fields and (2) H-bond donor and acceptor fields. Predictive power was analyzed using PLS statistical analysis, which was further evaluated using the LOO method. An external test set was also used to test CoMFA and CoMSIA models. The (M)-based model ($q^2 > 0.616$) was statistically better than the (A)-based model ($q^2 > 0.523$). In addition, the CoMFA models were statistically poor compared to CoMSIA for the same conformation and alignment. For its ability to predict, the simple statistical CoMFA models with less number of components based on automated (A) inhibitor alignment have better statistical significance ($q^2 = 0.647$) compared to that reported by Debnath ($q^2 = 0.629$) [108]. The CoMFA and CoMSIA contour maps confirmed the H-bonding interaction between an inhibitor and the backbone carbonyl oxygen of key amino acids.

Allen et al. [114] also analyzed the same dataset of cyclic urea derivatives as used by Debnath [108]. They developed a neural network-optimized molecular similarity-based QSAR model. Molecular similarity is a measure of the degree of overlap between a pair of molecules in some property space. Time constraints prevent the use of these techniques on large datasets or on large molecules. It was proposed that by reducing the molecular representation to a two-dimensional form, the alignment of the molecules could be speeded up greatly. The accuracy of the resulting similarity values can be further improved by using a neural network. Improvement of the results from a very poor correlation of 49% to a significant correlation of 86% was observed for the HIV protease inhibitor set. Overall, it was proposed that the neural network-based technique appeared to be fast with minimal loss of accuracy for molecular similarity calculation for larger sets.

Solovev and Varneck [115] applied the substructure molecular fragment (SMF) method to access anti-HIV activity of cyclic urea derivatives reported in the literature [74–76]. The SMF method uses molecular fragments (atom/bond sequences and augmented atoms) as variables in a multiple stepwise regression analysis. Forty-nine different types of fragments generated using the Trial program [116] were used with three linear and fitting equations building up to 147 models. The best-selected models for cyclic urea compounds involved atom/bond sequences from 2 to 4, 3 to 4 and exactly 4 atoms. Calculated for the best models, the average dataset was able to reproduce available experimental data at least as good as those from earlier QSAR studies. This paper demonstrated that the SMF method gives an interesting opportunity to build the optimal substituents whose attachment to the molecular core could result in increased activity. This feature of SMF was proposed to be useful in library design.

Avram et al. [117] conducted a 3D-QSAR study on the data reported by Nugiel et al. [70] to compare similarity and the differences between the electrostatic and the steric forcefields of the 3D-structure of the enzymes. Initially 45 inhibitor molecules were used to obtain the model with $q^2 = 0.575$ and $r^2 = 0.923$. Removal of three outliers yielded a better CoMFA model with good predictability ($q^2 = 0.670$ and $r^2 = 0.972$). The model showed that steric descriptors have a critical role for the determination of the predictable biological activity, while electrostatic interaction could compensate the steric repulsive effect. This observation was found to be true for the type of compounds analyzed in the article.

3.2.2.2
Cyclic Cyanoguanidines and Sulfamides

Avram et al. [118] used 3D-QSAR CoMFA techniques to study a dataset of symmetric and non-symmetric cyclic urea HIVPI. The anti-HIVPR inhibitory activity data was taken from the literature [72, 77]. The electrostatic and steric fields were calculated by default settings with sp^3 C-atom probe with a + 1 charge. The regression models derived with the PLS and LOO cross validation technique were developed. The best model with a correlation coefficient (r^2) of 0.981 and cross-validated correlation (q^2) of 0.525 showed a higher contribution from steric fields (58.6%) compared to electrostatic fields (41.1%). Two additional models were reported with $q^2 = 0.627$ and 0.536, respectively. All the models were improved further by omitting outliers.

Avram et al. [119, 120] also conducted CoMFA studies to compare the energetic and steric parameters of the wild type and mutant HIV-1 protease to explain the viral resistance. In their most recent work [120] they predicted the inhibition constants and inhibitor concentration of 127 symmetrical and unsymmetrical cyclic urea and cyanoguanidine derivatives. QSAR models with a correlation (r^2) of 0.70 and cross-validated correlation (q^2) of 0.63 were obtained indicating their better predictive ability. The biological activity predicted for 14 cyclic urea derivatives that inhibit the HIV protease mutants V82A, V82I and V82F was found to be in good agreement with experimental values. The contour plots of favorable and unfavorable steric and electrostatic fields were reported.

Schaal et al. [121] reported a flipped conformation of the cyclic sulfamide class of inhibitors [123] where a preferential binding of P_2/P_2' substituents in the S_2/S_2' subsite was unsuccessful unlike cyclic ureas. To test this hypothesis and to rationalize the SAR, 18 cyclic sulfamides were developed and a CoMFA study was performed. The protease inhibition data for this study was taken from Jadhav et al. [72]. The structure of the inhibitor was minimized using the AMBER forcefield with the GB/SA solvation model in Macromodel 5.5 [122]. The CoMFA model along with PLS and SAMPLS (sample-distance partial least square) were generated to obtain the best model with $q^2 = 0.54$ and a standard error of 0.69 with three components. The model showed a 66.1% contribution for the steric field and 39.9% for the electrostatic field. The model was used to predict the test-set of seven compounds that showed overall better prediction capability. This study showed that both types of compounds; cyclic urea and cyclic sulfamides, fit the CoMFA model well even though they are predicted to have opposite binding modes thus having a different SAR.

3.2.2.3
Tetrahydropyrimidinones (THPs)

Nair et al. [124] constructed 3D-QSAR models on PR inhibitory data of tetrahydropyrimidinone taken from the literature [87, 89]. The regression studies showed strong correlation between the observed pK_i values and the descriptors in terms of consistency (r^2) and internal predictive ability (r_{cv}^2). The important descriptors observed were E_{compl} (calculated enzyme-inhibitor complexation energies), E_{solv} (inhibitor solvation energies) and BSAs (inhibitor buried surface areas). The initial CoMFA model with 49 compounds in the training set has a strong correlation $(r^2 = 0.96)$ and a good cross-validation term $(r_{cv}^2 = 0.58)$, however, inclusion of inhibitor–solvent complexation energy, $E_{solv}[I]$ along with E_{compl} gave better predictive ability with a better cross-validation term $(r_{cv}^2 = 0.80)$. This study demonstrated the utility and potential of ligand and receptor-based descriptors in 3D-QSAR.

Senese and Hopfinger [125] used a set of 50 THP-based HIV inhibitors to construct a receptor independent 4D-QSAR model. They also used the dataset reported by Delucca et al. [87]. Out of 24 models, five unique models developed by a clustering algorithm with q^2 from 0.81–0.84 were chosen to map atom type morphology of the inhibitor and to predict the biological activity. The models identified the steric and non-polar characteristic of the receptor proving the success of a simple clustering technique. It also provided a discrete algorithm for model selection, as well as to predict the test set or unknown compounds. It has been long realized that there cannot be a single solution, or model, that can fully represent the multidimensional nature inherent to ligand–receptor binding. The multiple alignment method proposed in this 4D-QSAR study, was suggested to be one of the ways to approach the problem. It suggested a possible solution to a system that involves more than one binding mode or a dependence on different regions of the ligand molecule.

In another study Senese et al. [126] reported later on the derivation and validation of a potential set of universal descriptors referred to as 4D-fingerprints. One of the elusive goals in the field of cheminformatics and molecular modeling has been to generate descriptors that once calculated for a molecule may be used in a wide variety of applications. The assumption inherent in the generation of such descriptors is that such "universal descriptors" are generated free from external constraints, thus they are independent of the dataset in which they were employed.

The 4D-descriptors generated by Senese et al. [126] were derived from 4D molecular similarity analysis. It was validated on a dataset of THP inhibitors of HIVPI [87, 89]. They developed 4D-fingerprint QSAR models using PLSR (partial least square regression) and GFAR (genetic function approximation regression) techniques. The PLSR and GFAR models showed a correlation of $r^2 = 0.86$ & 0.92 and a quality of xv-$r^2 = 0.71$ & 0.88, respectively. Random scrambling analysis was performed to check the chance correlation for GFAR models. The prediction of the biological activity using PLSR and GFAR was poor but comparable to the other techniques. The 4D-fingerprint based QSAR models were comparable in quality to other 4D-QSAR and CoMFA studies based on statistical measures of fit. The authors suggested that these descriptors are independent of any alignment consideration and will be useful to develop much faster, easier, computationally simple and descriptive QSAR models.

3.2.3
Overview

In QSAR 1–30 the most common method adopted for developing QSAR models seems to be stepwise multiple linear regression (MLR) analysis except for a couple of new approaches tried in the derivation of QSAR 24 and 30. In comparing these QSAR, it is worth mentioning that 18 QSAR show the importance of hydrophobicity. Fourteen QSARs contain a $C \log P$ term, out of which four are positive, five negative and five non-linear (bilinear or parabolic). The non-linear QSARs (2, 3, 8, 10 and 11) identify a range of optimum $C \log P$ (5.69–6.96) for cyclic urea inhibitors. Garg et al. [14] in their seminal work reported a similar optimum $C \log P$ range (4.49 to 6.96) for HIVPI. This range was also confirmed later in another study [15]. It also seems that the optimum for antiviral activity is higher than that of enzyme inhibition (3 (IC_{90}) 6.96 vs. 2 (K_i) 6.53; 11 (IC_{90}) 6.93 vs. 10 (K_i) 6.40). This could be one of the reasons for the presence of a positive or negative $C \log P$ term in QSAR developed for antiviral or enzyme inhibitory activity of the same dataset. It is important to note that to observe the optimum value of a parameter, a sufficient spread in the data is required. In some of the recent papers [81, 92, 127] it has been mentioned that the investigation of the individual dataset shows insufficient spread in the range of $C \log P$ values to establish the optimum point for enzyme inhibition or antiviral activity for each dataset. A positive or a negative $C \log P$ term is observed in a QSAR because most of the data points fall on the positive or negative side of the optimum. These observations need to be investigated in more detail. Some of the QSAR studies showed both hydrophobic and hydrogen bond interactions to be equally significant (Figs. 5 and 7).

A significant steric interaction with the receptor seems to be important as shown by the presence of a volume term, Verloop's sterimol parameters (B1, B5 and L) and CMR in many of the QSARs. CMR is largely a measure of volume with a small correction for polarizability. Its presence in QSAR suggests involvement of polarizability. For a dataset that is difficult to parameterize for the electronic effect, it adds indirectly to the observation. It is not easy to generalize on these factors; however, a negative or positive coefficient indicates that reducing or increasing the size of that particular substituent may improve activity. Out of 30 QSAR, only four (6, 7, 12 and 13) have a negative σ term for X-substituents on the benzene ring of P_2/P_2' benzyl, indicating that electron-releasing substituents at that position would enhance inhibitory activity.

In molecular modeling studies, CoMFA appears to be the most popular approach used for developing 3D-QSAR models. The CoMFA results through the 3D contour plots of the steric and electrostatic fields provide valuable information for the design of a drug. Only in one of the CoMFA studies [108], the HINT method [109] was used to verify whether there was any additional effect from hydrophobicity. It revealed a significant contribution of hydrophobicity towards HIVPR inhibitory activity. It was interesting to note that the structure-activity data of some of the datasets [69, 72, 74–76, 87, 89, 119] were used extensively for these studies. Out of 12 molecular modeling studies reported on cyclic urea derivatives, four [108, 110, 114, 115] were based on Lam et al. [69] and Wilkerson et al. data [75, 76].

Other modeling approaches used for the study of cyclic urea derivatives involved neural network-optimized molecular similarity calculations [114], the substructure molecular fragment method [115] and 4D-QSAR. The neural network-optimized molecular similarity-based QSAR model [114] appears to be fast with minimal loss of accuracy for molecular similarity calculation of larger sets. The substructure molecular fragment (SMF) method applied on cyclic urea derivatives gives an interesting opportunity to build the optimal substituent whose attachment to the molecular core could result in increased activity. This feature of SMF was proposed to be useful in library design. The multiple alignment method proposed in a 4D-QSAR study [125] may provide a possible solution to a system that involves more than one binding mode or a dependence on different regions of the ligand/molecule. 4D-fingerprint-based QSAR models using PLSR and GFAR techniques were developed in another study [126]. They were found to be quiet comparable in quality to other 4D-QSAR and CoMFA studies based on statistical measures of fit. A potential set of universal descriptors referred to as 4D-fingerprints derived from 4D molecular similarity analyses were used in this analysis [126]. These descriptors can prove useful for developing computationally simple and descriptive QSAR models much faster and easier. Generation of a set of descriptors that once calculated for a molecule can be used in a wide variety of applications, will be very useful in HIVPR drug design. The assumption inherent in the generation of such descriptors is that such universal descriptors are generated free from external constraints, thus they are independent of the dataset in which they are investigated.

3.3
Pyranone Analogs

Early research on novel NPPIs resulted in the identification of warfarin (**25**) and 4-hydroxycoumarin (**26**) as weak inhibitors of HIVPR [128, 129]. Screen-

Warfarin
25

PD 099560
26

Phenprocoumon
27

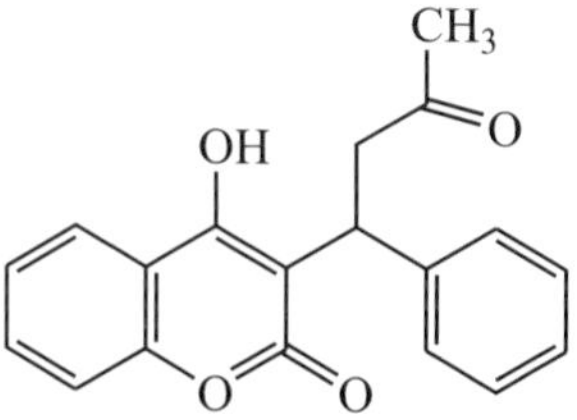

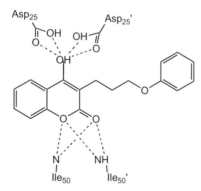

Scheme 5 Early screening hits on pyranone based on non-peptidic protease inhibitors [13]

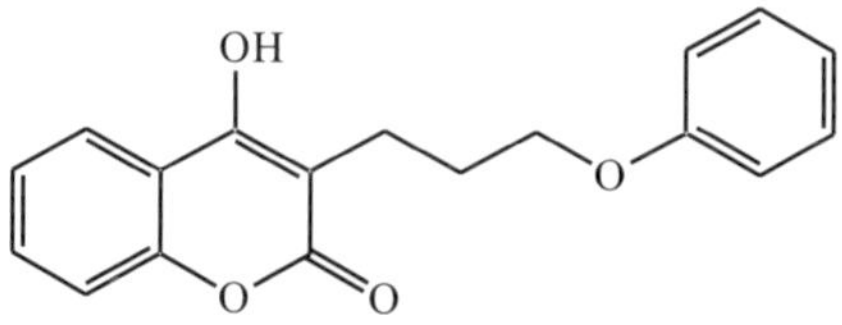

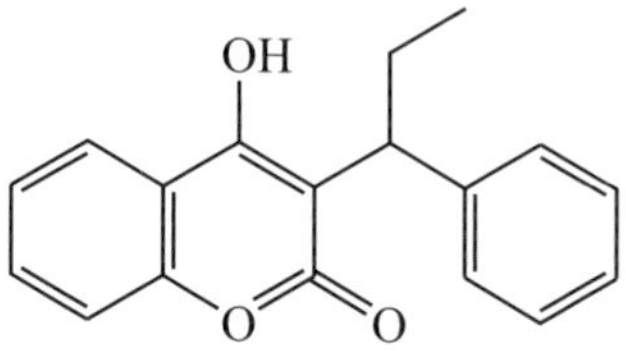

ing of large databases for structurally similar compounds identified phenprocoumon (**27**) as a potent HIVPI (Scheme 5) [13].

The X-ray crystal structure of **27** complexed with HIVPR revealed the absence of the structural water molecule in the complex and key hydrogen bonding interactions between the flap Ile amide proton and the lactone ring oxygen of the inhibitor. The enolic hydroxyl group of the coumarin ring was positioned within hydrogen bonding distance of the two catalytic amino acid residues (Fig. 9) [142]. This 4-hydroxycoumarin template was further

Fig. 9 A graphical representation of the hydrogen bonding interactions observed in the binding of PD099560 (Adapted with permission from [142]. Copyright 1994 American Chemical Society)

optimized using various structures based on drug design techniques. These intense research efforts resulted in the discovery of Tipranavir (**7**), the latest HIVPI approved by US-FDA [46, 47]. Several SAR studies on pyranone-based molecules were subsequently followed by a number of QSAR and molecular modeling papers. A discussion and analysis is presented in this section.

3.3.1
QSAR Studies

3.3.1.1
Cycloalkylpyranones

Various groups [130–135] extensively studied structure-activity data of cycloalkylpyranones. In a series of successive studies Romines et al. [131–133] reported various modifications in the structure (**28**), including the change in the size of the alkyl ring A and the effects of these changes on the inhibition constant (K_i) of HIVPR. The SAR data [133] reporting variation in the cycloalkyl ring along with the other substituent effects was modeled as QSAR 31 [14]. The enzyme inhibitory activity was found to correlate with a positive CMR term indicating that the bulky molecule would be good for increasing the activity.

$$\log 1/K_i = 0.98(\pm 0.13)\text{CMR} + 2.08(\pm 1.41)$$

$$n = 16, \quad r^2 = 0.948, \quad s = 0.52, \quad q^2 = 0.933 . \tag{31}$$

QSAR 32 was derived [14] for another series of sulfonamide-substituted cycloalkylpyranones (**29**) reported by Skulnick et al. [134]. Indicator variable, I, was used with a value of 1 for the cyclooctyl ring and 0 for cycloheptyl. Its positive coefficient indicated that the cyclooctyl ring is preferred over cyclo-

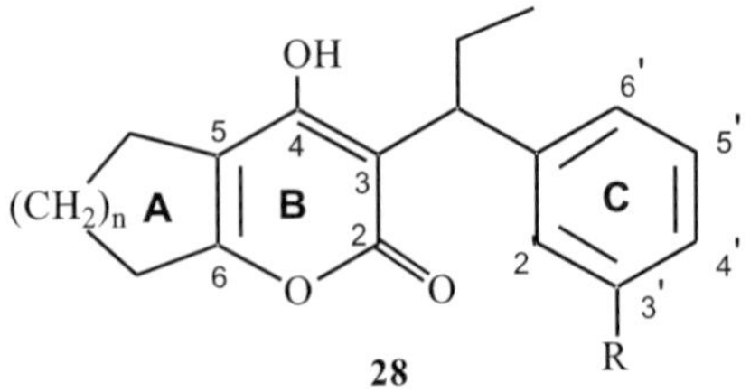

28

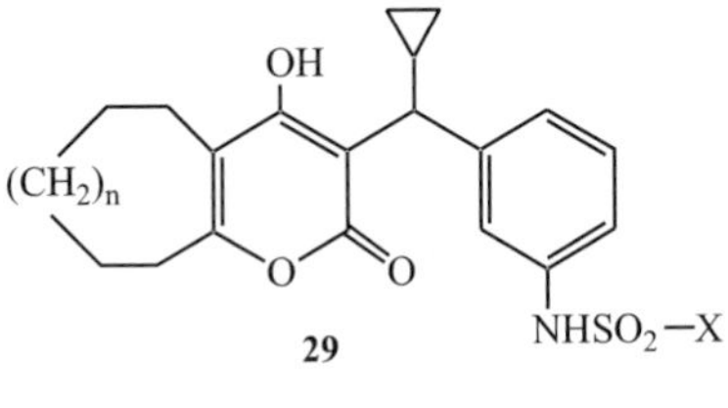

29

heptyl. The sterically bulky X-substituent was found to be useful for increased activity.

$$\log 1/K_i = 0.31(\pm 0.21)MR_X - 0.88(\pm 0.37)I + 7.69(\pm 0.37)$$
$$n = 9, \quad r^2 = 0.935, \quad s = 0.17, \quad q^2 = 0.883. \tag{32}$$

Once the cycloalkyl ring was optimized, Skulnick et al. [134] studied another series of cyclooctylpyranones (30). QSAR 33 and 34 were reported [14] for the enzyme inhibitory activity of *ortho/meta*-substituted and *para*-substituted analogs, respectively.

$$\log 1/K_i = - 0.35(\pm 0.28)B5_2 + 0.44(\pm 0.22)L_3$$
$$- 0.33(\pm 0.26)B5_3 + 8.07(\pm 0.75)$$
$$n = 14, \quad r^2 = 0.853, \quad s = 0.17, \quad q^2 = 0.627 \tag{33}$$

$$\log 1/K_i = - 0.57(\pm 0.28)B1 - 0.23(\pm 0.19)B5 + 0.55(\pm 0.26)\sigma^- + 9.87(\pm 0.60)$$
$$n = 17, \quad r^2 = 0.824, \quad s = 0.18, \quad q^2 = 0.706. \tag{34}$$

Both the QSAR seem to be governed by steric effects of the substituents. A *meta*-substituent of optimum length and small width was found to favor the activity whereas a large *ortho*-substituent appeared detrimental (QSAR 33). Electron-donating *para*-substituent also seems to interact with the steric binding sites of the receptor favorably.

Skulnick et al. [134] also investigated another series of cyclooctylpyranones (31), where the X-substituents were mostly 4-Cl/4-CN/4-F and the Y-substituents were mostly alkyl group. QSAR 35 was derived for that data [14]. Steric and electronic effects seem important as in previous QSAR. A negative coefficient of L and B1 parameters for the Y-substituent indicates that sterically large Y-groups are not suitable for enzyme inhibition activity. The positive σ^- term suggests that a more acidic H on N is significant.

$$\log 1/K_i = 0.41(\pm 0.19)\sigma_X^- - 0.16(0.14)L_Y$$
$$- 2.03(\pm 0.64)B1_Y + 12.31(\pm 1.45)$$
$$n = 14, \quad r^2 = 0.881, \quad s = 0.14, \quad q^2 = 0.780. \tag{35}$$

QSAR 36 was developed [14] for another dataset studied by Romines et al. (32) [132]. In place of sulfonamide groups, some carboxamide groups were also tried.

$$\log 1/K_i = 5.44(\pm 3.81)C \log P - 0.61(\pm 0.37)C \log P^2 - 3.84(\pm 9.58)$$
$$n = 9, \quad r^2 = 0.910, \quad s = 0.24, \quad q^2 = 0.770,$$
$$\log P_0 = 4.49(3.46 - 4.78). \tag{36}$$

The activity of the compounds of this series was found to be totally governed by the hydrophobicity of the molecule. An optimum $C \log P$ of 4.49 was identified for the compounds belonging to this class. It seems that the NH in these compounds is not acidic enough to benefit the activity. Also, note that QSAR 35 shows a positive electronic term for substituents attached to sulfonamide groups indicating that electron-attracting groups would favor activity whereas QSAR 36 derived for amide derivatives contains no electronic term.

QSAR 37 was reported for another series of cycloalkylpyranones where the substituents on the α-carbon and the cycloalkyl ring were studied (33) [132]. Indicator variables I_X and I_Y used with a value of unity for $X = C_2H_5$ and $Y = C_6H_5$ along with the size of the Z-substituent were found to be important

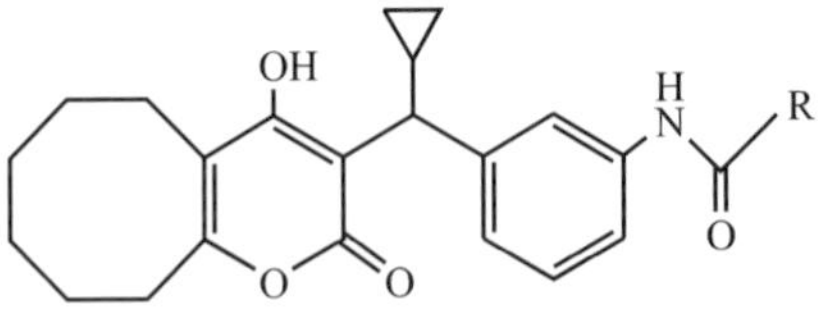

32

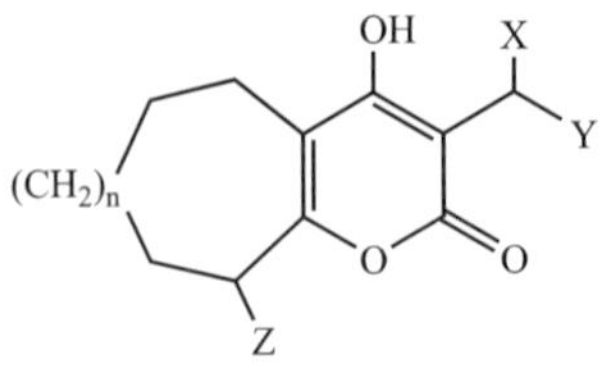

33

for the activity.

$$\log 1/K_i = -0.93(\pm 0.17)I_X + 0.88(\pm 0.20)I_Y - 1.09(\pm 0.33)B1_Z + 8.29(\pm 0.45)$$

$$n = 14, \quad r^2 = 0.954, \quad s = 0.11, \quad q^2 = 0.927 . \tag{37}$$

Gupta et al. [135] also conducted detailed QSAR analysis on cycloalkylpyranone data [131–133] and made similar observations.

3.3.1.2
Pyranones and Dihydropyranones

QSAR 38 was developed [14] for the SAR data on pyranone derivatives (34) reported by Vara Prasad et al. [136]. In 34, the X-groups were mostly phenyl/cycloalkyl and Y was mostly alkyl/cycloalkyl. Sterically bulky X- and Y-substituents were found to be favorable for the activity.

$$\log 1/IC_{50} = 2.42(\pm 1.66)B1_X + 0.37(\pm 0.25)B5_X$$
$$+ 0.58(\pm 0.20)B5_Y - 1.67(\pm 3.56)$$

$$n = 17, \quad r^2 = 0.848, \quad s = 0.32, \quad q^2 = 0.741 . \tag{38}$$

Gupta et al. [137] also studied the same data set [136] and reported QSAR 39 and 40.

$$\log 1/IC_{50} = 2.516(\pm 0.844)\pi_{R2} - 0.813(\pm 0.279)(\pi_{R2})^2$$
$$+ 0.591(\pm 0.309)I_{R2} + 0.586(\pm 0.403)I_{R1} + 4.150$$

$$n = 19, \quad r = 0.940, \quad s = 0.25, \quad F_{4,14} = 26.67(5.03),$$
$$(\pi_{R2})_{opt} = 1.55 \tag{39}$$

$$\log 1/IC_{50} = 6.238(\pm 2.186)V_{w,R2} - 4.404(\pm 1.726)(V_{w,R2})^2$$
$$+ 0.535(\pm 0.330)I_{R2} + 0.644(\pm 0.415)I_{R1} + 3.828$$

$$n = 19, \quad r = 0.937, \quad s = 0.26, \quad F_{4,14} = 25.16(5.03),$$
$$(V_{w,R2})opt = 0.708. \tag{40}$$

Hydrophobic as well as volume terms were found to be equally significant. Gupta et al. [137] noted that there was a high correlation between volume and hydrophobic term ($r = 0.865$). Since HIVPR has four hydrophobic pockets [31] and favorable interactions with these pockets are desirable for an

34

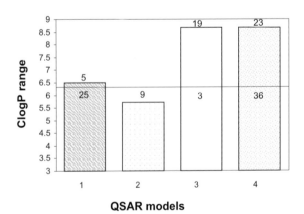

inhibitor to achieve nanomolar potency [138], it was suggested that in this case it is only the hydrophobic interaction of the substituents that are important for activity. Predicted activity of the most active compound in the dataset was the same as calculated experimentally.

QSAR 41 was developed by Bhhatarai and Garg [127] for the combined data taken from several series of 4-hydroxy-5,6-dihydropyranones PI (**35**) reported by Tait et al. [139].

$$\log 1/\mathrm{IC}_{50} = 0.82(\pm0.60)C\log P - 0.07(\pm0.06)C\log P^2$$
$$+ 0.47(\pm0.19)\mathrm{CMR} - 1.28(\pm2.93)$$
$$n = 57, \quad r^2 = 0.722, \quad q^2 = 0.676, \quad s = 0.436, \quad \log P_0 = 6.345. \tag{41}$$

This QSAR showed a parabolic dependence of antiviral activity on hydrophobicity. The activity first increases with increasing hydrophobicity up to an optimum value ($\log P_0 = 6.345$) and then decreases with further increase. The presence of CMR in the QSAR shows the importance of size dependent polarizability of these molecules for achieving good antiviral activity. The authors also analyzed individual datasets to study the effect of hydrophobicity of the molecules and reported three other QSAR [127]. It is of interest to note that out of the three QSAR, the first and third have a positive and negative $C\log P$

Fig. 10 The number of data points above and below the optimum $C\log P$ of 6.345 observed in QSAR 41 (Reprinted with permission from [127]. Copyright 2005 Elsevier)

term, respectively, whereas the second has no hydrophobic term. Only the combined dataset showed a parabolic $C \log P$ term (QSAR 41) bringing out the optimum value of $C \log P$ for the compounds belonging to this class. It was shown that only the combined dataset has sufficient spread in the range of parameter values to reveal the optimum value (Fig. 10) [127]. This study proposed the hypothesis that most of the data points in a dataset often fall either on the positive or negative side of the optimum value. This is why we do not observe an optimum and see either a positive or negative hydrophobic term in the QSAR. To observe the optimum value of a parameter, a sufficient spread in the data is required. More such studies are needed to fully establish this hypothesis.

QSAR 42 was reported [15] on 5,6-dihydropyranones possessing a substituted sulfonyl group (36) that were synthesized and tested against HIVPR by Boyer et al. [140].

$$\log 1/K_i = 0.92(\pm 0.47)C \log P - 1.21(\pm 0.53)C \log P(\beta.10^{C \log P} + 1)$$
$$+ 0.18(\pm 0.09)I_Z + 0.50(\pm 0.10)I_R + 3.37(\pm 2.46)$$
$$n = 23, \quad r^2 = 0.925, \quad q^2 = 0.892, \quad s = 0.095, \quad \log P_0 = 6.01 . \tag{42}$$

A bilinear $C \log P$ term brings out the optimum value of $C \log P$ for compounds of this dataset. Activity first increases in a linear fashion with an increase in the hydrophobicity up to an optimum of 6.01, then as hydrophobicity further increases the activity decreases linearly. Two indicator parameters were used, I_Z equals to 1 when $Z = NH$ and I_R equals to 1 for pyridyl or N-Me-imidazol-4-yl and 0 for R = thiophen-2-yl moieties. The presence of these parameters in QSAR showed the importance of specific substituents for which they were used.

QSAR 43 [15] was reported for another series (37) studied by the same group [140]. In these congeners the main difference was the presence of an

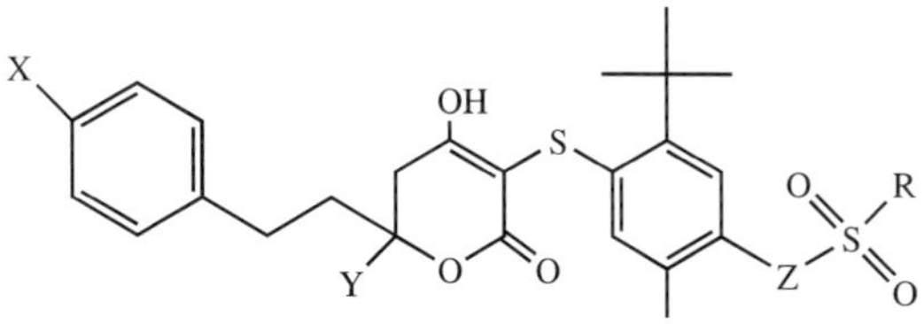

36

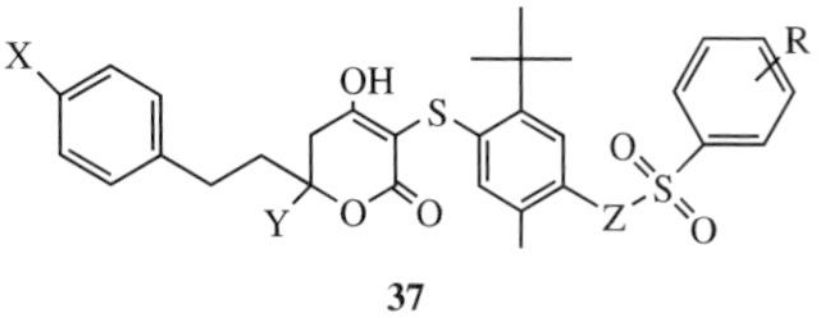

37

38

additional phenyl ring attached to the – SO_2 group.

$$\log 1/K_i = -0.44(\pm 0.10)C \log P - 0.63(\pm 0.61)\sigma_R^+$$
$$+ 0.34(\pm 0.22)L_{R,4} + 10.85(\pm 1.03)$$
$$n = 20, \quad r^2 = 0.851, \quad q^2 = 0.758, \quad s = 0.203 . \tag{43}$$

A very weak electronic effect was seen for the R-substituents. The length of the R-substituents at the *para* position was found to have a positive effect. A negative $C \log P$ term was found to be significant. Kurup et al. [15] noted that most of the compounds in this dataset have $C \log P \geq 6.61$ indicating that these congeners fall on the declining side of the bilinear relation. Note that the optimum $C \log P$ of 6.01 given by QSAR 42 was derived for another dataset of pyranones.

QSAR 44 was also developed [15] for another similar series of dihydropyranones (**38**) studied by Vara Prasad et al. [141].

$$\log 1/IC_{50} = -0.58(\pm 0.12)C \log P + 0.82(\pm 0.48)\sigma_X + 1.28(\pm 0.35)B1_Y$$
$$+ 0.45(\pm 0.10)B5_Y + 8.96(\pm 0.88)$$
$$n = 34, \quad r^2 = 0.839, \quad q^2 = 0.745, \quad s = 0.196 . \tag{44}$$

A negative $C \log P$ term and the range of $C \log P$ of the molecules (5.65–8.77) again suggested that it is desirable to have a compound having $C \log P$ near 6 (optimum value identified in QSAR 42) for good enzyme inhibitory activity. The electron-withdrawing X-substituent and steric interactions of the Y-substituent contribute additionally to the activity.

3.3.2
Molecular Modeling Studies

3.3.2.1
Cycloalkylpyranones

In an endeavor to identify a NPPI using the mass screening technique, Lunney et al. [142] synthesized several cycloalkylpyranones. They reported modeling studies using the Sybyl [107] software in a 4D-TG35 Silicon Graphics computer to understand the binding interaction of these compounds. The protein and H2O301 coordinates used were from the HIV-1/MVT-101 complex as in earlier reports [33]. Ten simulations each containing 50 cycles were carried out at a constant temperature. An evaluation was made of the interaction energy with the affinity grid. If the energy dropped, the move was accepted and if it rose, it was accepted depending upon the Boltzmann distribution. The program GRID [143, 144] was applied to HIV protease structure where both water and methyl probes were used to determine H-bonding and van der Waals interaction sites on the surface of the molecule. This study helped in identifying a potential binding position for a water molecule that could bridge an interaction between the 4-OH substitution of one of the pyranone inhibitors and Gly148 (CO) of HIV protease.

Skulnick et al. [145] synthesized and studied SAR of sulfonamide-based cycloalkylpyranone using a Monte Carlo/molecular mechanics procedure. The models were constructed from the crystal structure of the compound reported by Romines et al. [131]. Modeling based on the crystal structure of one of the compounds of this series bound to HIV-2 protease was conducted. The most important outcome of this study was the suggestion that an additional H-bond to Gly48 and Asp29 of protease is possible if the amide carbonyl group is replaced by an SO_2 group in the *meta* position of the C3-(α-benzyl) group. Also, a larger attachment such as the phenyl group to the SO_2 group was studied and approved.

Thaisirivongs et al. [146] designed a series of cycloalkylpyranones and conducted SAR and modeling studies. They used the GROW [147] program to compare and model the inhibitors under development with the X-ray crystal bound conformer of the reference inhibitor. The finding showed that the presence of the carboxamide functionality at the *meta*-position of benzene attached to the 3-(α-cyclopropyl) linker of the pyranone ring would yield a better inhibitor. Another two compounds without amino acid residues were also designed as an improved inhibitor. These studies were the basis for further exploration of non amino acid containing 4-OH-coumarine and 4-OH-2-pyranone analogs as HIVPI.

3.3.2.2
Pyranones and Dihydropyranones

Mass screening of the compounds from the Parke-Davis collection gave Vara Prasad et al. [148] low molecular weight pyran-2-one inhibitors. They used the Monte Carlo-based docking technique Autodock [149] to generate several low energy structures of pyran-2-one inhibitors (where the 4-hydroxyl group is enolic in nature) bound to HIV protease. The protein structure of the HIV-1/MVT101 crystal complex was used for this study. The selected inhibitor 4-OH-6-phenyl-3-(phenylthio)pyran-2-one and its analog were predicted to have H-bonding interaction with the two Asp residues present in the active site. Several other homologous compounds were synthesized which revealed the mode of binding of these inhibitors and their interactions with the active site.

Ellsworth et al. [150] synthesized and conducted modeling studies using the GA-based docking program GOLD [111]. Five docking runs were carried out resulting in identification of the most potent inhibitor with the highest scores. The most potent compound contained a 3-position benzothiophene ring system. It was proposed to fit the P2 pocket as well as aim for the P3 pocket of HIV protease. The benzothiophene moiety was within 4 Å distance of several residues in the protease cleft and formed hydrophobic contacts with key amino acids. This compound was found to demonstrate significant enzymatic and antiviral activity.

Schake [151] reported the binding pattern of a dihydropyranone Tipranavir, which has been recently approved by US-FDA for HIV treatment. In his report, he summarized the hydrophobic binding contacts of Tipranavir. Also, the I84V mutation caused by the hydrophobic interaction of the 3α-ethyl group with Ile84' was reported. It was mentioned that Tipranavir binds to the active site with fewer H-bonds than the peptidic PIs allowing for increased flexibility to adjust to amino acid changes in the active site [152].

Kulkarni and Kulkarni [153] predicted the binding affinity of HIVPI by developing 3D-QSAR models on the SAR data published by the Upjohn research group [130–132, 145, 146, 154–157]. The lower energy conformers developed by using the Insight-II package were further minimized by the conjugated gradient technique. Docking and molecular dynamics (MD) simulation were used to calculate molecular descriptors and develop ligand–enzyme interaction models. The PLS technique was used to derive a correlation between descriptors and biological activity. The correlation improved when a solvation term was included in the QSAR model ($q^2 = 0.649$). The QSAR was proposed to be useful for predicting binding affinities for a particular or homologous molecular targets.

3.3.3
Analysis

An analysis of the QSAR reported for pyranone-based HIVPI highlight the following important points. More studies were conducted on enzyme inhibitory activity than on antiviral. Out of 14 QSAR, six have the hydrophobic term (two negative and four non-linear). Similar to cyclic urea derivatives the range of optimum $C \log P$ observed in non-linear QSAR was 4.49–6.34 (see QSAR 36, 39, 41 & 42). This range is almost the same as reported in references 14 & 15 for HIVPI. Once again, the optimum for antiviral activity was found to be higher (QSAR 41, 6.34) in comparison to enzyme inhibitory activity (QSAR 42, 6.01). There seems to be a noticeable difference in the optimum value of $C \log P$ for enzyme inhibition by cycloalkylpyranone (QSAR 36, 4.49) versus dihydropyranone (QSAR 42, 6.01) inhibitors. Only one study (QSAR 39) noted the size of the hydrophobic binding pocket for a particular substituent. More such comparative studies will help in identifying the size and nature of binding pockets for different substituents of the ligand. Significant steric terms appeared in several QSAR models that indicate steric interactions of the substituent and/or group with the receptor. The Hammett σ constant describing the electronic effects of the substituent on the aromatic ring were found to be significant in few QSAR.

Several molecular modeling studies have been performed for the design and development of better synthetic analogs and to identify the full potential of different pyranone-based inhibitors of HIV protease. A collaborative approach involving chemical synthesis, X-ray crystallography and different molecular modeling techniques was adopted in most of these reports. Only one 3D-QSAR-based modeling study was conducted on pyranones [153]. Modeling studies by Lunney et al. [142] identified a potential binding position for a water molecule that could bridge an interaction between the 4-OH substitution of the pyranone inhibitor and Gly148 (CO) of HIV protease. Monte Carlo/molecular mechanics studies on sulfonamide-based cycloalkylpyranone suggested that an additional H-bond to Gly48 and Asp29 of protease is possible if the amide carbonyl group is replaced by the SO_2 group in the meta position of the C3α-benzyl group [145]. Modeling studies on cycloalkylpyranone using GROW showed that the presence of the carboxamide functionality at the *meta*-position of benzene attached to the 3-(α-cyclopropyl) linker of the pyranone-ring will give better inhibitors [146]. The use of the Monte Carlo-based docking technique Autodock to study pyranone derivatives predicted H-bonding interaction with the two Asp-residues present in the active site [148]. Studies using the GA-based docking program GOLD resulted in identification of the most potent compound carrying a 3-position benzothiophene ring system, which fits the P2 pocket as well as aims for the P3 pocket of HIV protease [150]. Schake [151] reported the hydrophobic binding contacts of Tipranavir and the I84V mutation caused by the hydrophobic interaction of the 3-(α-ethyl) group with Ile84′. This study also mentioned that Tipranavir (7) binds to the active site with fewer H-bonds than the peptidic PIs, which allows for increased flexibility to adjust to amino acid changes in the active site [152]. Kulkarni and Kulkarni [153] predicted the binding affinity of HIVPI by developing 3D-QSAR models that predicted binding affinities for a particular molecular target as well as homologous molecular targets. All of these studies helped in identifying key structural features of the pyranone derivatives that ultimately led to the development of Tipranavir (7), a recently approved non-peptidic dihydropyranone protease inhibitor.

4
Peptidic Protease Inhibitors (PPIs)

4.1
Definition

All the US-FDA approved PIs except Tipranavir (7) are substrate-based peptidic inhibitors. These inhibitors were designed using the "transition state peptidomimetic" principle, which means that in the inhibitors the hydrolyzable peptide linkage is replaced by a non-hydrolyzable transition-state isostere (Fig. 11a) [158]. A number of such isosteres were studied including statin, norstatin, hydroxyethylene, reduced amide, hydroxyethylamine, hydroxyethyl urea, monoalcohol, diol and aminodiols (Fig. 11b) [158]. Various classes of inhibitors containing dihydroxyethylene transition state isosteres were extensively developed. As an alternative to the peptide-based approach penicillin-derived C2 symmetric compounds were also pursued.

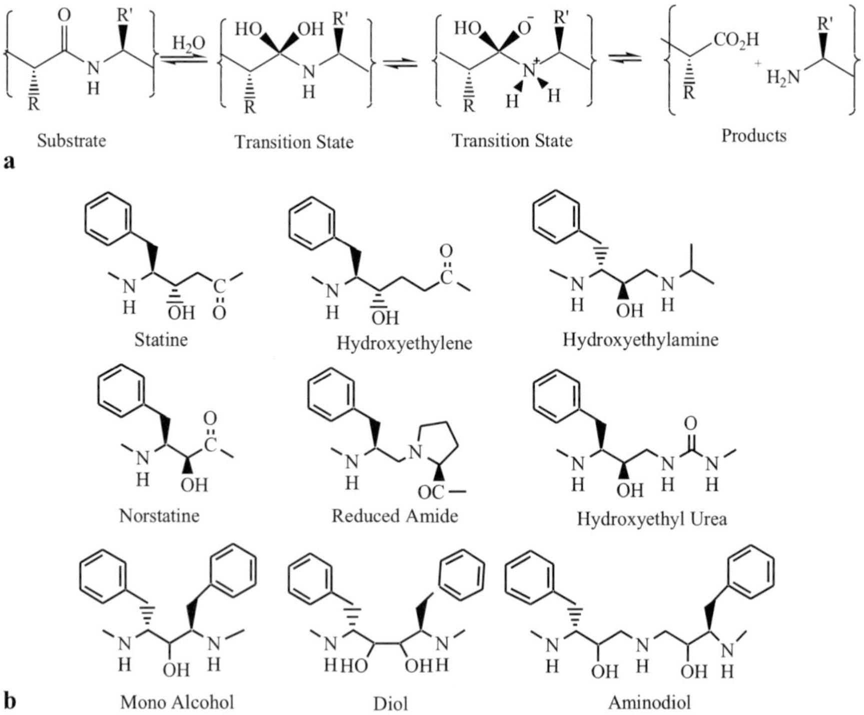

Fig. 11 a Hydrolysis of peptide linkage (Reprinted with permission from [158]. Copyright 1995 American Chemical Society). **b** Some non-hydrolyzable transition state isosteres employed to replace the P_1-P_1' amide bond of the substrate for the design of HIVPI (Reprinted with permission from [158]. Copyright 1995 American Chemical Society)

Subsequently, a number of QSAR and molecular modeling studies have been published on these inhibitors, which provide useful insight into the design and development of anti-HIV protease inhibitors.

4.2
QSAR Studies

4.2.1
Transition State Isostere

QSAR 45-47 were developed [14] for the biological activities of a series of 2-aminobenzylstatin as a novel scissile bond replacement (39) reported by Billich et al. [159]. Several structural modifications including the introduction of a benzimidazole heterocycle into the inhibitor were studied. QSAR 45 was derived for enzyme inhibitory data. Antiviral activity measured as the concentration required to reduce the P24 antigen level in the supernatant of infected cell cultures by 50% HIV was modeled as QSAR 46. The antiviral activity measured as the concentration required to reduce the virus-induced cytopathic effect by 50% in MT-4 cells was modeled in QSAR 47.

$$\log 1/K_i = -0.65(\pm 0.34)\sigma_Y + 0.34(\pm 0.16)I_Z + 7.93(\pm 0.12)$$
$$n = 9, \quad r^2 = 0.899, \quad s = 0.10, \quad q^2 = 0.799 \tag{45}$$

$$\log 1/IC_{50} = -0.35(\pm 0.40)\sigma_Y^+ + 1.02(\pm 0.31)I_Z + 6.80(\pm 0.23)$$
$$n = 11, \quad r^2 = 0.891, \quad s = 0.22, \quad q^2 = 0.791 \tag{46}$$

$$\log 1/IC_{50} = -0.77(\pm 0.55)\sigma_Y^+ + 0.69(\pm 0.39)I_Z + 6.45(\pm 0.28)$$
$$n = 10, \quad r^2 = 0.866, \quad s = 0.24, \quad q^2 = 0.727 . \tag{47}$$

Although a σ_Y^+ term of marginal value was observed in all the three models, it was suggested that an electron-donating 4-substituent (Y-substituent) at the aryl ring may be beneficial for improving antiviral and the protease inhibition activities. Indicator parameter, I_Z, was used with a value of 1 and 0 for Z = 2-benzimidazolyl and the phenyl group, respectively. The positive value for I_Z in all the models suggested that the benzimidazolyl group tried at the

39

terminal end of the chain would be better for activity. This superior effect of benzimidazolyl can be attributed to its nitrogen that can participate in the hydrogen bonding with the receptor. The X-substituents, being either H or CH_3, were not found to make any difference.

QSAR 48 was derived [14] for the antiviral data of hydroxyethylene isostere derivatives (40) for inhibition of HIV protease in H9 human T-lymphocytes cells, reported by De Solms et al. [160]. One of the prototype pentapeptides (L-682,679) was modified at the carboxy terminus and mostly variations of the P2′ amino acid and the elimination/replacement of the P3′ amino acid were studied [160]. This model showed that overall bulky molecules would favor the activity. However, the X-substituents would have a negative steric effect.

$$\log 1/IC_{50} = 2.56(\pm1.23)B1_X - 0.32(\pm0.29)B5_X$$
$$+ 1.36(\pm0.64)MgVol - 1.00(\pm2.81)$$
$$n = 31, \quad r^2 = 0.707, \quad s = 0.67, \quad q^2 = 0.623 . \tag{48}$$

Studies on IC_{50} data of hydroxyethylene-based isostere derivatives (41) reported by Thompson et al. [161] gave QSAR 49 [14]. Most of the compounds in this dataset were hydroxyethylene-based HIVPI containing heterocyclic P1′-P2′ amide bond isostere derivatives. X-substituents were either keto or hydroxyl group whereas Y were mainly alkyl.

$$\log 1/C_{50} = 1.26(\pm0.70)I + 0.49(\pm0.34)B5_Y + 3.94(\pm1.08)$$
$$n = 8, \quad r^2 = 0.885, \quad s = 0.33, \quad q^2 = 0.683 . \tag{49}$$

Indicator variable I was used with a value of 1 and 0 for X = keto and hydroxy group, respectively. Its large positive coefficient shows that the keto group is more favorable in comparison to hydroxyl. The positive B5 parameter for the

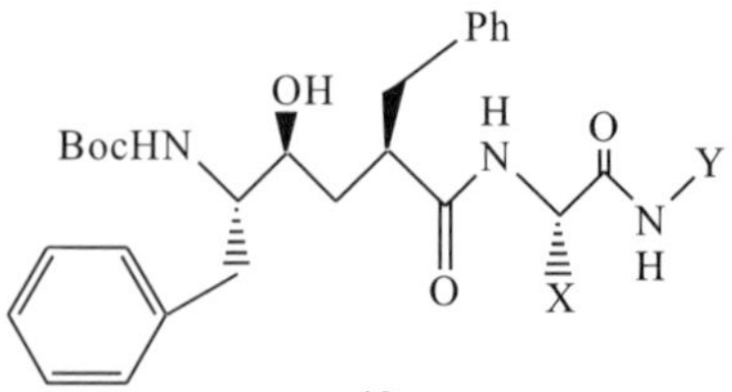

40

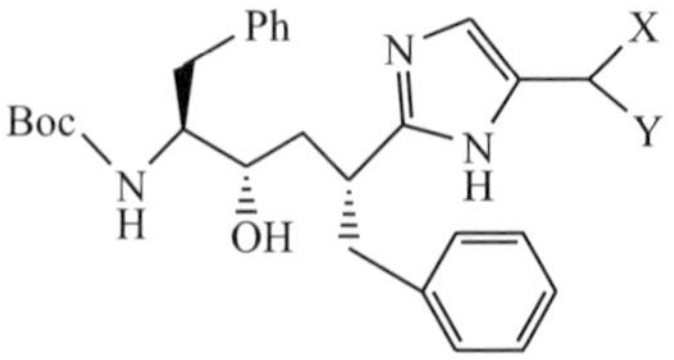

41

Y-substituent indicated steric interactions of these substituents with the receptor. X-ray crystallographic study of one of the heterocyclic isosteres (SB 206343) of this series revealed that this inhibitor participates in all polar and hydrophobic interactions that are commonly observed between enzyme and peptidic inhibitors [161].

QSAR 50 was reported [14] for a series of compounds containing R-hydroxyethyl urea isostere (42) studied by Getman et al. [162]. Replacement of the P1 chiral α-carbon center of the hydroxyethylene isostere with trigonal nitrogen gives a urea isostere (Fig. 11b) [158].

$$\log 1/C_{50} = 2.27(\pm 0.50)I_X - 0.75(\pm 0.34)I_R$$
$$+ 1.26(\pm 0.33)B1_Z + 2.70(\pm 1.09)$$
$$n = 18, \quad r^2 = 0.941, \quad s = 0.29, \quad q^2 = 0.912 . \tag{50}$$

In this model, I_X and $I_R = 1$ stands for X = Me and R = Cbz (carbobenzyloxy), respectively. It was shown that the width of the Z-substituent interacting with the S_2 site of the receptor is beneficial. A CHMe group as a Y-substituent interacting with the S_1 site has a negative effect relative to the CH_2 group. The carbobenzyloxy (Cbz) group, as an R-substituent, interacting with the S_3 site, appeared detrimental to the activity. It was noted by the authors that addition of a marginally significant hydrophobic term improves the correlation slightly ($r^2 = 0.96$). One of the representative compounds of this series was observed to interact with the hydrophobic regions of the receptor as shown in Fig. 12 [162].

One of the successful examples of employment of QSAR in the design and development of an HIV drug is the discovery of Indinavir (L-735,524) (2), one of the first HIVPI approved by the US-FDA. Holloway et al. [163] first reported this compound when they conducted SAR studies on a combined series of isostere derivatives of (43, 44). A high correlation between the intermolecular interaction energy (E_{int}) calculated for HIVPR inhibitor complexes and enzyme inhibition activity was observed. QSAR 51-53 were developed for native, acetylpepstatin-inhibited and L-689,502-inhibited HIVPR, respectively [163]. X-ray coordinates and the force field technique were employed in the calculation of E_{int} (intermolecular interaction energy). In these QSAR, r_{cv} is the cross-validated correlation coefficient.

42

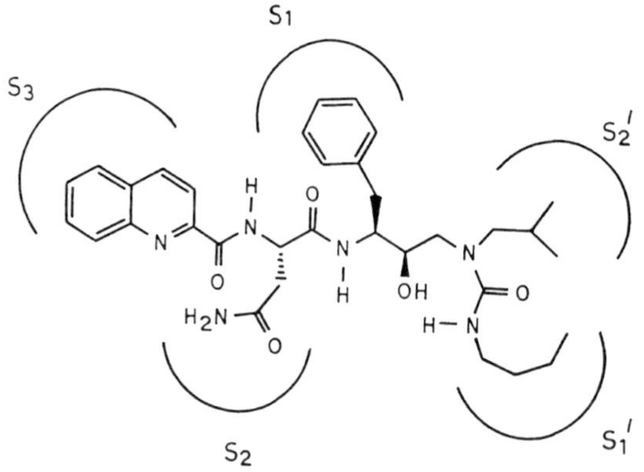

Fig. 12 Schematic representation of binding of a hydroxyethyl urea isostere with hydrophobic regions of HIV protease (Reprinted with permission from [162]. Copyright 1993 American Chemical Society)

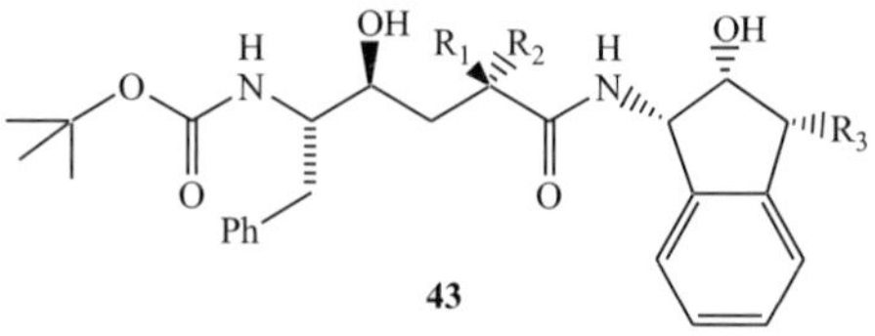

43

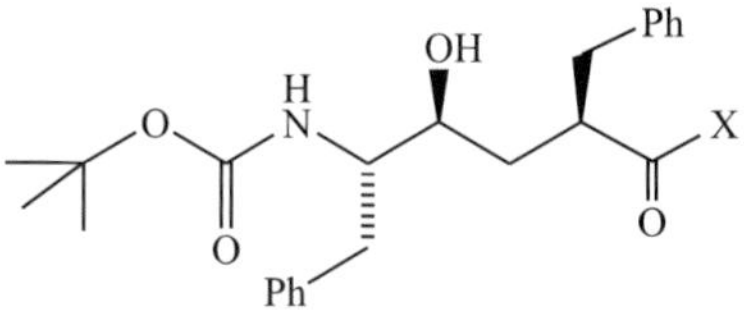

44

Native

$$\log 1/C_{50} = -0.15435\, E_{int} - 8.069$$

$$n = 33, \quad r^2 = 0.728, \quad r_{cv} = 0.831 \tag{51}$$

Acetylpepstatin-inhibited

$$\log 1/C_{50} = -0.17302\, E_{int} - 14.90$$

$$n = 33, \quad r^2 = 0.581, \quad r_{cv} = 0.724 \tag{52}$$

L-689, 502-inhibited

$$\log 1/C_{50} = -0.16946\, E_{int} - 15.707$$

$$n = 33, \quad r^2 = 0.783, \quad r_{cv} = 0.869 . \tag{53}$$

QSAR 54 was also developed [14] for the same series of compounds [163].

$$\log 1/C_{50} = 1.46(\pm 0.53)CMR - 1.56(\pm 1.99)\log(\beta.10^{CMR} + 1)$$
$$+ 2.38(\pm 0.64)I + 15.64(\pm 7.98)$$
$$n = 30, \quad r^2 = 0.816, \quad s = 0.69, \quad q^2 = 0.759, \quad (CMR)_0 = 16.87,$$
$$\log \beta = -15.72 .$$

(54)

This QSAR suggested that these molecules could be involved in dispersion interactions with the receptor, but very bulky molecules may be unfavorable. In this correlation, indicator variable $I = 1$ was used for an X moiety which has an OH group *cis* to its NH group as shown in **43**. A high positive coefficient with this variable suggested that the presence of the OH group *cis* to NH would increase the inhibition potency, possibly due to being in proper orientation to form a hydrogen bond with the receptor. The intermolecular interaction energy (E_{int}) used for correlating enzyme inhibitory activity in QSAR 51–53 corresponds to the sum of a volume and polar energy term ($E_{int} = E_{vdw} + E_{elec}$) [163]. E_{vdw} and E_{elec} stand for van der Waals and electrostatic interactions between the inhibitor and the enzyme when the inhibitor is minimized in the rigid enzyme active site [163]. The presence of CMR in a QSAR has also been found to hint towards involvement of size-dependent polarizability of the molecule in receptor-ligand interaction [164, 165]. It is of interest to note that QSAR 54 reconfirmed the observations of QSAR 51–53. In addition, this model also suggested an optimum size (16.87) of the molecules belonging to this series for achieving good inhibition.

In view of all these studies it is interesting to note that the determination of the relative free energies for the binding of peptide inhibitors of the type (**45**) with HIV protease led Ferguson et al. [166] to suggest that ethylamine hydroxyl group produces marked stabilization of the enzyme–inhibitor complex due to hydrogen bonding with aspartyl residues. This interaction was shown to induce a conformational change in the R diastereomer resulting in a decrease in binding affinity. The addition of a second hydroxyl group to the inhibitor might help avoid the conformational requirements for binding that depend on the configuration of the inhibitor. It had been shown that the binding of symmetric glycol-containing inhibitors is not dependent on the configuration of the two carbon centers with which the hydroxyl groups

R = H/OH, R' = H/OH

45

are attached. This may be due to flexibility of the inhibitors at the hydroxyl-carbon centers or a result of the availability of multiple binding modes for the diastereomers, afforded by the presence of a second hydroxyl group in the active site [167]. However, one must always question the hydrogen bonding effect since the free energy change in the OH binding to water and to the hydrogen bonding component in the receptor would probably be small. Multiple H-bonding would, of course, be more significant.

Maw and Hall [168] using E-state modeling developed another QSAR on the dataset reported by Holloway et al. [163]. In the E-state approach of structure representation information is developed for each atom and each hydride group in the molecule. The model was based upon topological descriptors, which do not distinguish among chiral compounds. The dataset was split between a training and test set. Statistical analysis was done using the SAS system [169]. Eight variables were selected as important based on the RSQUARE selection method [168]. The most prominent variables were atom type and atom level E-state indices [170–172] that emphasize H-bond donating ability, non-polar structure features, hydrogen accessibility at a few atomic sites and skeletal structures. The best four variable QSAR model ($r^2 = 0.86$, $q^2 = 0.79$) included $HS^T(HBd)$ sum of hydrogen E-state values for hydrogen bond donors, $HS^T(other)$ sum of hydrogen E-state values for non-polar CH bonds, $^1\chi^\nu$ first order molecular connectivity index and $^2\kappa\alpha$ second order shape index.

k-Nearest neighbors (kNN)-based 2D-QSAR COMBINE model was also developed by Golbraikh and Tropsha [173] to predict activities of the compounds reported by Holloway et al. [163] and Perez et al. [174]. A combination of chirality-based descriptors with conventional (chirality insensitive) topological descriptors were used and it was proposed that these new descriptors will help to circumvent the problem of stereoisomers, which limit the application of the conventional QSAR. Two models of highest predictive ability were obtained. Model 1 was generated using non-chiral and Id chirality descriptors, chirality correction = 2.5 with $q^2 = 0.77$, $R^2 = 0.85$, $R_0^2 = 0.79$, $k = 0.95$, $F = 81.1$ and $\alpha = 1 - 3.3633 \times 10^{-7}$. Model 2 contained non-chiral and all chirality descriptors, chirality correction = 0.5 with $q^2 = 0.79$, $R^2 = 0.85$, $R_0^2 = 0.77$, $k = 0.92$, $F = 76.4$ and $\alpha = 1 - 4.8142 \times 10^{-7}$. A model using Molconn-Z descriptors was found statistically less significant than other models ($q^2 = 0.79$, $R^2 = 0.7188$, $R_0^2 = 0.77$, $k = 0.96$, $F = 47.4$, and $\alpha = 1 - 7.5045 \times 10^{-6}$).

IC_{50} and K_i data of the hydroxyethyl sulfonamide isostere (**46**) measured for the inhibition of recombinant HIV protease in CEM cells by Vazquez et al. [175] was modeled as QSAR 55 and 56 [14]. A sulfonamido moiety in place of the P_1/P_1 amide linkage was studied in this series for the first time. These models showed that the width of the X-substituent is important for activity. The authors [14] mentioned that no effect of Y-substituent was ob-

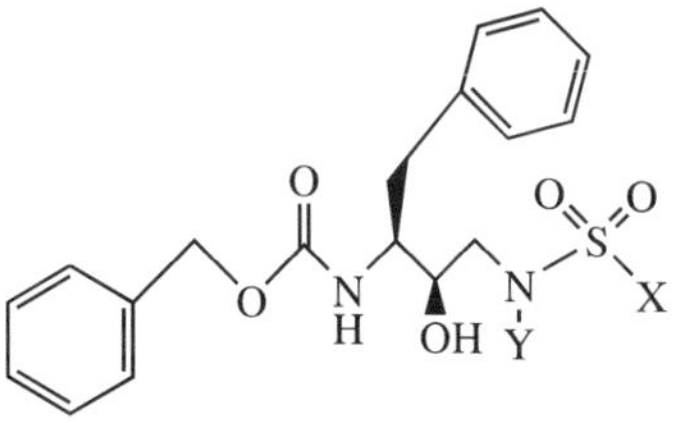

46

served on activity due to insufficient variation.

$$\log 1/IC_{50} = 11.12(\pm 5.82)B1_X - 11.55(\pm 9.72)$$
$$n = 7, \quad r^2 = 0.829, \quad s = 0.59, \quad q^2 = 0.763 \tag{55}$$

$$\log 1/K_i = 0.62(\pm 0.30)L_X + 4.06(\pm 1.75)$$
$$n = 6, \quad r^2 = 0.894, \quad s = 0.40, \quad q^2 = 0.819 . \tag{56}$$

New analogs of Palinavir (**47**) were synthesized by incorporating $2',6'$-dimethylphenoxyacetyl as the P3-P2 ligand by Beaulieu et al. [176]. Palinavir is a hydroxyyethylamine transition state isostere with a 4-hydroxypipecolic acid fragment that spans the S1-S3' pockets of the protease. QSAR 57 was developed [15] for the IC_{50} data of recombinant HIV protease [176]. Here the negative coefficient of CMR indicates that the activity decreases with an increase in size of the molecules. However, positive $L_{Y,O}$ suggests that the length of the *ortho*-Y-substituents improve the activity.

$$\log 1/C_{50} = -2.42(\pm 0.80)CMR + 1.39(\pm 0.51)L_{Y,O} + 41.62(\pm 12.42)$$
$$n = 11, \quad r^2 = 0.872, \quad q^2 = 0.741, \quad s = 0.332 . \tag{57}$$

Beaulieu and his group [176] also studied a second series of Palanavir analogs (**48**) for further improving the activity. QSAR 58 was reported for this data [15]. Although a very small dataset, a highly significant correlation with $C \log P$ indicated that the hydrophobicity of the molecule is conducive to

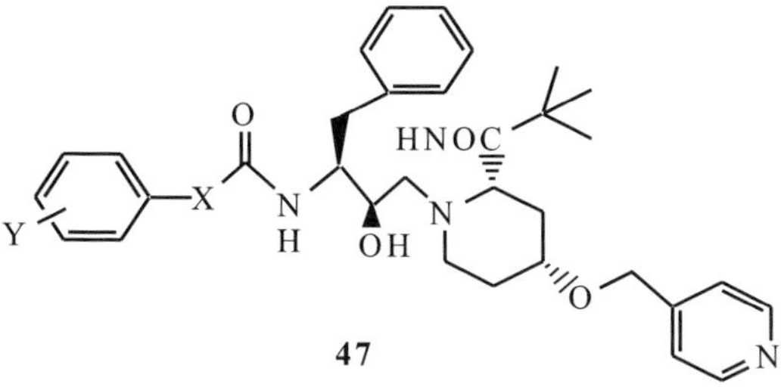

47

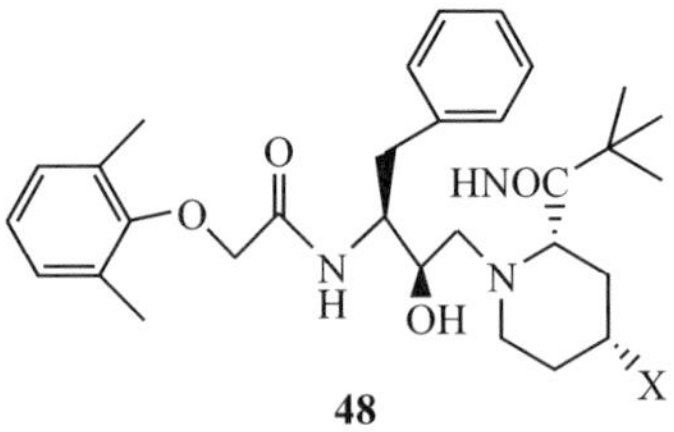

48

activity.

$$\log 1/C_{50} = 0.61(\pm0.18)C\log P + 5.56(\pm0.91)$$
$$n = 4, \quad r^2 = 0.991, \quad q^2 = 0.668, \quad s = 0.037 \,. \tag{58}$$

Takashiro et al. [177–179] reported structure-activity relationship studies of HIVPI containing AHPBA (3-amino-2-hydroxy-4-phenylbutanoic acid). Various modifications at different sites were studied. Detailed QSAR analysis of the data was conducted by Kurup et al. [15].

QSAR 59 [15] was developed for the derivatives of (**49**) [177]. In this series, small-sized dipeptide HIVPI containing cyclic urethane at the P2 site were designed to improve the oral bioavailability.

$$\log 1/IC_{50} = -2.70(\pm0.79)\text{MgVol} + 18.32(\pm3.04)$$
$$n = 10, \quad r^2 = 0.887, \quad q^2 = 0.840, \quad s = 0.131 \,. \tag{59}$$

This QSAR indicated that very bulky molecules do not bind well to the active site. It was noted that there exist a high mutual correlation between CMR and MgVol ($r^2 = 0.943$) [15]. However, the QSAR with the MgVol term was more significant.

In the same paper, Takashiro et al. [177] also reported inhibitory data for AHPBA derivatives bearing substituted benzamides at the P2 site (**50**). QSAR 60 was derived for the data [15].

$$\log 1/IC_{50} = 0.40(\pm0.25)\sigma^+ + 3.99(\pm1.07)\text{MgVol} - 7.27(\pm4.14)$$
$$n = 14, \quad r^2 = 0.869, \quad q^2 = 0.815, \quad s = 0.195 \,. \tag{60}$$

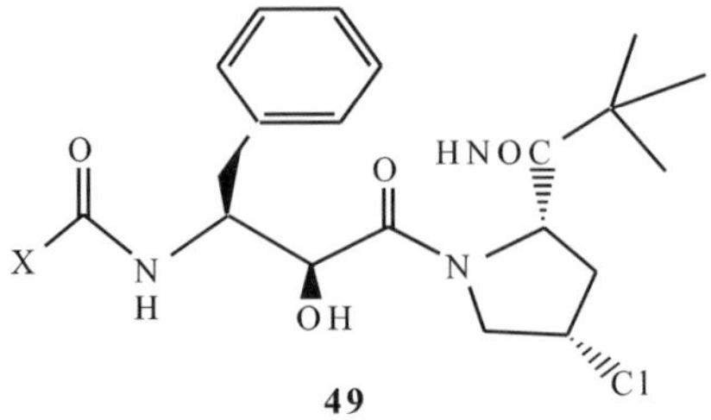

49

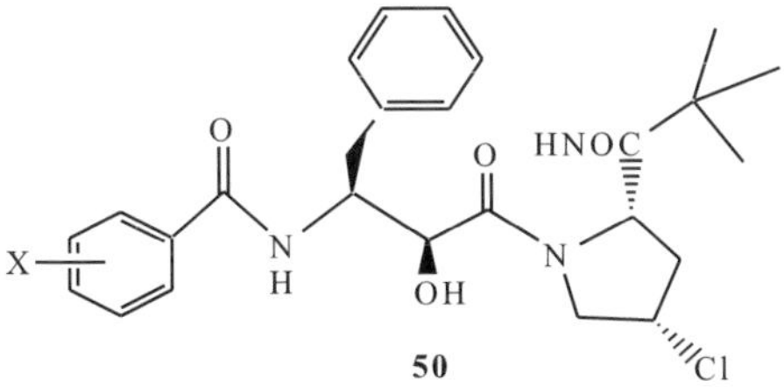

50

The electron-attracting X-substituents were found to be conducive to the activity. Also, unlike urethane analogs the MgVol has a positive effect on the activity of benzamide derivatives.

Takashiro et al. [178] later on studied the substitution on the P1 aromatic rings of the 2-Me-3-OH benzamide bearing AHPBA derivatives (51). QSAR 61 was derived for their data [15].

$$\log 1/IC_{50} = 0.26(\pm 0.18)\sigma_Y - 1.17(\pm 0.36)B1_{Y,2} + 10.24(\pm 0.39)$$

$$n = 15, \quad r^2 = 0.850, \quad q^2 = 0.700, \quad s = 0.097. \tag{61}$$

In QSAR 61, the presence of the positive σ_Y implied that electron-attracting groups attached to phenyl ring enhance the activity. However, an *ortho* Y-substituent decreases the activity in terms of B1-sterimol, which primarily measures the width of the first atom of the substituent. In this series all the molecules except one had X = CH$_2$. The only derivative with X = $-$ (CH$_2$)$_2$ $-$ was found to be an outlier which suggested that a long chain by which the substituted phenyl is attached is detrimental to the activity.

Takashiro et al. [178] also tested another series of AHPBA analog (52) for their inhibitory activity against HIV protease for which QSAR 62 gave the best correlation [15].

$$\log 1/IC_{50} = 0.80(\pm 0.27)B1_{Y,5} - 0.23(\pm 0.14)I_{Y,4}$$
$$- 0.94(\pm 0.18)I_{X,2} + 7.65(\pm 0.35)$$

$$n = 18, \quad r^2 = 0.955, \quad q^2 = 0.925, \quad s = 0.109. \tag{62}$$

A positive B1$_{Y,5}$ term in this QSAR showed that meta Y-substituents would have a positive steric interaction with the active site. Two indicator variables,

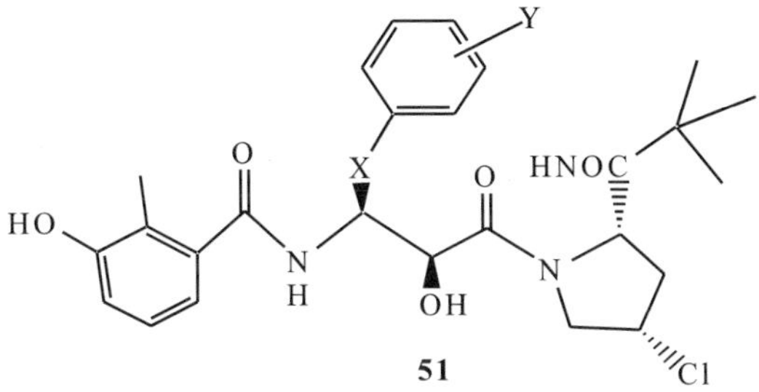

51

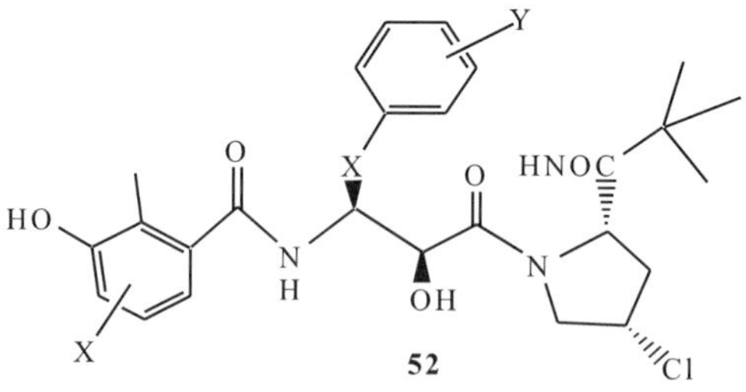

52

$I_{Y,4}$ and $I_{X,2}$ were used with a value of 1 for Y = 4-Me and X = 2-Me, respectively. The negative coefficient of both these parameters indicated that the presence of methyl at the 4-position as the Y-substituent and at the 2-position as X-substituents would be detrimental to the activity.

In another study Takashiro et al. [179] further studied the antiviral activity of AHPBA inhibitors (**53**). To reduce the peptide-like characteristics of the inhibitors, AHPBA and proline components were connected without using peptide bonds. QSAR 63 developed for the data indicated that electron-releasing Z-groups would enhance the activity [15]. A strong correlation with σ^+ was observed even though it was a very small dataset.

$$\log 1/IC_{50} = -0.43(\pm 0.20)\sigma_Z^+ + 5.70(\pm 0.12)$$
$$n = 5, \quad r^2 = 0.940, \quad q^2 = 0.860, \quad s = 0.038. \tag{63}$$

One more series comprised of AHPBA derivatives (**54**) was also tested by Takashiro et al. [179]. QSAR 64 derived for the data brought out the role of

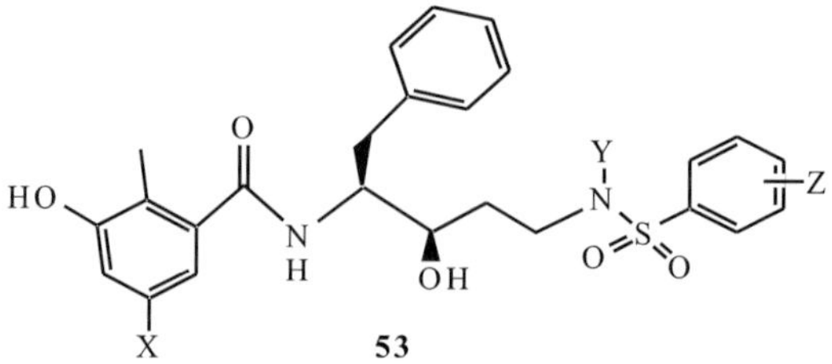

53

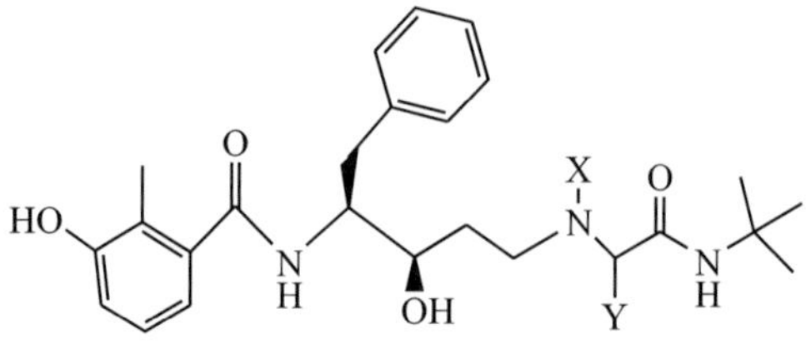

54

hydrophobicity and polarity of the Y- and X-substituents [15].

$$\log 1/IC_{50} = -0.65(\pm 0.54)C_{\pi Y} + 1.21(\pm 0.52)\sigma_X^* + 5.29(\pm 1.45)$$

$$n = 9, \quad r^2 = 0.844, \quad q^2 = 0.694, \quad s = 0.357 . \tag{64}$$

$C_{\pi,Y}$ is the calculated π for the Y substituent and measures its hydrophobicity. Its negative coefficient suggested that the hydrophobic Y-substituents would have a negative influence on activity. However, electron-attracting X-substituents seem to have a positive contribution to the activity as implied by the positive σ_X^*.

Ghosh et al. [180] developed cyclic sulfolanes (**55**) as novel and high affinity P2 ligands for HIV protease inhibition. Gupta et al. [137] conducted detailed QSAR studies on this data and gave QSAR 65.

$$\log 1/IC_{50} = 1.524(\pm 0.489)\pi_2 - 0.575(\pm 0.240)\pi_2^2 + 0.479(\pm 0.292)I_m$$
$$+ 0.426(\pm 0.256)I_5 - 0.893(\pm 0.420)I_6$$
$$+ 0.287(\pm 0.190)I_s + 6.470$$

$$n = 32, \quad r = 0.951, \quad s = 0.23, \quad F_{6,25} = 39.16(3.63), \quad \pi_{opt} = 1.32 . \tag{65}$$

The inhibitory activity was found to have a good parabolic correlation with the hydrophobic constant π of the 2-substituent in the X-group of the compounds. Several indicator variables were used to account for the specific effects of substituents in QSAR 65. Variable, I_m, was given a value of 0 for a five-member A-ring and 1 for a six-member A-ring. A second variable, I_5, was used for a five member sulfolane-ring in the X-substituent with a value of 1 and zero for others. A third variable, I_6, equals to 1 for a six-member ring in the X-substituent and zero for others. A fourth variable $I_s = 1$ was used for describing the effect of a 3(S)-configuration of the ring in X relative to a 3(R)-configuration. From this QSAR model, it is evident that a five member A-ring and a sulfolane-ring in the 3(S)-configuration with a lipophilic 2-substituent (*cis* to the 3-substituent) will be highly beneficial. It was suggested that compounds possessing the 2-substituent with a $\pi = 1.32$ and having all other positive factors can be the most potent compound of this series with a predicted IC_{50} value of 8.67 [137]. On the basis of these studies Gupta et al. [137] proposed a hypothetical model for the binding of a sulfolane with HIVPR

55

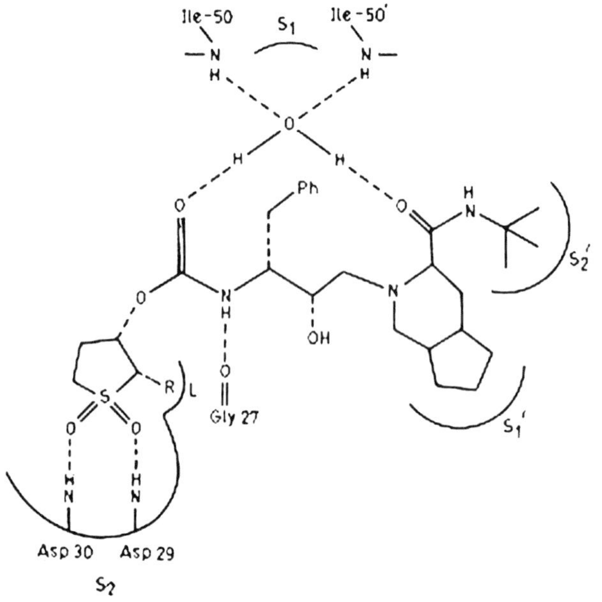

Fig. 13 A model proposed for the binding of a sulfolane with HIV protease. S_1', S_2' are the hydrophobic sites in the enzyme. S_2 site is shown to participate in the hydrogen bondings but also to contain a small hydrophobic cavity L, able to accommodate the hydrophobic 2-substituent (Reprinted with permission from [137]. Copyright 1998 Elsevier)

(Fig. 13). Hydrophobic and hydrogen bonding interactions were found to be important for inhibition of protease.

In search of novel HIVPI, Di Santo et al. [181] reported the design, synthesis and QSAR study of heteroarylisopropanolamines, a new class of HIVPI. It is one of the very few studies where QSAR was part of the design and synthetic efforts. The isopropanolamine unit of the US-FDA approved peptidic PIs was modified by using heteroaryl rings in a very systematic manner (Fig. 14) [181]. Substituted indole (**56**), 3-arylpyrrole (**57**) and a number of substituents on these indole, aryl pyrrole and aryl groups were tested along with other typical moieties shown to improve anti-PR activity in known drugs. QSAR 66 was developed for the PR inhibitory activity [181].

$$pIC_{50} = -4.284 - (0.659)^* Sfit + (0.010)^* Dip\text{-}mom - (0.340)^* HOMO$$
$$+ (0.008)^* MW + (0.008)^* volume - (0.053)^* G\_CDS\_aq. \qquad (66)$$

In this model, S fit was used to calculate steric fit between ligand and receptor, Dip-mom was for dipole moment, HOMO was for HOMO ligand energy, G_CDS_aq was for cavity-dispersion-solvent free energy. The most important parameters in the QSAR model were steric descriptors (MW and volume) since their contribution to the regression equation was > 66% of the total.

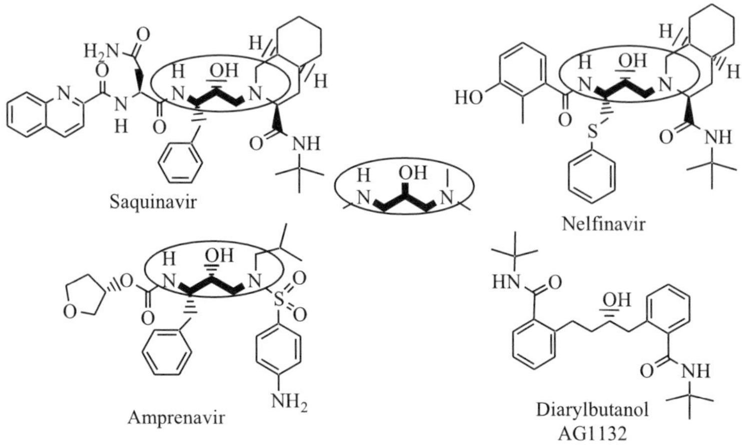

Fig. 14 Design of new protease inhibitor with an isopropanolamine unit (Reprinted with permission from [181]. Copyright 2002 Elsevier)

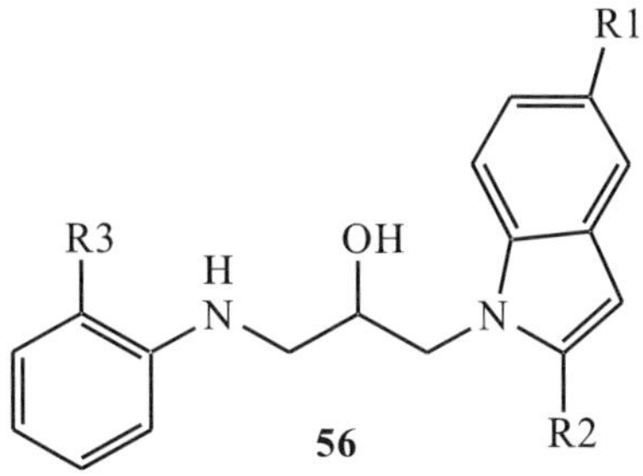

$R_1 = H, Cl$
$R_2 = COOEt, CONH\text{-}tert\text{-}Bu$
$R_3 = COOMe, CONH\text{-}tert\text{-}Bu$

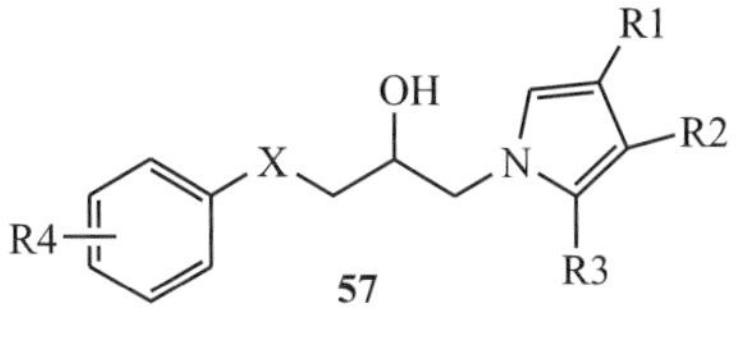

$X = S, NH; R_1 = H, Ph, 4\text{-}Cl\text{-}Ph$
$R_2 = H, COOEt, CONH\text{-}tert\text{-}Bu$
$R_3 = H, COOEt$
$R_4 = H, Me, Cl, NH_2, OMe, COOMe,$
    $COOEt, CONH\text{-}tert\text{-}Bu$

This QSAR was derived using a training set of diarylbutanols taken from the literature [182–184]. The QSAR method used 3D-calculated parameters compiled by mean of VALIDATE and VALIDATE II as implemented in SYBYL [107]. VALIDATE is a hybrid approach (QSAR and scoring functions) that makes the maximal use of the three-dimensional information from the 3D-coordinates of known ligand–receptor complexes. Chemical and physical-chemical parameters/scores were calculated to account for the free energy contribution. A PLS-based QSAR model using an optimal number of PC was derived. The calculated parameters correlated with the reported IC_{50} of the training set in a PLS model ($q^2 = 0.751$, SEP = 0.743 using only two principal components). The non-cross-validated PLS model resulted in an r^2 of 0.812 and a standard error of estimation (SEE) of 0.646 log unit of IC_{50} (F-test = 58.252).

4.2.2
C2 Symmetric Diols and Aminodiols

C2 symmetric diols having a L-mannaric acid backbone were developed as shown in Fig. 15 [185]. A series of fluoro-substituted P_1/P_1' analogs (58) of these symmetrical diols were synthesized and tested with the aim of improving their anti-HIV activity. QSAR 67 [15] reported for the data showed that electron-releasing substituents on the two phenyl rings and the hydrophobicity of the molecule have a positive effect on the activity. Similar to QSAR 66,

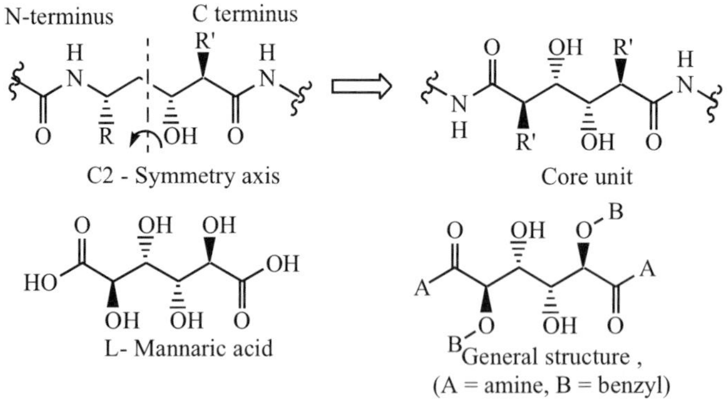

Fig. 15 Design of C2-symmetric diols (L-Mannaric acid) (Reprinted with permission from [185]. Copyright 2001 American Chemical Society)

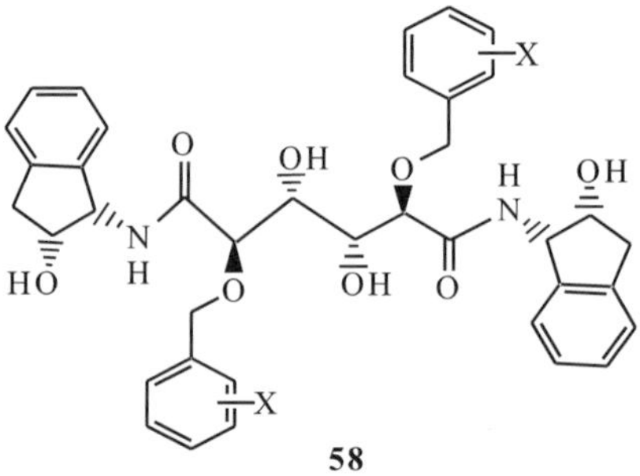

58

the unsubstituted compound did not fit this model.

$$\log 1/K_i = 1.48(\pm0.56)C\log P - 0.40(\pm0.18)\sigma_{sum} + 6.16(\pm0.92)$$
$$n = 9, \quad r^2 = 0.899, \quad q^2 = 0.783, \quad s = 0.080. \tag{67}$$

A variety of peptidic inhibitors was developed based on the C2 symmetric structure. An early study by Kempf et al. [186–188] based on a symmetric core diamine and a pseudo symmetric core amine led to the identification of A-77003 and A-80987 compounds possessing adequate anti-HIV activity. Further studies on analogs of A-80987 led to the discovery of Ritonavir (3) (ABT-538). A systematic QSAR study on analogs of A-80987 (59) and Ritonavir (3) [189] was conducted by Mekapati et al. [190] using a Fujita–Ban type analysis [191].

In this analysis, the activity contribution of each substituent or moiety was obtained [191]. The anti-HIV activity (EC$_{50}$ and CCIC$_{50}$) under study was evaluated in terms of the ability of the compound to block the spread of HIV in the immortalized human T-cell line MT4 by measuring the cytopathic effect of the virus in those cells by uptake of tetrazolium dye. A number of observations were noted regarding the most suitable substituent for each position in all the series under investigation. One important observation was the involvement of hydrophobic interaction and hydrogen bonding for inhibitory activity. The authors [190] noted that since there was not much variation in the substituent at any position in any of the series of compounds, the dataset was not ideal for QSAR analysis using physico-chemical or structural properties.

C2 symmetric aminodiols and their tyrosine-derived analogs were some of the early leads on diol derivatives. Chen et al. [192] reported a detailed SAR study on P$_1$/P$_1'$-substituted aminodiols (60). A high degree of correlation was observed between lipophilicity, measured by reverse-phase HPLC constant k' and the cytotoxicity (CC$_{50}$) of the compounds (QSAR 68) [192]. It was also found that appropriate substitution at the para position of the P1' phenyl group of 60 resulted in the identification of compounds that possess good antiviral and enzyme inhibitory activity and significantly decreased cytotoxicity [192].

$$\log 1/CC_{50} = 1.1886 \log k' - 2.7466$$
$$n = 174, \quad r^2 = 0.702. \tag{68}$$

Bisacchi et al. [193] studied BOC (t-butyloxycarbonyl) modified analogs of the C2 symmetric aminodiols (61) for HIV protease inhibition. The inhibitory

59

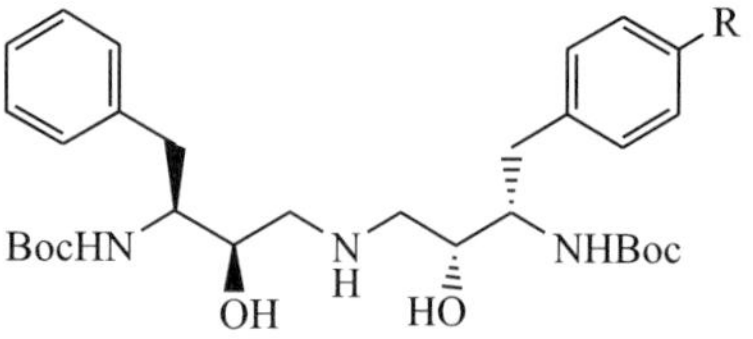

60

activity (IC_{50}) tested to inhibit the cleavage of V-S-Q-N-(b-naphthylalanine)-P-1-V by 50% was found to be significantly correlated with Verloop's B5 width parameter of X-substituents as shown in QSAR 69 [15]. It was a very small dataset and the variation in the congeners was brought out by X-substituents only. The negative coefficient of B5 indicated that an increase in size of the X-substituents would produce negative steric hindrance in binding of the molecules to the enzyme. The unsubstituted molecule was found to be unfit in the final QSAR.

$$\log 1/IC_{50} = -0.99(\pm 0.12)B5_x + 9.37(\pm 0.35)$$
$$n = 5, \quad r^2 = 0.996, \quad q^2 = 0.991, \quad s = 0.049 . \tag{69}$$

Leonard and Roy [194] recently reported QSAR 70–73 on the HIV protease inhibitory data of 1,2,5,6-tetra-o-benzyl-D-mannitols (**62**) studied by Bouzide et al. [195]. Several statistical techniques such as stepwise regression, multiple linear regression with factor analysis as the data preprocessing step (FA–MLR), principal component regression analysis (PCRA) and partial least square (PLS) analysis were applied to identify the structural and physicochemical requirements for HIV protease inhibitory activity.

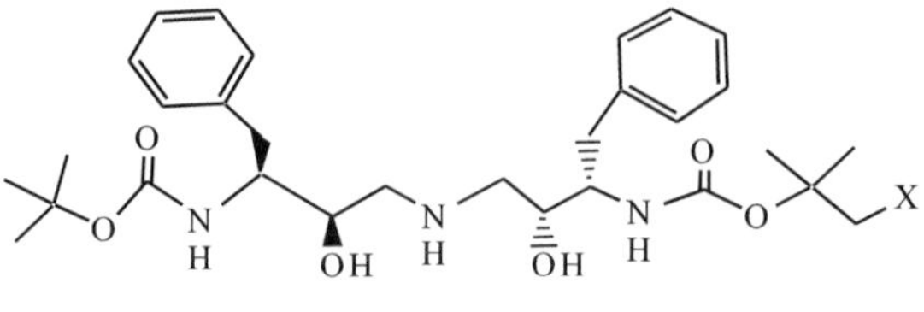

61

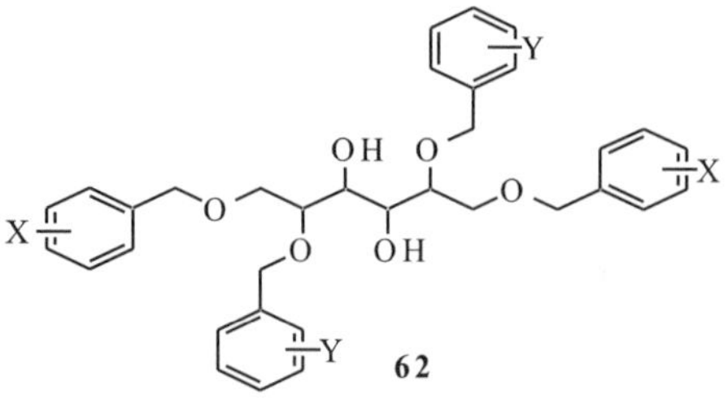

62

Stepwise regression

$$pC = -0.311(\pm 0.249)\pi_{X\_p} - 0.961(\pm 0.721)\sigma^2_{Y\_p}$$
$$- 0.804(\pm 0.272)I_{Y\_Hbond\_do\_p} + 0.523(\pm 0.170)N_{X\_O}$$
$$+ 2.600(\pm 0.114)$$

$n = 35,\quad R^2_a = 0.764,\quad R^2 = 0.792,\quad R = 0.890,\quad F = 28.5(df\,4, 30),$

$s = 0.247,\quad \text{SDEP} = 0.261,\quad S_{PRESS} = 0.282,\quad Q^2 = 0.730,$

$\text{PRESS} = 2.390$ \hfill (70)

FA-MLR

$$pC = -0.020(\pm 0.012)MR + 0.464(\pm 0.176)N_{X\_O}$$
$$- 0.253(\pm 0.231)I_{Y\_F\_O}$$
$$- 0.715(\pm 0.274)I_{Y\_Hbond\_do\_p} + 5.941(\pm 1.997)$$

$n = 35,\quad R^2_a = 0.770,\quad R^2 = 0.797,\quad R = 0.893,\quad F = 29.4(df\,4, 30),$

$s = 0.244,\quad \text{SDEP} = 0.255,\quad S_{PRESS} = 0.276,\quad Q^2 = 0.741,$

$\text{PRESS} = 2.281$ \hfill (71)

PCRA

$$pC = -0.273(\pm 0.080)fs1 + 0.132(\pm 0.080)fs3 + 0.278(\pm 0.080)fs4$$
$$- 0.141(\pm 0.080)fs7 + 0.117(\pm 0.080)fs9$$
$$- 0.119(\pm 0.080)fs12 + 2.527(\pm 0.078)$$

$n = 35,\quad R^2_a = 0.805,\quad R^2 = 0.839,\quad R = 0.916,\quad F = 24.3(df\,6, 28),$

$s = 0.225,\quad \text{SDEP} = 0.259,\quad S_{PRESS} = 0.289,\quad Q^2 = 0.734,$

$\text{PRESS} = 2.341$ \hfill (72)

PLS

$$pC = -0.246\pi_{X\_p} - 0.198B1_{X\_p} + 0.148\pi_{Y\_p}$$
$$- 0.252mr_{Y\_p} - 0.156B5_{Y\_p} + 0.191N_{X\_F}$$
$$+ 0.151N_{Y\_F} + 0.246N_{X\_O} + 0.116I_{Y\_p}$$
$$- 0.461I_{Y\_Hbond\_do\_p} - 0.189I_{Y\_di\_F}$$
$$- 0.219I_{Y\_F\_O} + 2.997$$

$n = 35,\quad R^2_a = 0.791,\quad R^2 = 0.865,\quad R = 0.873,\quad Q^2 = 0.771,$

$\text{PRESS} = 2.022 \,.$ \hfill (73)

The equations were developed using several descriptors: hydrophobicity (π), electronegativity (σ), molar refractivity (mr), steric terms (B1–B5), H-bond donor (Hbond_do) and indicator variables (I, N) for *ortho* (O) and *para* (p/P) positions of the X- and Y-substituted benzene ring (**62**). The positive and negative coefficient of parameters, respectively, shows the positive and negative effect to the biological activity (C) with acceptable statistical range (R^2

from 0.792 to 0.865). The significance of the original variable for modeling was observed using factor scores (*fs*) that contained contributions of different descriptors (QSAR 72). In addition, the presence of fluorine (F) and the di-fluoro group (di-F) was found to be detrimental to the activity.

4.2.3
Other Classes

In an effort to identify inhibitors that possess high potency, reduced molecular weight and lipophilicity, a series of novel unsymmetrical hydrid anthranilamide-containing HIVPI (**63**) were designed by Randad et al. [196]. Two datasets were reported: (a) P3–P2′ inhibitors: possessing a pyridyl-anthranilamide group at the P3–P2 and a benzamide at the P2 position; (b) P2–P3′ inhibitors: possessing a pyridyl-anthranilamide group at the P3′–P2′ positions. The data for both sets were combined and QSAR 74 was developed [15].

$$\log 1/K_i = 3.24(\pm1.88)\text{CMR}_Z - 0.47(\pm0.33)\text{CMR}_Z^2$$
$$- 0.94(\pm0.35)I + 1.24(\pm0.43)I_Z + 5.42(\pm2.44)$$
$$n = 23, \quad r^2 = 0.850, \quad q^2 = 0.764, \quad s = 0.331, \quad (\text{CMR})_0 = 3.44. \tag{74}$$

The QSAR model showed a parabolic correlation for size in terms of CMR_Z (calculated MR for the Z-substituents). It appeared that the activity increases with an increase in size (i.e. CMR value up to an optimum of 3.44) and decreases with a further increase. The indicator, I, was used with a value of 1 for P3–P2′ inhibitors. Its negative coefficient shows that these analogs will have lower activity. Another indicator variable, I_Z, was used with a value of 1 for Z = substituted phenyl. Its positive coefficient indicated that a substitution in the Z-phenyl ring is desirable for good activity.

QSAR 75 was derived [15] for a series of macrocyclic peptidomimetic (**64**) tested for activity against HIV protease by Glenn et al. [197].

$$\log 1/K_i = 0.53(\pm0.29)I_Z - 0.51(\pm0.16)B5_Y + 10.35(\pm0.77)$$
$$n = 8, \quad r^2 = 0.964, \quad q^2 = 0.921, \quad s = 0.149. \tag{75}$$

63

64

A bulkier Y-substituent was found to be detrimental to the activity. The positive coefficient of the indicator variable, I_Z, used with a value of unity for Z = H hinted that unsubstituted Z is better for the activity. There was not much variation for the X-substituent. It was noted that the dataset is too small to include a third parameter in the QSAR.

4.3
Molecular Modeling Studies

Many molecular modeling and 3D-QSAR studies have been conducted on PPIs. In these studies a diverse dataset containing various classes of transition state isostere were used. It was difficult to group them according to different classes of peptidomimetic HIVPI.

Waller et al. [198] reported the very first CoMFA study on a diverse set of transition state isosteres (TSI). The SAR data was taken from a few papers. The availability of crystal structure data for at least one compound from each class helped in identifying the active conformation and position of each ligand. These molecules were used as a template for field-fit minimization of other compounds. The training set of 59 molecules of five different TSI classes (hydroxyethylamine, statine, norstatine, keto amide and dihydroxyethylene) was minimized using the Tripos force field within Sybyl [107] software. The charges were assigned using the AM1 method in MOPAC 5.0. The default CoMFA setting was used and fields were calculated. A classical PLS technique with LOO cross-validation was used to derive predictive models. The predictive ability of each model was evaluated using a test set of 18 hydroxyethylamine [162]. Force field minimization of the steric and electrostatic field of the most closely related crystal structures and subsequent relaxation of the conformation in the active site (without field-fit option) produced the best predictive model ($r_{cv}^2 = 0.778$, $r^2 = 0.984$ and standard error = 0.146) using six principal components. This model was used to create the contour maps to gain visual understanding of the steric and electrostatic interactions. These results were used in designing third generation peptidic HIVPIs.

Oprea et al. [199] developed a semi-automated procedure "NewPred" to evaluate alternate binding modes and assist 3D-QSAR studies in predictive power. Five CoMFA models developed earlier for 59 HIVPI by the same group [198] were evaluated. There were three sets in the test set. Set A has 18 different TSI hydroxyethyl urea derivatives for investigating the binding mode of P1' and P2. Set B having 12 dihydroxy ethylene was used to study the binding mode of P2 and P3 as well as P2' and P3'. Set C comprising of six other compounds was part of the total sets but not investigated with New-Pred. Each compound was aligned; geometrically different conformers were generated to compare energies. The r^2 value of the five CoMFA models varied from 0.1 to 0.7 after omitting outliers. The predictive abilities for set A, B and the entire set with r^2 of 0.670, 0.711 and 0.661, were observed. The model correctly predicted the poor inhibitory activity of set B, which was explained and interpreted from 3D-QSAR perspectives.

Doweyko [200] applied hypothetical active site lattice (HASL) [201–203] methodology on a diverse set of 84 peptidic protease inhibitors. The experimental pK_i values used to generate a putative 3D-pharmacophore were taken from the literature. It was proposed that this method is capable of quantitatively predicting the binding activity. There are two fundamental steps in this technique. First is the creation of a 3D-QSAR HASL model in which molecular structures are converted to a set of points in a 3D grid with each point retaining a fourth dimensional variable describing the type of atom present at that point. The second step includes trimming the model to the smallest subset of points that can retain the reasonable predictive properties. The model generated at 2 Å and containing 899 lattice points identified an 11-point pharmacophore predicting the pK_i activity with a correlation (r^2) of 0.827. The predictivity was determined by equally dividing the data into a training set (odd-numbered inhibitors) and a test set (even-numbered inhibitors). Using previously determined optimum resolution of 2.0 Å, a 784-point HASL yielded r^2 of 1.00 from the 42-odd inhibitor and r^2 of 0.726 from the 42-even inhibitors test set. The pharmacophore was found to be consistent with known H-bonding sites between the inhibitor and the active site.

Kroemer et al. [204] conducted a CoMFA study for successful design and prediction of a set of 100 heterosubstituted statine derivatives [205]. The receptor-based approach used the X-ray structure of a HIV proteinase/MVT-101 complex [33] where the ligand was replaced with the prototype compound SDZ-282870. The DISCOVER force field was used and the conjugate gradient minimization technique applied [206]. CoMFA studies were performed with a training/test set of 100/75 compounds. Several models were developed using both "no volume averaging" and "volume averaging" that showed steric descriptors to be dominant in both cases with minor differences. Since the cross-validation technique with LOO may not provide models with general predictivity, authors decided to use two validation groups each consisting 50% of the compounds randomly. The best model with both steric and electrostatic interaction energy had r^2_{cv} of 0.574 and r^2 of 0.841 with 77.2% steric contribution. The chance correlation was avoided and the models were compared with receptor topology to observe high correlation between calculated fields and biological activities.

Cho and Tropsha [207] reported cross-validated r^2-guided region selection (q²-GRS)-based CoMFA models that were orientation-independent, for the same 59 HIVPIs as studied by Waller et al. [198]. It was found that the q^2 value is sensitive to the overall orientation of superimposed molecules on a computer terminal and can vary by as much as 0.5 q^2 units when the orientation is varied by systematic rotation. The authors hypothesized that the low q^2 value obtained from classical CoMFA might be caused by poor orientation of the molecular aggregates rather than by poor alignment. To validate their hypothesis rectangular lattice obtained by classical CoMFA were subdivided into 125 small boxes for reorientation of aggregates. One hundred and twenty five independent analyses were performed using probe atoms placed within each box with the step size of 1 Å. Only those small boxes for which q² was higher than a specified optimal cutoff value were selected. Finally, CoMFA was repeated with the union of small boxes selected at the previous step. It was proposed that the reorientation of the aggregates significantly improved the results.

Viswanadhan et al. [208] used a new approach for rapid estimation of relative binding affinities. It was based on variables representing enthalpy of binding and strength of hydrophobic interaction on 11 peptidomimetic HIVPIs taken from the literature. 3D-structures were constructed based on crystal structures of HIV protease complex with known inhibitors. Energy minimization was performed using the BORN module of the AMBER v.3 program [209, 210]. The relative differences in the free energy of binding measured experimentally correlated well ($r = 0.94$) with the calculated scores of relative enthalpy differences for analogous inhibitors. Addition of hydrophobic interaction strength used to compute and display a color-coded "molecular hydrophobicity map" [211] did not improve the correlation significantly ($r = 0.92$). The lack of influence of hydrophobic interaction was attributed to a smaller dataset.

Perez et al. [174] used molecular mechanics (MM)-based comparative binding energy (COMBINE) analysis on HIVPIs reported by Holloway et al. [163]. In COMBINE methodology a data-matrix is produced containing a large number of energy descriptors and PLS analysis is used to generate predictive CoMFA models. The COMBINE analysis produces a set of weights indicative of relative importance of each residue for activity, whereas CoMFA only gives information about the interaction properties of ligands. The COMBINE model was developed using a training set of 33 hydroxyethylene moiety-based inhibitors and a test set of 16 inhibitors containing hydroxyethylene, hydroxyethylamine, statine and symmetrical-diol isosteres. An AMBER forcefield was used to observe the high correlation between ligand–receptor interaction energies and inhibitory activities ($r^2 = 0.81$). The final regression model has acceptable predictive ability both in the internal validation test set ($q^2 = 0.79$, $SDEP_{cv} = 0.61$) and the external set ($SDEP_{ex} = 1.08$). Analysis of interaction energies using the COMBINE improved correlation ($r^2 = 0.89$) with better estimation of the activity of the external dataset ($SDEP_{ex} = 0.83$) was carried out. Further incorporation of the solvent-screened residue-based electrostatic interactions and two additional descriptors representing the electrostatic energy contributions to the partial desolvation of both the ligands and the receptor resulted in a better model. Remarkable predictive ability for both internal ($q^2 = 0.73$, $SDEP_{cv} = 0.69$) and external validation tests ($SDEP_{ex} = 0.59$) was achieved. The final model developed for the combined set of all the inhibitors has the best variance in biological activity ($r^2 = 0.91$) and the highest predictive ability ($q^2 = 0.81$, $SDEP_{cv} = 0.66$) revealing the importance of MM-based QSAR studies.

Pastor et al. [212] reported a structure-based QSAR to improve the model given by Perez et al. [174] to study the alternative binding modes. Only one binding orientation was observed for each compound in the training set, which has structural variability only in one-half of the pseudo symmetrical binding cavity of HIV protease (pockets S_1' and S_2'). The improved model was proposed to give accurate predictions for new compounds exhibiting structural variation in both halves. The new model was derived by manipulating the data matrix by duplicating and swapping the variables, which describe the electrostatic and steric interaction energies of every residue in the subunit. The interactions involving residues Asp25 and Asp25' were not duplicated since they were considered different due to their different protonation state. An AMBER force field [210] and Delphi [213] were used to calculate the steric and electrostatic contributions of the ligand–receptor interactions, respectively. The new model, termed as C_{duplo} was built in a similar way to C_{single} (C stands for COMBINE). Golpe 3.0 [214] was used for building the model and validation. The new model (C_{duplo}) has better predictive ability especially when the external test set molecules were considered in dual binding orientations compared to the single orientation model.

Vedani et al. [215] reported a quasi-atomistic receptor modeling technique. This technique bridges 3D-QSAR and receptor fitting by generating receptor surface models populated with atomistic properties such as H-bond, salt bridges, aromatic and aliphatic regions, etc. This approach allows for H-bond flip-flop particles, which can simultaneously act as H-bond donor and H-bond acceptor when interacting with different ligand molecules. Genetic algorithm (GA) combined with cross validation protocols were used for generating a family of receptor models. The relative free energies of ligand binding was calculated for the training/test set of 15/8 HIV protease inhibitors. The cross-validated correlation (q^2) of 0.950 and RMS deviation of 0.173 was obtained for the training set and an RMS deviation of 0.545 was obtained for the test set (prediction). The models were able to reduce the influence of random errors and scan the receptor space exhaustively.

Jayatilleke et al. [216] also performed a theoretical study on the inhibitory data [163, 174] of compounds that are structurally similar to Indinavir (2). They derived CoMFA-based 3D-QSAR models. Predicted biological activity (pIC_{50}) of compounds was linearly correlated with experimentally determined values ($r^2 = 0.82$, $q^2 = 0.64$). These models were used for screening large databases for candidate inhibitors of HIV protease.

Di Santo et al. [181] also conducted CoMFA studies for the same dataset for which the 2D-QSAR model was developed (QSAR 66). A cross-validated model with two principal component ($r^2 = 0.607$) and a non-cross validated model with r^2 of 0.869 were obtained. The model showed a major contribution of steric fields (56.7%) over electrostatic fields (43.3%), which was in agreement with the 2D-QSAR model. The 3D-QSAR models were able to predict the anti-protease activity but were unable to discriminate anti-protease active compounds from the inactive compounds. The model also predicted the biological activity of a set of 20 newly synthesized arylpropanolamines and indicated their binding mode to be similar to the diarylbutanols used to derive the CoMFA model.

Nair et al. [217] conducted a molecular modeling study to understand the binding pattern of the mutant resistance pattern for data published [218–220] on Ritonavir (3) derivatives. Comparison of the complexation energies of 12 mutants of HIV (V82I, M46I and others) was reported. Results showed good agreement between the calculated and the experimentally measured complexation energies. The analysis showed that resistance is principally due to a decrease in complexation energy between the mutant protease and the inhibitor. Using these models, a significant improvement in calculated complexation energies for mutant as well as wild-type protease was obtained for several new analogs.

Huang et al. [221] reported docking studies on 27 AHPBAs derivatives [177, 178, 222, 223] into the active site of HIV protease using the Lamarckian genetic algorithm (LGA) of Autodock 3.0 [224]. The binding mode demonstrating inhibitors' conformation, sub-site interaction and H-bonding was observed and compared with a known inhibitor. The regression equation correlating antiviral activity of 27 analogs with the total binding free energies, ΔG, was highly significant ($r^2 = 0.860$). In the same study consistent and highly predictive CoMFA, CoMSIA and HQSAR models were developed with reasonable r^2_{cross} values of 0.613, 0.530 and 0.717, respectively for AHPBAs derivatives. The stability and predictive ability of these models was validated by a test set with a known inhibitor and a set of nine compounds that were not included in the training set. The predicted values were found to be in good agreement with experimental data. Automated molecular docking studies were performed to investigate structural differences of binding mode between test and training sets. A successful combination of structure-based design elucidating enzyme–substrate interactions and 3D-QSAR models showing characteristics of the binding conformations was applied to demonstrate AHPBA–HIVPR interactions.

Visco Jr. et al. [225] introduced the concept of signature of an atom as a molecular descriptor and various topological indices for developing an inverse QSAR. The effectiveness of the method was measured by correlating activities of 121 HIVPI in a training set [142, 148, 172, 226–228]. The test set of nine compounds was chosen with at least one test set compound coming from each of the literature sources. Standard forward-stepping MLR-based QSAR models using atomic signatures were compared with the one developed using Molconn-Z [229]. Overall, the QSAR based on signature calculation were able to correlate training set data with errors slightly larger than Molconn-Z parameters. The optimum sets of descriptors were used to generate the focused library of candidate structures using inverse-QSAR approach. Faulon et al. [230] improved the model given by Visco Jr. et al. [225] by avoiding use of redundant Molconn-Z descriptors. QSAR and QSPR models were developed and compared. Topological indices from atomic signatures for a large dataset of compounds containing octanol-water partition coefficients were also calculated.

Jenwitheesuk and Samudrala [231] reported an improved prediction method to calculate the binding energy on a set of 25 HIVPI obtained from the PDB database by molecular dynamics simulations. The experimental binding energies used were computed from the experimental inhibition constants (K_i) using a simple rate equation ($\Delta G^0 = - RT \ln K_i$) and were plotted against ΔG^0 to obtain a good fit. The MD simulations were carried out using NAMD software [232] and the X-PLOR forcefield [233]. The structures at different time intervals were recorded and used in the docking step. The binding energy was calculated using AutoDock 3.0.5 [224]. Cluster analysis was performed on multiple docking runs to cluster ligands with all-atom RMSD within 1 Å of each other and ranked by the lowest-energy. Protein-rigid docking without any dynamics produced a low correlation of 0.38 between the experimental and calculated binding energies. Correlation improved significantly when the calculated binding energy from 0.1 picoseconds of dynamic simulation was used with a highest correlation coefficient of 0.87.

Kiralj and Ferreira [234] reported on an a priori molecular modeling approach for pseudo C2 symmetry-based 2-hydroxy-1-aminoindane and its derivatives. They calculated molecular volumes, established electron density-distance relations and modeled the inhibitor inside the active site to understand the binding interaction. The electron density and HOMO-LUMO iso-surfaces were generated using MOPAC 6.0 Hamiltonian [235, 236]. The intermolecular interaction was established in terms of projected surface area (S), molecular height (H) and length (L). Molecular modeling was performed using the Weblab viewer [237]. The most active compound and its inhibitor-environment interaction energy at a 10 Å cut-off ($E_{10.0}$) were found to be comparable to the results of Holloway et al. [163] and Perez et al. [174]. In addition, the predicted biological activities of compounds correlated highly with empirically predicted activities ($r > 0.93$). The calculated log P values of the inhibitors were found to be in the range reported by Hansch and co-workers [14] revealing the importance of the lipophilicity. The molecular graphics analysis revealed that the enzyme–inhibitor intermolecular interaction cut-offs are at the interval 1.6–5.5 Å for defining the physical boundary for protease-water-inhibitor interaction.

Skalova et al. [238] reported the analysis of conformational interaction energies and compared the wild-type and mutated HIV protease-inhibitor complexes based on an ethylenamine isostere. A triple mutant HIV protease (A71V, V82T and I84V) was complexed with isostere to analyze the interactions. Tight binding with wild-type ($K_i = 1.5$ nM) and mutant ($K_i = 4.1$ nM) protease was observed with H-bond formation. The bifurcated H-bonds from the isosteric NH group to both catalytic aspartates were observed. Binding to mutant and wild-type HIV protease revealed the difference in the conformation of the peptide bond isostere and the orientation of the phenyl ring in the P1 position. Weaker van der Waals interactions of the mutated residues Val84 and Val184 were found to be compensated by new aromatic H-bond between the phenyl ring of the inhibitor and mutated residue Thr182. Comparison of X-ray and computational structures showed a similar pattern.

Kiralj and Ferreira [239] developed QSAR models on 48 peptidic HIVPIs by using 14 a priori molecular descriptors. Hierarchical cluster analysis (HCA), principal component analysis (PCA) and partial least squares (PLS) regression were employed. PLS models with 32/16 (model I) and 48/0 (model II) molecules in the training/external validation set were constructed. The a priori molecular descriptors were related to two energetic variables using PLS. HCA and PCA on data from model II classified the inhibitors as slightly, moderately and highly active. Three principal components: (1) bulk, electronic and hydrophobic properties, (2) stereochemical fit to enzyme (steric and electrotopological properties) and (3) distribution of electron density (polarity and hydrogen bonding) were found to be enough for describing the enzyme–inhibitor binding. The study showed that a good peptidic inhibitor should have four aromatic groups and/or ring substituents rich in polar and hydrophobic groups. The authors noted that Model I ($r^2 = 0.91$, $q^2 = 0.84$) was comparable to literature models obtained by various QSAR software, justifying the use of a priori descriptors.

Ungwitayatorn et al. [240] reported the 3D-QSAR CoMFA/CoMSIA studies for a series of 30 Chromone derivatives of HIVPI. The dataset was divided into a training/test set of 30/5 compounds based on the distribution of biological activity and the variety of substitution pattern. Superposition and field fit alignment criteria were used for model development in CoMFA. The best predictive CoMFA model with steric (46%) and electrostatic (54%) fields gave cross-validated r^2 of 0.763 and non-cross-validated r^2 of 0.967 with a standard error of estimate (S) of 5.092. The PLS protocol and stepwise procedure were used in CoMSIA. The best CoMSIA model with steric (27.3%), electrostatic (27.9%), hydrophobic (21.7%) and H-bond donor (23.1%) fields has a cross-validated r^2 of 0.707 and a non-cross-validated r^2 of 0.943 with a standard error of estimate (S) of 7.018.

Senese and Hopfinger [241] reported on a receptor-independent 4D-QSAR study of 27 norstatine-derived HIVPI based on an AHPBA core. The QSAR was developed using a reference grid cell lattice of 1 Å cubes in which the 3D structure of a training set of compounds was kept. The preferred geometry was determined using molecular mechanics with the MM+ forcefield. Partial charges were assigned using a semiemperical AM1 method. Each training set compound generated from the 4D-QSAR analysis was not restricted to a single conformation. Grid cell-occupancy descriptors (GCOD) were generated based on conformational ensemble profile (CEP) of every compound as the pool of trial descriptors in model building and optimization. Five QSAR models were developed from two different alignments to map the atom type morphology of the inhibitor–binding site at the HIV protease as well as to predict the inhibitory potencies of seven test set compounds. The models correctly identified the hydrophobic nature of the HIV protease receptor site and helped in structural modification to improve the potency of the AHPBA inhibitors. This set of unique and equally individual models were referred to as a "manifold model". This study demonstrated that there could be more than one way to fit structure-activity data within a QSAR methodology.

Senese et al. [126] also reported a 4D-fingerprint-based QSAR model using the approach discussed earlier to analyze a dataset of AHPBA inhibitors. 445 4D-fingerprints were used to develop a model that was compared with other QSAR techniques such as CoMFA, CoMSIA, H-QSAR and normal 4D-QSAR. The partial least square regression (PLSR) and genetic function approximation regression (GFAR) models were generated independent of any receptor structure or alignment information. These models exhibited comparable statistical data with CoMFA, CoMSIA and H-QSAR approaches. The 4D-fingerprints model predicted well the biological activity of test set inhibitors. This study proved that genuine representation of 3D and conformational properties of compounds is possible using this approach.

Radestock et al. [242] used a "reverse" protein-based CoMFA (AFMoC) approach to study 66 peptidic HIVPIs. Protein-specific objective functions were adapted to observe binding mode prediction in docking. Shannon entropy-based column filtering of the descriptors matrix and the capping of adaptive repulsive potentials within the binding site were shown to be successful. Use of protein-specific adapted potential fields in a modified AFMoC approach (AFMoC^{obj}) as an objective function led to 14% improvement in docking accuracy as compared to non-adapted DrugScore or AutoDock fields [224].

Boutton et al. [243] reported a genotype-dependent QSAR for HIV protease inhibition. A computational structure-based approach was used to predict the resistance of the HIVPI strain to Amprenavir (5) by calculating the interaction energy of the drug with the enzyme. Two thousand nine hundred and eighty homology models were developed by modifying amino acids according to point mutation found in corresponding HIV protease strains of 2980 patients. To accommodate the geometry difference of mutated amino acids the protein structure was optimized first and then minimized. Six different interaction energies such as coulomb, van der Waals and H-bond contributions were calculated per residue to identify the mutation that contributes to drug-resistance. The dataset was divided 80 : 20 into a training and test set. The improved prediction model was developed with only 40 variables instead of 136 variables to avoid bias. The best model gave an r^2 value of 0.794 and RMSE of 0.345 for the training set and r^2 of 0.783 and RMSE of 0.348 for the test set. Out of the 40 variables used, 22 were used for backbone interaction and 18 involved side chain interactions. This research suggested that the changes in energy values at the mutated positions and their environment should be considered first instead of side-chain mutations. The asymmetry in the prediction model observed was correlated with the asymmetry of HIV protease after drug binding.

4.4
Summary

The crystal structures of the protease complexed with four structurally different peptide isosteres exhibited that all four inhibitors were bound in an extended conformation, spanning the P4 to P3' site [33, 166, 244–247]. An extensive network of hydrogen bonds was also proposed between the enzyme and the polar atoms in the inhibitor. These postulated hydrogen bonds were formed primarily with backbone atoms of the floor and flap regions of HIV proteases (Fig. 3a) [31]. The binding pockets discernable from P_2 to P'_2 were comprised almost entirely of hydrophobic residues in the enzyme [33, 244, 245]. According to Huff [247] the inhibitor–enzyme binding is dominated by hydrophobic interactions. In some protease-inhibitor binding studies it has also been found that the protease is singly protonated [166, 174]. Although hydrogen bonding plays a crucial role in the stabilization of protease-inhibitor complexes, adequate treatment of the enzyme active site protonation state is important for their accurate molecular simulations. Calculations have shown that in HIVPR-inhibitor complexes, only one catalytic aspartic acid residue is protonated [166, 174].

In view of all this information, a survey of QSAR 45–75 on peptidic inhibitors reveals that only four models (58, 64, 65 and 68) contain a hydrophobic term. Also, note that QSAR 65 points toward the optimum size of the hydrophobic binding pocket for a specific substituent. This is surprising since the protease receptor does have hydrophobic binding sites. Hydrophobic and H-bond interactions were found to be important in some studies (Figs. 12 and 13) [137, 162]. Several studies have noted the importance of hydrophobic interactions for protease inhibition by peptidic inhibitors [137, 161, 162, 190]. It is possible that due to some spatial restrictions these molecules are unable to bind in hydrophobic space.

Holloway et al. [163] SAR data was studied in many QSAR studies [14, 163, 168, 171]. It was found that the enzyme inhibition activity of peptide isosteres could be correlated with the interaction energies (QSAR 51–53) [163]. However, QSAR 54 suggested the involvement of dispersion interactions in the binding of some isosteres with the enzyme [14]. H-bonding and hydrophobic interactions were found to play an important role in the inhibition of protease in another QSAR study [168]. Chirality-based descriptors were proposed to solve the problem of stereoisomers in QSAR [171]. Significant steric and polar interactions with the receptor seem to be important as shown by the presence of steric and electronic terms in many of the QSAR. Results of some 3D-QSAR studies [198, 204, 207, 216] using the CoMFA methodology on different kinds of peptide isosteres supported almost a common mode of binding and stressed the involvement of steric and electrostatic interactions.

QSAR 67–75 developed for C2 symmetric diols and aminodiols highlighted the hydrophobic, steric and polar interactions of the specific substituents/groups with the specific binding site of the receptor. Studies [248, 249] on the binding modes of a series of penicillin-based C2 symmetric dimer inhibitors suggested that these inhibitors would bind in a symmetrical fashion, tracing an S-shaped course through the active site, with good hydrophobic interactions in the S_1/S_1' and S_2/S_2' pockets and hydrogen bonding of inhibitor amide groups (Fig. 16) [248]. Interactions with the cata-

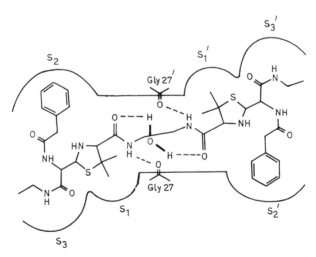

Fig. 16 Model proposed for the binding of penicillin derived C2 symmetric dimer inhibitor with HIV protease (Reprinted with permission from [248]. Copyright 1993 American Chemical Society)

lytic aspartates were found to be poor and the protein conformation to be very similar to that observed in complexes with peptidomimetics, in spite of the major differences in ligand structure [248].

Besides stepwise multiple linear regression (MLR) analysis, other methods used for deriving QSAR models were E-state modeling [168], кNN based COMBINE [173] and VALIDATE [181]. The VALIDATE method makes use of 3D-coordinates of known ligand–receptor complexes to calculate physicochemical parameters. It was one of the very few studies where QSAR was part of the design and synthetic efforts [181]. Several statistical techniques such as stepwise regression, FA-MLR, PCRA and PLS analysis were applied in a recent study [194] to identify the structural and physicochemical requirements for HIV protease inhibitory activity.

Many molecular modeling and 3D-QSAR studies have been conducted on PPIs. In these studies biological data of diverse classes of transition state isostere were used. The CoMFA methodology was extensively used for developing 3D-QSAR models. As mentioned earlier CoMFA results through the 3D contour plots of the steric and electrostatic fields provide valuable information for the design of a drug. Few new approaches in CoMFA were introduced to improve the predictive power of the model. In one study, a semi-automated procedure "NewPred" was developed to evaluate alternate binding modes and assist 3D QSAR studies in predictive power [199]. Another study used superposition and field fit alignment criteria to improve the model [240]. Orientation-independent CoMFA models using cross-validated r^2-guided region selection (q^2-GRS) were also studied [207]. Molecular mechanics (MM)-based comparative binding energy (COMBINE) analysis on Holloway et al. data [163] was used to generate better predictive CoMFA models [172]. This method produces a set of weights indicative of relative importance of each residue for activity, whereas CoMFA only gives information about the interaction properties of ligands. In another study, CoMFA-based 3D-QSAR models developed on Holloway et al. [163] and Perez et al. [174] data were used for screening large databases for candidate inhibitors of HIV protease [216].

A successful combination of structure-based design elucidating enzyme-substrate interactions and 3D-QSAR models showing characteristics of the binding conformations was applied to demonstrate AHPBA–HIVPR interactions [221]. In another interesting study the AFMoC (reverse protein-based CoMFA) approach was adapted to observe improved binding mode prediction in docking [242]. Hypothetical active site lattice (HASL) methodology was used to generate putative 3D-pharmacophore and it was proposed that this method is capable of quantitatively predicting the binding activity [200]. The pharmacophore was found to be consistent with known H-bonding sites between the inhibitor and the active site.

A quasi-atomistic receptor modeling technique was proposed to bridge receptor fitting by generating receptor surface models populated with atomistic properties [215]. This was an interesting approach since it allows for

H-bond flip-flop particles, which can simultaneously act as H-bond donor and H-bond acceptor when interacting with different ligand molecules. An a priori molecular modeling approach modeled the inhibitor inside the active site to understand the binding interaction and revealed the importance of the lipophilicity [234]. In another study, an approach for rapid estimation of relative binding affinities based on variables representing enthalpy of binding and strength of hydrophobic interaction was developed [208]. The lack of influence of hydrophobic interaction was attributed to the smaller dataset. A structure-based QSAR study was proposed to study the alternative binding modes [212]. Use of external test set molecules in dual binding orientations compared to the single orientation model showed better predictability [212]. Molecular dynamics simulation-based cluster analysis was also proposed to improve the correlation [231]. Concepts of signature-based inverse QSAR were used to generate the focused library of candidate structures [225, 230].

4D-QSAR studies demonstrated that there could be more than one way to fit structure-activity data within a QSAR methodology. A receptor-independent 4D-QSAR study identified the hydrophobic nature of a HIV protease receptor site and helped in structural modification to improve the potency of the AHPBA inhibitors [241]. A 4D-fingerprint-based QSAR model developed for AHPBA inhibitors of HIV was generated independent of any receptor structure or alignment information [126]. These models exhibited comparable statistical data with CoMFA, CoMSIA and H-QSAR approaches. This study proved that genuine representation of 3D and conformational properties of compounds is possible using this approach.

A molecular modeling study to understand the binding pattern of the mutant resistance pattern showed that resistance is principally due to a decrease in complexation energy between the mutant protease and the inhibitor [217]. A genotype-dependent QSAR was used to predict the resistance of the HIVPI strain [243]. Homology models were developed to accommodate the geometry difference of mutated amino acids. Different types of interaction energies were calculated per residue to identify the mutation that contributes to drug-resistance. This research suggested that the changes in energy values at the mutated positions and their environment should be considered first instead of side-chain mutations [243]. Not many studies have been reported which compare the binding pattern of mutant vs. wild-type protease. More such studies will provide useful insight for designing drugs targeting mutant virus.

5
QSAR Studies on Mutant Protease

Protease crystal complexes with Lopinavir and DMP 450 exhibited completely different binding modes [250]. Another study of high resolution crystal structures of the HIVPR V82A mutant with a potent non-peptide inhibitor (UIC-94017) active against multi-drug-resistant clinical strains was reported [251]. This study showed differences in the main-chain atoms of residue 82 compared to wild-type protease and an unusual distribution of electron density for the catalytic aspartate residues. Perryman et al. [252] in their recent molecular dynamics study of a wild-type and the drug-resistant V82F/I84V mutant of HIVPR suggested the possibility of a new allosteric binding site in the mutant protease.

Analysis of the antiviral data of 3-aminoindazole cyclic urea HIVPR inhibitors (21) tested against HIV mutant I84V resulted in QSAR 76 [81].

$$\log 1/\text{IC}_{90} = 0.082(\pm 0.020)C \log P + 5.90(\pm 0.14)$$

$$n = 4, \quad r^2 = 0.994, \quad q^2 = 0.977, \quad s = 0.008 \,. \tag{76}$$

Analysis of the antiviral data of similar compounds tested against second wild-type viral isolate HXB$_2$ gave QSAR 77 [81].

$$\log 1/\text{IC}_{90} = 8.46(\pm 3.17)C \log P - 0.58(\pm 0.22)C \log P^2 - 23.23(\pm 11.54)$$

$$n = 8, \quad r^2 = 0.906, \quad q^2 = 0.721, \quad s = 0.119, \quad \log P_0 = 7.32 \,. \tag{77}$$

QSAR 77 showed a parabolic dependence of antiviral activity on hydrophobicity, with an optimum of 7.32 [81]. The range of optimum $C \log P$ for structurally diverse classes of protease inhibitors derived for wild-type protease has been found to be from 4.49 to 6.96 [14, 15]. The optimum of 7.32 observed in the QSAR 77 is higher than values observed so far. It may be attributed to the type of viral isolate HXB2 that was used for studies. This second wild-type viral isolate HXB2 seems to have a larger hydrophobic binding pocket as compared to the more common wild-type HIV. More such studies on mutant HIVPR biological data will help in delineating the difference in size of wild-type and mutant hydrophobic binding pockets.

A series of mutations, first identified in protease inhibitor–resistant HIV viral isolate, studied as individual substitutions in HIVPR exhibited wild-type preference for large hydrophobic residues, especially in the P1′ substrate position [49]. QSAR 76 and 77 also emphasize the importance of the hydrophobicity of the substituent of ligands in the design of new drugs active against mutant viruses.

6
Concluding Remarks and New Approaches

A number of QSAR models describing biological activity as a function of physico-chemical parameters and molecular descriptors have been established and described. Table 1 lists a summary of the type of inhibitors studied and methods used by different research groups for developing 2D-QSAR

Table 1 2D-QSAR publications on HIV protease inhibitors

Class	Sub-class	Inhibitor type	QSAR methods	Refs.
Non-peptides	Cyclic ureas	Unsym/Sym cyclic urea	MLR	Garg R et al. 1999 [14]
		Benzamide[a]	MLR	Wilkerson WW et al. 1996 [75]
		Unsym benzamide[a]	MLR	Wilkerson WW et al. 1996 [76]
		Cyclic cyanoguanidine[a]	MLR	Gupta SP et al. 1999 [78]
		Aminoindazole[b]	MLR	Garg R, Bhhatarai B 2004 [81]
		Tetrahydropyrimidinone	MLR	Gayathri P et al. 2001 [86]
		Tetrahydropyrimidinone	Codessa Pro	Katritzky AR et al. 2002 [88]
		Tetrahydropyrimidinone	MLR	Garg R, Patel D 2005 [92]
		Tricyclic urea	MLR	Kurup A et al. 2003 [15]
		Other classes	Scoring function	Frecer V et al. 2005 [94]
	Pyra-nones	Cycloalkylpyranone[a]	MLR	Gupta SP et al. 1999 [135]
		Pyranone[a]	MLR	Gupta SP et al. 1998 [137]
		Dihydropyranone[b]	MLR	Bhhatarai B, Garg R 2005 [127]
		Dihydropyranone[a]	MLR	Kurup A et al. 2003 [15]
Peptides	Transition state isosteres (TSIs)	Diverse hydroxyethyl isostere, statin, AHPBA, hybrid, macrocyclic derivatives	MLR	Garg R et al. 1999 [14] Kurup A et al. 2003 [15]
		Indinavir analog[a,b]	MLR/ Forcefield	Holloway MK et al. 1995 [163]
		Indinavir analog	E-state modeling	Maw HH, Hall LH 2002 [168]
		Indinavir analog	k-NN	Golbraikh A, Tropsha A 2003 [173]
		Palinavir analog	MLR	Kurup A et al. 2003 [15]

[a] Garg R et al. 1999 [14],
[b] Kurup A et al. 2003 [15]

Table 1 (continued)

Class	Sub-class	Inhibitor type	QSAR methods	Refs.
		Cyclic sulfolane	MLR	Gupta SP et al. 1998 [137]
		N-Diarylbutanol	VALIDATE/PLS	Di Santo R et al. 2002 [181]
	C2 symmetric diols & aminodiols	Ritonavir analog	Fujita-Ban	Mekapati SB et al. 2001 [190]
		Aminodiol[b]	MLR	Chen P et al. 1996 [192]
		Mannitol derivative	LFER/PCRA/PLS	Leonard TJ, Roy K 2005 [194]

[a] Garg R et al. 1999 [14],
[b] Kurup A et al. 2003 [15]

models. Table 2 provides an overview of the 3D-QSAR and molecular modeling studies. It is clear that some methods have been used extensively.

Good statistical quality models and their relevance to possible modes of action provide better understanding to ligand–receptor interactions and structure-based drug design. These models lead to assessment of the specific effects of various substituent reducing trial experiments. Analysis of these QSAR studies suggests that the hydrophobic, steric and polar interactions play a key role in the inhibition of the viral protease. Various substituents designed to target individual pockets need to be designed cautiously to achieve better inhibition.

The three important factors, which describe the physicochemical properties of the molecules and are used in developing QSAR, are hydrophobic, steric and electronic. One needs variation in these properties of the substituent at each position of the parent structure to be sure that these properties are considered. In addition, the test sets should be large enough to be able to include the three factors to see their influence on activity. Very often either of the two aspects is not considered while designing a series for investigation. Many times there is a rather high correlation between two parameters and one can be replaced with the other giving very similar correlation. Comparative study of models derived for similar systems/datasets can provide valuable guidance in the choice of the right parameter.

Some of the QSAR were developed omitting the compounds that do not fit in the final QSAR (outliers). This problem of "misfit" of the congeners in the final QSAR has been associated with several reasons such as

Table 2 Molecular modeling (mostly 3D-QSAR) publications on HIV protease inhibitors

Class	Subclass	Inhibitor type	Modeling methods	Refs.
Non-peptides	Cyclic ureas	Unsym/Sym benzamide	CoMFA	Debnath AK 1998 [99]
		Sym benzamide	CoMFA	Debnath AK 1999 [108]
		Unsym/Sym benzamide	CoMFA/CoMSIA	Tervo AJ et al. 2004 [110]
		Unsym/Sym benzamide	Neural network/	Allen BCP et al. 2001 [114]
			Molecular mechanics/	
			QSAR	
		Unsym/Sym benzamide	QSPR (SMF)	Solov'ev VP, Varnek A 2003 [115]
		Sym cyclic urea	CoMFA	Avram S et al. 2001 [117]
		Unsym/sym cyclic urea	CoMFA	Avram S et al. 2003 [118]
		Cyclic urea and Cyanoguinadine	CoMFA	Avram S et al. 2005 [120]
		Cyclicsulfamides	CoMFA	Schaal W et al. 2004 [121]
		Tetrahydropyrimidinone	CoMFA	Nair AC et al. 2002 [124]
		Tetrahydropyrimidinone	4D-QSAR	Senese CL et al. 2003 [125]
		Tetrahydropyrimidinone	4D-fingerprint based QSAR	Senese CL et al. 2004 [126]
	Pyranones	Cycloalkylpyranone	Modeling (GRID)	Lunney EA et al. 1994 [142]
		Cycloalkylpyranone	Modeling/ Molecular mechanics	Skulnick HI et al. 1995 [145]
		Cycloalkylpyranone	Modeling (GROW)	Thaisrivongs S et al. 1995 [146]
		Pyranone	Modeling (Autodock)	VaraPrasad JVN et al. 1994 [148]
		Dihydropyranone	Modeling (GOLD)	Ellsworth EL et al. 1999 [150]
		Dihydropyranone	Modeling	Schake D et al. 2004 [151]
		Pyranone	3D-QSAR/ Molecular dynamics/ Docking	Kulkarni SS, Kulkarni VM 1999 [153]

Table 2 (continued)

Class	Subclass	Inhibitor type	Modeling methods	Refs.
Peptides	Transition state isosteres (TSIs)	Diverse	CoMFA	Waller CL et al. 1993 [198]
		Diverse	CoMFA	Oprea TI et al. 1994 [199]
		Diverse	HASL	Doweyko AM 1994 [200]
		Statin derivative	CoMFA	Kroemer RT et al. 1995 [204]
		Diverse	CoMFA (q^2 GRS)	Cho SJ, Tropsha A 1995 [207]
		Peptidomimetic	Binding studies (AMBER)	Viswanadhan VN et al. 1996 [208]
		Indinavir analog	COMBINE/Molecular mechanics	Perez C et al. 1998 [174]
		Indinavir analog	Structure based QSAR	Pastor M et al. 1997 [212]
		Diverse	Quasi-atomistic receptor modeling	Vedani A et al. 1998 [215]
		Indinavir analog	CoMFA	Jayatilleke PRN et al. 2000 [216]
		N-Diarylbutanol	CoMFA	Di Santo R et al. 2002 [181]
		Ritonavir derivative	CoMFA	Nair AC et al. 2002 [217]
		AHPBA	CoMFA/CoMSIA	Huang X et al. 2002 [221]
		Diverse	Inverse QSAR/Atomic signature	Visco DP Jr et al. 2002 [225]
		Diverse	QSAR/QSPR	Faulon JL et al. 2003 [230]
		Diverse	Molecular dynamics/Cluster analysis	Jenwitheesuk E, Samudrala R 2003 [231]
		C2 sym aminoindane	A Priori modeling	Kiralj R, Ferreira MMC 2003 [234]
		Ethylamine	Modeling	Skalova T et al. 2003 [238]
		Diverse	HCA/PCA/PLS/QSAR	Kiralj R, Ferreira MMC 2003 [239]
		Chromone derivative	CoMFA/CoMSIA	Ungwitayatorn J et al. 2004 [240]
		Norstatine/AHPBA	4D-QSAR	Senese CL, Hopfinger AJ 2003 [241]
		AHPBA	4D-fingerprint QSAR	Senese CL et al. 2004 [126]
		Diverse	AFMoC	Radestock S et al. 2005 [242]
		Amprenavir	Genotype dependent/ Homology modeling/QSAR	Boutton CW et al. 2005 [243]

- Outliers due to what seem to be "congeners" but are not.
- Mathematical form of the equation may be off the mark.
- Different rates of metabolism of the members of a set.
- The quality of the experimental data.
- Finally, the parameters used may not be the best. Sometimes, experimentally obtained parameters are better than those calculated and vice versa.

Results of QSAR and molecular modeling studies supported a common mode of binding and stressed the involvement of steric and electrostatic interactions. In many QSAR models hydrophobic terms are not observed although there are certainly hydrophobic binding sites at the receptor. An overview of the QSAR derived for the protease inhibitors revealed several QSAR with a positive, negative and parabolic/bilinear $C\log P$ term. Optimum value of $C\log P$ ($\log P_0$) shown by parabolic or bilinear QSAR ranged from 4.49 to 6.96. Closer inspection of all QSAR showing optimum value revealed that they were developed for different classes of HIVPR inhibitors and for different biological end points. It is possible that ligand-induced conformational changes modify the size of the hydrophobic cavity at a receptor binding site differently. Identification of an optimum value for various classes of inhibitors for different biological endpoints is desirable. Comparative study of different datasets showed that in many of them there was insufficient spread in the range of $C\log P$ values to firmly establish the optimum point. The optimum hydrophobicity of the molecule is important for its translation across the cell membrane, which is essential for a compound to achieve good potency and bioavailability.

The majority of HIV research is done with cells and these studies tend to over estimate the value of $\log P_0$ (i.e., 1 or 2 log unit higher) for whole organisms [253]. A study of the $C\log P$ value of approved HIVPI drugs emphasizes this observation.

US-FDA Approved Protease Inhibitors [a]	$C\log P$ [b]
1. Saquinavir	4.73
2. Ritanovir	4.94
3. Indinavir	3.68
4. Nelfinavir	5.84
5. Amprenavir	3.29
6. Lopinavir	5.54
7. Tipranavir	7.76
Recently approved Kaletra® is a combination of Lopinavir and Ritonavir	

[a] All except Tipranavir are peptidic in nature
[b] Calculated using CQSAR program [253]

In conclusion, a combination of several approaches is required to understand and represent the multidimensional nature inherent to ligand–receptor binding. The physicochemical parameters delineating the characteristics of the substituent of the ligands need to be investigated thoroughly. Further lateral validation of these models with molecular modeling studies provides useful insight for the design of new HIVPR inhibitors. 3D-QSAR models as derived using CoMFA can be useful in selecting areas of the molecules for adjusting the lipophilicity for improved in vivo activity. The 3D steric contour plots can assist in identifying and avoiding any possible steric clashes with the receptor site. 3D electrostatic contour plots can be useful in identifying the positions where more negative or positive charges will increase or decrease the activity. Chirality-based molecular descriptors can be useful to account for the stereochemistry.

In the study of HIVPI, QSAR methodology seems to evolve from 2D to 3D to 4D-QSAR. The ensemble averaging the conformational behavior in constructing a 3D-QSAR using 4D-QSAR analyses permits an estimation of the effect of conformational entropy on activity. 4D-QSAR not only appears to yield 3D-QSAR models at least as good as can be generated using other methods but also provide added value information not realized by other methods.

A number of peptidic protease inhibitors are currently used in clinical practice. Although these inhibitors represent a major advance in HIV chemotherapy, adverse side effects and viral resistance is a cause of constant concern. A further limitation of current protease inhibitors is their complexity and difficult synthetic pathway with high cost of production. In order to overcome all these difficulties, it is desirable to identify new inhibitors of simpler structure. Up to now, only one non-peptide compound (Tipranavir, 7) has been approved for HIV treatment. Some reached clinical trials but could not be developed further, signifying the urgent need for development of new inhibitors.

Except for a very few studies most of the work so far has been done using SAR data of wild-type protease. Further QSAR and molecular modeling studies on prodrug derivatives, wild-type vs. mutant SAR data and lateral validation of these models via comparative analysis can provide useful insight. Such studies can highlight the differences and similarity, if any, in their mechanism of interaction with wild-type and mutant protease receptor.

The main goals that are currently being pursued by many laboratories around the world in the design of new HIVPI are increased efficacy, enhanced specificity and minimizing adverse side effects. We have reported in this contribution several examples that point in these directions. An attempt has been made to point out to the reader the important factors that may affect the development of new, more efficient and more specific anti-HIV agents. Much work, however, remains to be done in order to put these new and exciting results to practical use.

Acknowledgements This work was supported by grant 2 R15 GM 069323-02 from NIH/NIGMS. We are thankful to Disha Patel and Raghava Kasara for their help in the preparation of the MS. Every attempt has been made to make this work as complete as possible. However, sincere apology is extended if any pertinent work(s) was inadvertently omitted from this review.

References

1. Gottlieb MS, Schroff R, Schanker HM, Weisman JD, Fan PT, Wolf RA, Saxon A (1981) N Engl J Med 305:1425
2. UNAIDS, Aids Epidemic Update, 2005 http://www.unaids.org (accessed 2/9/2006)
3. Gallo RC, Sarin PS, Gelmann EP, Robert-Guroff M, Richardson E, Kalyanaraman VS, Mann D, Sidhu GD, Stahl RE, Zolla-Pazner S, Leibowitch J, Popovic M (1983) Science 220:865
4. Barre-Sinoussi F, Chermann JC, Rey F, Nugeyre MT, Chamaret S, Gruest J, Dauguet C, Axler-Blin C, Vezinet-Brun F, Rouzioux C, Rozenbaum W, Montagnier L (1983) Science 220:868
5. Sarngadharan MG, Popovic M, Bruch L, Schupbach J, Gallo RC (1984) Science 224:506
6. Levy JA, Hoffman AD, Kramer SM, Landis JA, Shimabukuro JM, Oshiro LS (1984) Science 225:840
7. Coffin J, Haase A, Levy JA, Montagnier L, Oroszlan S, Teich N, Temin H, Toyoshima K, Varmus H, Vogt P (1986) Science 232:697
8. Clavel F, Guetard D, Brun-Vezinet F, Chamaret S, Rey MA, Santos-Ferreira MO, Laurent AG, Dauguet C, Katlama C, Rouzioux C (1986) Science 233:343
9. Fauci AS (1988) Science 239:617
10. De Clercq E (1995) Clin Microbiol Rev 8:200
11. De Clercq E (2002) Biochim Biophys Acta 258:1587
12. Gullick RM (2003) Clin Microbiol Infect 9:186
13. Chrusciel RA, Strohbach JW (2004) Curr Top Med Chem 4:1097
14. Garg R, Gupta SP, Gao H, Mekapati SB, Debnath AK, Hansch C (1999) Chem Rev 99:3525
15. Kurup A, Mekapati SB, Garg R, Hansch C (2003) Curr Med Chem 10:1819
16. Debnath AK (2005) Curr Pharma Design 11:3091
17. Ratner L, Haseltine W, Patarca R, Livak KJ, Starcich B, Josephs SF, Doran ER, Rafalski JA, Whitehorn EA, Baumeister K (1985) Nature 313:277
18. Toh H, Miyata T (1985) Nature 316:21
19. Tang J, James MN, Hsu IN, Jenkins JA, Blundell TL (1978) Nature 271:618
20. Huff JR, Kahn J (2001) Adv Protein Chem 56:213
21. Pearl LH, Taylor WR (1987) Nature 329:351
22. Kohl NE, Emini EA, Schleif WA, Davis LJ, Heimbach JC, Dixon RA, Scolnick EM, Sigal IS (1988) Proc Natl Acad Sci USA 85:4686
23. Krausslich HG, Schneider H, Zybarth G, Carter CA, Wimmer E (1988) Virol 62:4393
24. Nutt RF, Brady SF, Darke PL, Ciccarone TM, Colton CD, Nutt EM, Rodkey JA, Bennett CD, Waxman LH, Sigal IS (1988) Proc Natl Acad Sci USA 85:7129
25. Lapatto R, Blundell T, Hemmings A, Overington J, Wilderspin A, Wood S, Merson JR, Whittle PJ, Danley DE, Geoghegan KF (1989) Nature 342:299
26. Navia MA, Fitzgerald PM, McKeever BM, Leu CT, Heimbach JC, Herber WK, Sigal IS, Darke PL, Springer JP (1989) Nature 337:615

27. Wlodawer A, Miller M, Jaskolski M, Sathyanarayana BK, Baldwin E, Weber IT, Selk LM, Clawson L, Schneider J, Kent SB (1989) Science 245:616
28. Toh H, Ono M, Saigo K, Miyata T (1985) Nature 315:691
29. Miller M, Jaskolski M, Rao JKM, Leis J, Wlodawer A (1989) Nature 337:576
30. Schechter I, Berger A (1967) Biochem Biophys Res Commun 27:157
31. Babine RE, Bender SL (1997) Chem Rev 97:1359
32. Spinelli S, Liu QZ, Alzari PM, Hivel PH, Poljak RJ (1991) Biochimie 73:1391
33. Miller M, Schneider J, Sathyanarayan BK, Toth MV, Marshall GR, Clawson L, Selk L, Kent SBH, Wlodawer A (1989) Science 246:1149
34. Wlodawer A, Ericson JW (1993) Annu Rev Biochem 62:543
35. Kramer RA, Schaber MD, Skalka AM, Ganguly K, Wong-Staal F, Reddy EP (1986) Science 231:1580
36. McQuade TJ, Tomasselli AG, Liu L, Karacostas V, Moss B, Sawyer TK, Heinrikson RL, Tarpley WG (1990) Science 247:454
37. Boehme RE, Borthwick AD, Wyatt PG (1995) Annu Rep Med Chem 30:139
38. Winslow DL, Otto MJ (1995) AIDS 9 Suppl A:183–192
39. Prasad JV, Lunney EA, Para KS, Tummino PJ, Ferguson D, Hupe D, Domagala JM, Erickson JW (1996) Drug Des Discov 13:15
40. Deeks SG, Volberding PA (1997) AIDS Clin Rev 145:85
41. Korant BD, Rizzo CJ (1997) Adv Exp Med Biol 421:279
42. Flexner C (1998) N Engl J Med Chem 338:1281
43. De Clerq E (2004) Inter J Biochem Cell Bio 36:1800
44. Olson GL, Bolin DR, Bonner MP, Bos M, Cook CM, Fry DC, Graves BJ, Hatada M, Hill DE, Kahn M, Madison VS, Rusiecki VK, Sarabu R, Sepinwall J, Vincet GP, Voss ME (1993) J Med Chem 36:3039
45. Maligres PE, Upadhyay V, Rossen K, Cianciosi SJ, Purick RM, Eng KK, Reamer RA, Askin D, Volante RP, Reider PJ (1995) Tetrahedron Lett 36:2195
46. http://www.niaid.nih.gov/daids/dtpdb/FDADRUG.HTM (accessed 12/25/2005)
47. www.fda.gov/oashi/aids/virals.html (accessed 2/9/2006)
48. Drug cocktails fight HIV (1997) Chemistry in Britain 38–40
49. Ridky T, Leis J (1995) J Biol Chem 270:29621
50. Hansch C, Maloney PP, Fujita T, Muir RM (1962) Nature 194:178
51. Hansch C, Fujita T (1964) J Am Chem Soc 86:1616
52. Topliss JG (ed) (1983) Quantitative structure-activity relationships of drugs. Academic Press, New York
53. Hansch C, Fujita T (1994) Classical and three-dimensional QSAR in agrochemistry. Am Chem Soc, Washington, DC
54. Fujita T (ed) (1995) QSAR and drug design: new developments and applications. Elsevier, Amsterdam
55. Devillers J (ed) (1997) Comparative QSAR. Taylor & Francis, Washington, DC
56. Parrill AL, Reddy MR (1999) Rational drug design novel methods and practical applications. Am Chem Soc, Washington, DC
57. Debnath AK (2006) Quantitative structure-activity relationship (QSAR)—a versatile tool in drug design. In: Ghose AK, Viswanadhan VN (eds) Combinatorial library design and evaluation: principles, software tools, and applications. Mercel Dekker Inc, New York
58. Hansch C, Leo A (1995) Exploring QSAR, vol 1: Fundamentals and applications in chemistry and biology. Am Chem Soc, Washington, DC
59. Hansch C, Leo A, Hoekman D (1995) Exploring QSAR vol 2: Hydrophobic electronic and steric constants. Am Chem Soc, Washington, DC

60. Debnath AK (2001) Mini Rev Med Chem 1:187
61. Kellogg GE, Semus SF (2003) 3D-QSAR in modern drug design EXS 223
62. Mason JS, Good AC, Martin EJ (2001) Curr Pharm Des 7:567
63. Podlogar BL, Ferguson DM (2000) Drug Des Discov 17:4
64. Selassie CD, Mekapati SB, Verma RP (2002) Curr Top Med Chem 2:1357
65. Wlodawer A (2002) Vox Sang 83(Suppl 1):23
66. Wlodawer A, Vondrasek J (1998) Annu Rev Biophys Biomol Struct 27:249
67. Hardy LW, Malikayil A (2003) Curr Drug Discov 15:230
68. Gupta SP (2002) Prog Drug Res 58:223
69. Cramer RD III, Bunce JD, Patterson DE, Frank IE (1988) Quant Struct-Act Relat 7:18
70. Nugiel DA, Jacobs K, Worley T, Patel M, Kaltenbach RF III, Meyer DT, Jadhav PK, DeLucca GV, Smyser TE, Klabe RM, Bacheler LT, Rayner MM, Seitz SP (1996) J Med Chem 39:2156
71. Lam PYS, Ru Y, Jadhav PK, Aldrich PE, DeLucca GV, Eyerman CJ, Chang CH, Emmett G, Holler ER, Danekar WF, Li L, Confalone PN, McHugh RJ, Han Q, Li R, Markwalder JA, Seitz SP, Harpe TR, Bacheler LT, Rayner MM, Klabe RM, Shum L, Winslow DL, Kornhauser DM, Jackson DA, Erickson-Viitanen S, Hodge CN (1996) J Med Chem 39:3514
72. Jadhav PK, Woerner FJ, Lam PYS, Hodge CN, Eyermann CJ, Man HW, Daneker WF, Bacheler LT, Rayner MM, Meek JL, Erickson-Viitanen S, Jackson DA, Calabrese JC, Schadt M, Chang CH (1998) J Med Chem 41:1446
73. Gupta SP, Babu MS, Garg R, Sowmya S (1998) J Enzyme Inhib 13:399
74. Lam PYS, Jadhav PK, Eyermann CJ, Hodge CN, Ru Y, Bacheler LT, Meek JL, Otto MJ, Rayner MM, Wong NY, Chang CH, Wever PC, Jackson DA, Sharpe TR, Erickson-Viitanen S (1994) Science 263:380
75. Wilkerson WW, Akamike E, Cheatham WW, Hollis AY, Collins RD, DeLucca I, Lam PYS, Ru Y (1996) J Med Chem 39:4299
76. Wilkerson WW, Dax S, Cheatham WW (1997) J Med Chem 40:4079
77. Han Q, Chang CH, Li R, Ru Y, Jadhav PK, Lam PYS (1998) J Med Chem 41:2019
78. Gupta SP, Babu MS (1999) Bioorg Med Chem 7:2549
79. Rodgers JD, Johnson BL, Wang H, Erickson-Viitanen S, Klabe RM, Bachelor L, Cordova BC, Chang CH (1998) Bioorg Med Chem Lett 8:715
80. Patel M, Bachelor LT, Rayner MM, Cordova BC, Klabe RM, Erickson-Viitanen S, Seitz SP (1998) Bioorg Med Chem Lett 8:823
81. Garg R, Bhhatarai B (2004) Bioorg Med Chem 12:5819
82. Rodgers JD, Johnson BL, Wang H, Greenberg RA, Erickson-Viitanen S, Klabe RM, Cordova BC, Rayner MM, Lam GN, Chang CH (1996) Bioorg Med Chem Lett 6:2919
83. Kaltenbach RF III, Klabe RM, Cordova BC, Seitz SP (1999) Bioorg Med Chem Lett 9:2259
84. Rodgers JD, Johnson BL, Wang H, Erickson-Viitanen S, Klabe RM, Bachelor L, Cordova BC, Chang CH (1998) Bioorg Med Chem Lett 8:715
85. Kaltenbach RF III, Patel M, Waltermire RE, Harris GD, Stone BRP, Klabe RM, Garber S, Bacheler LT, Cordova BC, Logue K, Wright MR, Erickson-Viitanen S, Trainor GL (2003) Bioorg Med Chem Lett 13:605
86. Gayathri P, Pande V, Sivakumar R, Gupta SP (2001) Bioorg Med Chem 9:3059
87. De Lucca GV, Liang J, Aldrich PE, Calabrese J, Cordova B, Klabe RM, Rayner MM, Chang CH (1997) J Med Chem 40:1707
88. Katritzky AR, Oliferenko A, Lomaka A, Karelson M (2002) Bioorg Med Chem Lett 12:3453
89. De Lucca GV, Liang J, De Lucca I (1999) J Med Chem 42:135

90. Karelson M (2000) Molecular Descriptors in QSAR/QSPR. Wiley, New York
91. Zefirov NS, Kirpichenok MA, Izmailov FF, Trofimov MI (1987) Dokl Akad Nauk (Engl Transl) 296:883
92. Garg R, Patel D (2005) Bioorg Med Chem Lett 15:3767
93. Han W, Pelletier JC, Hodge CN (1998) Bioorg Med Chem Lett 8:3615
94. Frecer V, Burello E, Miertus S (2005) Bioorg Med Chem 13:5492
95. Rodgers JD, Lam PY, Johnson BL, Wang H, Li R, Ru Y, Ko SS, Seitz SP, Trainor GL, Anderson PS, Klabe RM, Bacheler LT, Cordova B, Garber S, Reid C, Wright MR, Chang CH, Erickson-Viitanen S (1998) Chem Biol 5:597
96. Patel M, Kaltenbach RF, Nugiel DA, McHugh RJ, Jadhav PK, Bacheler LT, Cordova BC, Klabe RM, Erickson-Viitanen S, Garber S, Reid C, Seitz SP (1998) Bioorg Med Chem Lett 8:1077
97. Ala PJ, DeLoskey RJ, Huston EE, Jadhav PK, Lam PY, Eyermann CJ, Hodge CN, Schadt MC, Lewandowski FA, Weber PC, McCabe DD, Duke JL, Chang CH (1998) J Biol Chem 273:12325
98. Böhm HJ (1994) J Comput Aided Mol Des 8:243
99. Debnath AK (1998) J Chem Inf Comput Sci 38:761
100. Corelli F, Manetti F, Tafi A, Campiani G, Nacci V, Botta M (1997) J Med Chem 40:125
101. Kim KH (1997) Med Chem Res 7:45
102. Kaminski JJ, Doweyko AM (1997) J Med Chem 40:427
103. Cho SJ, Serrano Gracia ML, Bier J, Tropsha A (1996) J Med Chem 39:5064
104. Steinmetz WE (1995) Quant Struct-Act Relat 14:19
105. Debnath AK, Jiang S, Strick N, Lin K, Haberfield P, Neurath AR (1994) J Med Chem 37:1099
106. Martin YC, Lin CT, Wu J (1993) Application of CoMFA to D1 dopaminergic agonists: A case study. In: Kubinyi H (ed) 3D QSAR in drug design: theory methods and applications. ESCOM, Leiden, p 643
107. SYBYL molecular modeling system version 6.4, Tripos Associates Inc, 1699 South Hanley Road, St. Louis, MO
108. Debnath AK (1999) J Med Chem 42:249
109. HINT 2.25S, EduSoft, LC, P.O. Box 1811, Ashland, VA 23005
110. Tervo AJ, Nyrönen TH, Rönkkö T, Poso A (2004) J Chem Inf Comput Sci 44:807
111. Jones G, Willett P, Glen RC, Leach AR, Taylor R (1997) J Mol Biol 267:727
112. Clark M, Cramer RD III, Van Opdenbosch N (1989) J Comput Chem 10:982
113. Weiner SJ, Kollman PA, Nguyen DT, Case DA (1986) J Comput Chem 7:230
114. Allen BCP, Grant GH, Richaards WG (2001) J Chem Inf Comput Sci 41:330
115. Solov'ev VP, Varnek A (2003) J Chem Inf Comput Sci 43:1703
116. Forsythe GE, Malcolm MA, Moler CB (1977) Computer methods for mathematical computations. Prentice Hall Inc, Englewood Cliffs, NJ
117. Avram S, Bologa C, Banda M, Flonta ML (2001) Romanian J of Biophy 11:11
118. Avram S, Svab I, Bologa C, Flonta ML (2003) J Cell Mol Med 7:287
119. Avram S, Movileanu L, Mihailescu D, Flonta ML (2002) J Cell Mol Med 6:251
120. Avram S, Bologa C, Flonta ML (2005) J Mol Mod 11:105 (Online)
121. Schaal W, Karlsson A, Ahlsen G, Lindberg J, Andersson HO, Danielson UH, Classon B, Unge T, Samuelsson B, Hulten J, Hallberg A, Karlen A (2004) J Med Chem 44:155
122. Mohamadi F, Richards NGJ, Guida WC, Liskamp R, Lipton M, Caufield C, Chang G, Hendrickson T, Still WC (1990) J Comput Chem 11:440
123. Hulten J, Andersson HO, Schaal W, Danielsson UH, Classon B, Kvarnstrom I, Karlen A, Unge T, Samuelsson B, Hallberg A (1999) J Med Chem 42:4054
124. Nair AC, Jayatilleke P, Wang X, Miertus S, Welsh WJ (2002) J Med Chem 45:973

125. Senese CL, Hopfinger AJ (2003) J Chem Inf Comput Sci 43:2180
126. Senese CL, Duca J, Pan D, Hopfinger AJ, Tseng YJ (2004) J Chem Inf Comput Sci 44:1526
127. Bhhatarai B, Garg R (2005) Bioorg Med Chem 13:4078
128. Hagen S, Vara Prasad JVN, Tait BD (2000) Adv Med Chem 5:159
129. Romines KR, Chrusciel RA (1995) Curr Med Chem 2:825
130. Thaisrivongs S, Tomich PK, Watenpaugh KD, Chong KT, Howe JW, Yang CP, Strohbach JW, Turner SR, McGrath JP, Bohanon MJ, Lynn JC, Mulichak AM, Spinelli PA, Hinshaw RR, Pagano PJ, Moon JB, Ruwart MJ, Wilkinson KF, Rush BD, Zipp GL, Dalga RJ, Schwende FJ, Howard GM, Padbury GE, Toth LN, Zhao Z, Koeplinger KA, Kakuk TJ, Cole SL, Zaya RM, Piper RC, Jeffrey P (1994) J Med Chem 37:3200
131. Romines KR, Watenpaugh KD, Tomick PK, Howe WJ, Morris JK, Lovasz KD, Mulichak AM, Finzel BC, Lynn JC, Horng MM, Schwende FJ, Ruwart MJ, Zipp GL, Chong K-T, Dolak LA, Toth LN, Howard GM, Rush BD, Wilkinson KF, Possert PL, Dalga RJ, Hinshaw RR (1995) J Med Chem 38:1884
132. Romines KR, Watenpaugh KD, Howe WJ, Tomich PK, Lovasz KD, Morris JK, Janakiraman MN, Lynn JC, Horng MM, Chog KT, Hinshaw RR, Dolak LA (1995) J Med Chem 38:4463
133. Romines KR, Morris JK, Howe WJ, Tomich PK, Horng MM, Chong KT, Hinshaw RR, Anderson DJ, Strohbach JW, Turner SR, Mizsak SA (1996) J Med Chem 39:4125
134. Skulnick HI, Johnson PD, Aristoff PA, Morris JK, Lovasz KD, Howe WJ, Watenpaugh KD, Janakiraman MN, Anderson DJ, Reischer RJ, Schwartz TM, Banitt LS, Tomich PK, Lynn JC, Horng MM, Chong KT, Hinshaw RR, Dolak LA, Seest EP, Schwende FJ, Rush BD, Howard GM, Toth LN, Wilkinson KR, Kakuk TJ, Johnson CW, Cole SL, Zaya RM, Zipp GL, Possert PL, Dalga RJ, Zhong WZ, Williams MG, Romines KR (1997) J Med Chem 40:1149
135. Gupta SP, Babu MS, Kaw N (1999) J Enzyme Inhib 14:109
136. Vara Prasad JVN, Para KS, Tummino PJ, Ferguson D, Mcquade EJ, Lunney EA, Rapundalo ST, Batley BL, Hingorani G, Domagala JM, Gracheck SJ, Bhat TN, Liu B, Baldwin ET, Erickson JW, Sawyer TK (1995) J Med Chem 38:898
137. Gupta SP, Babu MS, Sowmya S (1998) Bioorg Med Chem 6:2185
138. Bernstein FC, Koetzle TF, William GTB, Meyer EF Jr, Brice MD, Rodgers JR, Kennard O, Shimanouchi T, Tasumi M (1978) Arch Biochem Biophys 185:584
139. Tait BD, Hagen S, Domagala J, Ellsworth EL, Gajda C, Hamilton HW, Vara Prasad JVN, Ferguson D, Graham N, Hupe D, Nouhan C, Tummino PJ, Humblet C, Lunney EA, Pavlovsky A, Rubin J, Gracheck SJ, Baldwin ET, Bhat TN, Erickson JW, Gulnik SV, Liu B (1997) J Med Chem 40:3781
140. Boyer FE, Vara Prasad JVN, Domagala JM, Ellsworth EL, Gajda C, Hagen SE, Markoski LJ, Tait BD, Lunney EA, Palovsky A, Ferguson D, Graham N, Holler T, Hupe D, Nouhan C, Tummino PJ, Urumov A, Zeikus E, Zeikus G, Gracheck SJ, Sanders JM, VanderRoest S, Brodfuehrer J, Iyer K, Sinz M, Gulnik SV, Erickson JW (2000) J Med Chem 43:843
141. Vara Prasad JVN, Boyer FE, Domagala JM, Ellsworth EL, Gajda C, Hamilton HW, Hagen SE, Markoski LJ, Steinbaugh BA, Tait BD, Humblet C, Lunney EA, Pavlovsky A, Rubin JR, Ferguson D, Graham N, Holler T, Hupe D, Nouhan C, Tummino PJ, Urumov A, Zeikus E, Zeikus G, Gracheck SJ, Saunders JM, VanderRoest S, Brodfuehrer J, Iyer K, Sinz M, Gulnik SV, Erickson JW (1999) Bioorg Med Chem 7:2775
142. Lunney EA, Hagen SE, Domagala JM, Humblet C, Kosinski J, Tait BD, Warmus JS, Wilson M, Ferguson D, Hupe D, Tummino PJ, Baldwin ET, Bhat TN, Liu B, Erickson JW (1994) J Med Chem 37:2664

143. Goodford PJ (1985) J Med Chem 28:849
144. GRID Software Program, Molecular Discovery Ltd, West Way House, Elms Parade, Oxford OX2 9LL, England
145. Skulnick HI, Johnson PD, Howe WJ, Tomich PK, Chong KT, Watenpaugh KD, Janakiraman MN, Dolak LA, McGrath JP, Lynn JC, Horng MM, Hinshaw RR, Zipp GL, Ruwart MJ, Schwende FJ, Zhong WZ, Padbury GE, Dalga RJ, Shiou L, Possert PL, Rush BD, Wilkinson KF, Howard GM, Toth LN, Williams MG, Kakuk TJ, Cole SL, Zaya RM, Lovasz KD, Morris JK, Romines KR, Thaisrivongs S, Aristoff PA (1995) J Med Chem 38:4968
146. Thaisrivongs S, Watenpaugh KD, Howe WJ, Tomich PK, Dolak LA, Chong KT, Tomich CSC, Tomasselli AG, Turner SR, Strohbach JW, Mulichak AM, Janakiraman MN, Moon JB, Lynn JC, Horng MM, Hinshaw RR, Curry KA, Rothrock DJ (1995) J Med Chem 38:3624
147. Moon JB, Howe WJ (1991) Proteins: Struct Funct Gene 11:314
148. Vara Prasad JVN, Para KS, Lunney EA, Ortwine DF, Dunbar JB Jr, Ferguson D, Tummino PJ, Hupe D, Tait BD, Domagala JM, Humblet C, Bhat TN, Liu B, Guerin DMA, Baldwin ET, Erickson JW, Sawyer TK (1994) J Am Chem Soc 116:6989
149. Goodsell DS, Olson AJ (1990) Proteins: Struct Funct Genet 8:195
150. Ellsworth EL, Domagala JM, Vara Prasad JVN, Hagen S, Ferguson D, Holler T, Hupe D, Graham N, Nouhan C, Tummino PJ, Zeikus G, Lunney EA (1999) Bioorg Med Chem Lett 9:2019
151. Schake D (2004) AIDS 18:579
152. Larder BA, Hertogs K, Bloor S, vanden Eynde CH, Decian W, Wang Y, Freimuth WW, Tarpley G (2000) AIDS 14:1943
153. Kulkarni SS, Kulkarni VM (1999) J Chem Inf Comp Sci 39:1128
154. Thaisrivongs S, Janakiraman MN, Chong KT, Tomich PK, Dolak LA, Turner SR, Strohbach JW, Lynn JC, Horng MM, Hinshaw RR, Watenpaugh KD (1996) J Med Chem 39:2400
155. Schwartz TM, Bundy GL, Strohbach JW, Thaisrivongs S, Johnson PD, Skulnick HI, Tomich PK, Lynn JC, Chong KT, Hinshaw RR, Raub TJ, Padbury GE, Toth LN (1997) Bioorg Med Chem Lett 7:399
156. Thaisrivongs S, Romero DL, Tommasi RA, Janakiraman MN, Strohbach JW, Turner SR, Biles C, Morge RR, Johnson PD, Aristoff PA, Tomich PK, Lynn JC, Horng KT, Hinshaw RR, Howe WJ, Finzel BC, Watenpaugh KD (1996) J Med Chem 39:4630
157. Thaisrivongs S, Skulnick HI, Turner SR, Strohbach JW, Tommasi RA, Johnson PD, Aristoff PA, Judge TM, Gammill RB, Morris JK, Romines KR, Chrusciel RA, Hinshaw RR, Chong KT, Trapley WG, Poppe S, Slade DE, Lynn JC, Horng MM, Tomich PK, Seest EP, Dolak LA, Howe WJ, Howard GM, Schwende FJ, Toth LN, Padbury GE, Wilson GJ, Shiou L, Zipp GL, Wilkinson KF, Rush BD, Ruwart MJ, Koeplinger KA, Zhao Z, Cole S, Zaya RM, Kakuk TJ, Janakiraman MN, Watenpaugh KD (1996) J Med Chem 39:4349
158. De Clercq E (1995) J Med Chem 38:2491
159. Billich A, Charpiot B, Fricker G, Gstach H, Lehr P, Peichl P, Scholz D, Rosenwirth B (1994) Antiviral Res 25:215
160. De Solms SJ, Giuliani EA, Guare JP, Vacca JP, Sanders WM, Graham SL, Wiggins JM, Darke PL, Sigal IS, Zugay JA, Emini EA, Schleif WA, Quintero JC, Anderson PS, Huff JR (1991) J Med Chem 34:2852
161. Thompson SK, Murthy KHM, Zhao B, Winborne E, Green DW, Fisher SM, Desjarlais RL, Tomaszek TA Jr, Meek TD, Gleason JG, Abdel-Meguid SS (1994) J Med Chem 37:3100

162. Getman DP, DeCrescenzo GA, Heintz RM, Reed KL, Talley JJ, Bryant ML, Clare M, Houseman KA, Marr JJ, Mueller RA, Vazquez ML, Shieh H-S, Stallings WC, Stegeman RA (1993) J Med Chem 36:288
163. Holloway MK, Wai JM, Halgren TA, Fitzgerald PMD, Vacca JP, Dorsey BD, Levin RB, Thompsom WJ, Chen J, DeSolms SJ, Gaffin N, Ghosh AK, Giuliani EA, Graham SL, Guare JP, Hungate RW, Lyle TA, Sanders WM, Tucker TJ, Wiggins M, Wiscount CM, Woltersdorf OW, Young SD, Darke PL, Zugay JA (1995) J Med Chem 38:305
164. Hansch C, Garg R, Kurup A, Mekapati SB (2003) Bioorg Med Chem 11:2075
165. Hansch C, Garg R, Kurup A (2001) Bioorg Med Chem 9:283
166. Ferguson DM, Radmer RJ, Kollman PA (1991) J Med Chem 34:2654
167. Kempf DJ, Norbeck DW, Codacovi L, Wang XC, Kohlbrenner WE, Wideburg NE, Paul DA, Knigge MF, Vasavanonda S, Graig-kennard A, Saldivar A, Rosenbrook W Jr, Clement JJ, Plattner JJ, Erickson J (1990) J Med Chem 33:2687
168. Maw HH, Hall LH (2002) J Chem Inf Comput Sci 42:290
169. SAS, ver 8.0, SAS Institute, Cary, NC 27513
170. Maw HH, Hall LH (2000) J Chem Inf Comput Sci 40:1270
171. Maw HH, Hall LH (2001) J Chem Inf Comput Sci 41:1248
172. Hall LH, Kier LB (2000) J Chem Inf Comput Sci 40:784
173. Golbraikh A, Tropsha A (2003) J Chem Inf Comput Sci 43:144
174. Perez C, Pastor M, Ortiz AR, Gago F (1998) J Med Chem 41:836
175. Vazquez ML, Bryant ML, Clare M, Decrescenzo GA, Doherty EM, Freskos JN, Getman DP, Houseman KA, Julien JA, Kocan GP, Mueller RA, Shieh HS, Stallings WC, Stegeman RA, Talley JJ (1995) J Med Chem 38:581
176. Beaulieu PL, Anderson PC, Cameron DR, Croteau G, Gorys V, Grand-Maitre C, Lamarre D, Liard F, Paris W, Plamondon L, Soucy F, Thibeault D, Wernic D, Yoakim C (2000) J Med Chem 43:1094
177. Takashiro E, Watanabe T, Nitta T, Kasuya A, Miyamoto S, Ozawa Y, Yagi R, Nishigaki T, Shibayama T, Nakagawa A, Iwamoto A, Yabe Y (1998) Bioorg Med Chem 6:595
178. Takashiro E, Hayakawa I, Nitta T, Kasuya A, Miyamoto S, Ozawa Y, Yagi R, Yamamoto I, Shibayama T, Nakagawa A, Yabe Y (1999) Bioorg Med Chem 7:2063
179. Takashiro E, Nakamura Y, Miyamoto S, Ozawa Y, Sugiyama A, Fujimoto K (1999) Bioorg Med Chem 7:2105
180. Ghosh AK, Lee HY, Thompson WJ, Culberson C, Holloway K, McKee SP, Munson PM, Duong TT, Smith AM, Darke PL, Zugay JA, Emini EA, Schleif WA, Huff JR, Anderson PS (1994) J Med Chem 37:1177
181. Di Santo R, Costi R, Artico M, Massa S, Ragno R, Marshall GR, La Colla P (2002) Bioorg Med Chem 10:2511
182. Melnick M, Reich SH, Lewis KK, Mitchell LJ Jr, Nguyen D, Trippe AJ, Dawson H, Davies JF II, Appelt K, Wu BW, Musick L, Gehlhaar DK, Webber S, Shetty B, Kosa M, Kahil D, Andrada D (1996) J Med Chem 39:2795
183. Reich SH, Melnick M, Pino MJ, Fuhry MA, Trippe AJ, Appelt K, Davies JF II, Wu BW, Musick L (1996) J Med Chem 39:2781
184. Reich SH, Melnick M, Davies JF II, Appelt K, Lewis KK, Fuhry MA, Pino M, Trippe AJ, Nguyen D, Dawson H, Wu B, Musick L, Kosa M, Kahil D, Webber S, Gehlhaar DK, Andrada D, Shetty B (1995) Proc Natl Acad Sci USA 92:3298
185. Pyring D, Lindberg J, Rosenquist A, Zuccarello G, Kvarnstrom I, Zhang H, Vrang L, Unge T, Classon B, Hallberg A, Samuelsson B (2001) J Med Chem 44:3083
186. Kempf DJ (1994) Methods Enzymol 241:334
187. Kempf DJ, Marsh JF, Denissen E, McDonald S, Vasavanonda S, Flentge CA, Green BE, Fino L, Park CH, Kong X-P, Wideburg NE, Saldivar A, Ruiz L, Kati WM, Sham HL,

Robinn T, Stewart KD, Hau A, Plattner JJ, Leonard JM, Norbeck DW (1995) Proc Natl Acad Sci USA 92:2484

188. Kempf DJ, Codacovi L, Wang XC, Kohlbrenner WE, Wideburg NE, Saldivar A, Vasavanonda S, Marsh KC, Bryant P, Sham HL, Green BE, Betebenner DA, Erickson J, Norbeck DW (1993) J Med Chem 36:320

189. Kempf DJ, Sham HL, Marsh JF, Flentge CA, Betebenner DA, Green BE, McDonald S, Vasavanonda S, Saldivar A, Wideburg NE, Kati WM, Ruiz L, Zhao C, Fino L, Patterson J, Molla A, Plattner JJ, Norbeck DW (1998) J Med Chem 41:602

190. Mekapati SB, Sivakumar R, Gupta SP (2001) J Enzyme Inhib 16:185

191. Fujita T, Ban T (1971) J Med Chem 14:148

192. Chen P, Cheng PTW, Alam M, Beyer BD, Bisacchi GS, Dejneka T, Evans AJ, Greytok JA, Hermsmeier MA, Humphreys WG, Jacobs GA, Kocy O, Lin P-F, Lis KA, Marella MA, Ryono DE, Sheaffer AK, Spergel SH, Sun C-Q, Tino JA, Vite G, Colonno RJ, Zahler R, Barrish JC (1996) J Med Chem 39:1991

193. Bissachi GS, Alam SAM, Ashfaq A, Barrish J, Cheng PTW, Greytok J, Hermsmeier M, Lin P-F, Merchant Z, Skoog M, Spergel S, Zahler R (1995) Bioorg Med Chem Lett 5:459

194. Leonard JT, Roy K (2006) Bioorg Med Chem 14:1039

195. Bouzide A, Sauvé G, Sévigny G, Yelle J (2003) Bioorg Med Chem Lett 13:3601

196. Randad RS, Lubkowska L, Eissenstat MA, Gulnik SV, Yu B, Bhat TN, Clanton DJ, House T, Stinson SF, Erickson JW (1998) Bioorg Med Chem Lett 8:3537

197. Glenn MP, Pattenden LK, Reid RC, Tyssen DP, Tyndall JDA, Birch CJ, Fairlie DP (2002) J Med Chem 45:371

198. Waller CL, Oprea TI, Giolitti A, Marshall GR (1993) J Med Chem 36:4152

199. Oprea TI, Waller CL, Marshall GR (1994) J Med Chem 37:2206

200. Doweyko AM (1994) J Med Chem 37:1769

201. Doweyko AM (1988) J Med Chem 31:1396

202. Doweyko AM (1989) New tool for the study of structure-activity relationships in three dimensions. In: Magee P, Henry DR, Block JH (eds) Probing bioactive mechanisms. ACS Symposium Series 413. Am Chem Soc, Washington, DC, p 82

203. Doweyko AM (1991) J Math Chem 7:273

204. Kroemer RT, Peter E, Peter H (1995) J Med Chem 38:4917

205. Scholz D, Billich A, Charpiot B, Ettmayer P, Lehr P, Rosenwirth B, Schreiner E, Gstach H (1994) J Med Chem 37:3079

206. DISCOVER, Biosym Technologies, 9685 Scranton Rd, San Diego, CA 92121

207. Cho SJ, Tropsha A (1995) J Med Chem 38:1060

208. Viswanadhan VN, Reddy MR, Wlodawer A, Varney MD, Weinstein JN (1996) J Med Chem 39:705

209. Weiner SJ, Kollman PA, Case DA, Singh UC, Ghio C, Alagona G, Profeta S, Weiner PK (1984) J Am Chem Soc 106:765

210. Singh UC, Weiner PK, Caldwell JK, Kollman PA (1986) AMBER University of California at San Francisco, San Francisco

211. Furet P, Sele A, Cohen NC (1988) J Mol Graph 6:182

212. Pastor M, Pérez C, Gago F (1997) J Mol Grap Mod 15:364

213. Nicholls A, Honig B (1991) J Comput Chem 12:435

214. Pastor M (1996) GOLPE version 3.0. Multivariate Infometric Analysis (MIA). Perugia, Italy

215. Vedani A, Zbinden P (1998) Pharm Acta Helvet 73:11

216. Jayatilleke PRN, Nair AC, Zauhar R, Welsh WJ (2000) J Med Chem 43:4446

217. Nair AC, Bonin I, Tossi A, Welsh WJ, Miertus S (2002) J Mol Grap Mod 21:171

218. Gulnick SV, Suvorov LI, Liu B, Yu B, Anderson B, Mitsuya H, Erickson JW (1995) Biochemistry 34:9282
219. Klabe RM, Bacheler LT, Ala PJ, Erickson-Viitanen S, Meek JL (1998) Biochemistry 37:8735
220. Wilson SI, Lowri PH, Mills JS, Gulnick SV, Erickson JW, Dunn BM, Kay J (1997) Biochim Biophys Acta 1339:113
221. Huang X, Xu L, Luo X, Fan K, Ji R, Pei G, Chen K, Jiang H (2002) J Med Chem 45:333
222. Sakurai M, Higashida S, Sugano M, Komai T, Yagi R, Ozawa Y, Handa H, Nishigaki T, Yabe Y (1994) Bioorg Med Chem 2:807
223. Komai T, Higashida S, Sakurai M, Nitta T, Kasuya A, Miyamaoto S, Yagi R, Ozawa Y, Handa H, Mohri H, Yasuoka A, Oka S, Nishigaki T, Kimura S, Shimada K, Yabe Y (1996) Bioorg Med Chem 4:1365
224. Morris GM, Goodsell DS, Halliday RS, Huey R, Hart WE, Belew RK, Olson AJ (1998) J Comput Chem 19:1639
225. Visco DP Jr, Pophale RS, Rintoul MD, Faulon J-L (2002) J Mol Grap Model 20:429
226. Young SD, Payne LS, Thompson WJ, Gaffin N, Lyle TA, Britcher SF, Graham SL, Schultz TH, Deana AA, Darke PL, Zugay J, Schleif WA, Quintero JC, Emini EA, Anderson PS, Huff JR (1992) J Med Chem 35:1702
227. Thompson WJ, Fitzgerald PMD, Holloway MK, Emini EA, Darke PL, McKeever BM, Schleif WA, Quintero JC, Zugay J, Tucker TJ, Schwering JE, Homnick CF, Nunberg J, Springer JP, Huff JR (1992) J Med Chem 35:1685
228. Beaulieu PL, Wernic D, Abraham A, Anderson PC, Bogri T, Bousquet Y, Croteau G, Guse I, Lamarre D, Liard F, Paris W, Thibeault D, Pav S, Tong L (1997) J Med Chem 40:2164
229. Hall LH, Hall M-Z (1991) Associates Consulting, Quincy, MA
230. Faulon J-L, Visco DP Jr, Pophale RS (2003) J Chem Inf Comp Sci 43:707
231. Jenwitheesuk E, Samudrala R (2003) BMC structural biology 3:2
232. Kale L, Skeel R, Bhandarkar M, Brunner R, Gursoy A, Krawetz N, Phillips J, Shinozaki A, Varadarajan K, Schulten K (1999) J Comput Phys 151:283
233. Brunger AT (1992) X-PLOR version 3.1, A system for X-ray crystallography, NMR. Yale University Press, New Haven, CT
234. Kiralj R, Ferreira MMC (2003) J Mol Grap Model 21:499
235. Stewart JJP (1989) J Comput Chem 10:209
236. Stewart JJP (1989) J Comput Chem 10:221
237. WebLab Viewer version 2.01 (1997) Molecular Simulations Inc., San Diego, CA
238. Skalova T, Hasek J, Dohnalek J, Petrokova H, Buchtelova E, Duskova J, Soucek M, Majer P, Uhlikova T, Konvalinka J (2003) J Med Chem 46:1636
239. Kiralj R, Ferreira MMC (2003) J Mol Grap Model 21:435
240. Ungwitayatorn J, Samee W, Pimthon J (2004) J Mol Str 689:99
241. Senese CL, Hopfinger AJ (2003) J Chem Inf Comput Sci 43:1297
242. Radestock S, Bohm M, Gohlke H (2005) J Med Chem 48:5466
243. Boutton CW, De Bondt HL, De Jonge MR (2005) J Med Chem 48:2115
244. Fitzgerald PMD, McKeever BM, Van Middlesworth JF, Springer JP, Heimbach JC, Leu C-T, Herber WK, Dixon RAF, Darke PL (1990) J Biol Chem 265:14209
245. Erickson J, Neidhart DJ, VanDrie J, Kempf DJ, Wang XC, Norbeck DW, Plattner JJ, Rittenhouse JW, Turon M, Wideburg N, Kohlbrenner WE, Simmer R, Helfrich R, Paul DA, Knigge M (1990) Science 249:527
246. Swain AL, Miller MM, Green J, Rich DH, Schneider J, Kent SBH, Wlodawer A (1990) Proc Natl Acad Sci USA 87:8805
247. Huff JR (1991) J Med Chem 34:2305

248. Wonacott A, Cooke R, Hayes FR, Hann MM, Jhoti H, McMeekin P, Mistry A, Murray-Rust P, Singh OMP, Weir MP (1993) J Med Chem 36:3113

249. Humber DC, Bamford MJ, Bethell RC, Cammack N, Cobley K, Evans DN, Gray NM, Hann MM, Orr DC, Saunders J, Shenoy BEV, Storer R, Weingarten GG, Wyatt PG (1993) J Med Chem 36:3120

250. Logsdon BC, Vickery JF, Martin P, Proteasa G, Koepke JI, Terlecky SR, Wawrzak Z, Winters MA, Merigan TC, Kovari LC (2004) J Virol 78:3123

251. Tie Y, Boross PI, Wang Y-F, Gaddis L, Hussain AK, Leshchenko S, Ghosh AK, Louis JM, Harrison RW, Weber IT (2004) J Mol Biol 338:341

252. Perryman AL, Lin J-H, McCammon JA (2004) Protein Sci 13:1108

253. CQSAR Program, Biobyte Corp., Claremont, CA, USA (www.biobyte.com)

Author Index Volumes 1–3

The volume numbers are printed in italics

Subject Index